NUCLEAR ARMS CONTROL

BACKGROUND AND ISSUES

Committee on International Security and Arms Control
National Academy of Sciences

NATIONAL ACADEMY PRESS
Washington, D.C.   1985

NATIONAL ACADEMY PRESS   2101 CONSTITUTION AVENUE, N.W.   WASHINGTON, D.C. 20418

NOTICE: The National Academy of Sciences was established in 1863 by Act of Congress as a private, nonprofit, self-governing membership corporation for the furtherance of science and technology for the general welfare.

The Committee on International Security and Arms Control is a committee of the National Academy of Sciences.

This work was supported by special grants from the John D. and Catherine T. MacArthur Foundation and by grants from the Carnegie Corporation of New York, the Charles E. Culpeper Foundation, the William and Flora Hewlett Foundation, the Andrew W. Mellon Foundation, the Rockefeller Foundation, and the Richard Lounsbery Foundation.

Library of Congress Catalog Card Number 84-62287

International Standard Book Number 0-309-03491-4

Contents

Foreword

The advent of the nuclear age after World War II profoundly changed the nature of warfare and the strategic relationship of the superpowers. The scientific developments that produced this revolution in warfare also created a new, special relationship between the scientific community and the government. Scientists were not only partners in the rapid evolution of military technology but were also major participants in the formulation of military and foreign policy reflecting the new technology. Conscious of the terrible consequences of nuclear war, scientists played a central role in developing approaches to control nuclear weapons and reduce the likelihood that they would ever be used. Over the years many U.S. scientists have served in important government positions and as influential advisors on these matters.

In this tradition the National Academy of Sciences has an important role to play. It has undertaken many studies relating to matters of national security, and currently several committees of the National Research Council advise branches of the military on questions of scientific research. One committee of experts is evaluating the impact of a major nuclear war on the earth's atmosphere and climate. Another is advising the government on issues related to scientific communication and national security.

The Committee on International Security and Arms Control reflects the Academy's deep interest in international security and the potential of arms control to reduce the threat of nuclear war. I believe this is as expert a group of individuals as one could assemble to consider these critical problems. Its members have been deeply involved in many aspects of military technology and arms control. They have advised several presidents and served in senior governmental posts; they have been involved in military research since the days of the Manhattan Project; they have headed universities and research centers; they have been involved with important arms control negotiations. The members of this committee have thought long and hard about these issues.

The committee has pursued a number of activities in response to its broad charter. Twice each year it has met with its counterparts from the Soviet Academy of Sciences to explore problems of international security and arms control. In response to the widely expressed interest of members of our Academy in learning more about issues and opportunities in arms control, it has also served an important educational role, holding a number of meetings and sessions on arms control for the Academy's membership. This educational role culminated in the spring of 1984 in a major tutorial that brought together over 200 Academy members for two days of meetings and discussions prior to the Academy's annual meeting.

The response to the background materials prepared for the tutorial was so positive that I asked that they be expanded and refined for a broader audience. I believe that the result is a unique volume—timely, comprehensive, authoritative. It thoroughly describes the history and status of the arms control debate. At the same time, it presents a wide diversity of views on the underlying issues in a nontechnical, nonpartisan fashion. I believe that it will prove to be a valuable resource for our national leaders, for students and researchers, and for the growing number of people who are concerned about this issue of vital importance to our future.

Frank Press
President
National Academy of Sciences

Preface

There is no more important challenge in our time than how to prevent the unprecedented catastrophe of nuclear war. But despite almost universal agreement on the overriding imperative of averting such a disaster, there are fundamental differences between the United States and the Soviet Union and between groups within our own country on how best to accomplish this goal. In particular, serious observers differ strongly over the appropriate role for arms control in this process and over the formulation of specific approaches to arms control.

The Committee on International Security and Arms Control was created by the National Academy of Sciences in 1980 to study these issues and to advance understanding of them both in the United States and abroad. In the course of our study, we have been impressed by the extensive literature dealing with the appalling consequences of nuclear war, with nuclear arsenals and strategic doctrine, and with the detailed diplomatic and bureaucratic politics of particular efforts to achieve specific nuclear arms control agreements. At the same time, we have sensed the lack of an objective overview of current arms control agreements and proposals that brought into focus the evolution of these concepts and the issues underlying the often confusing domestic and international debate on them. We concluded that there was a useful role to be filled in sharing our collective background on these subjects with our colleagues in the Academy.

This book had its immediate origins in a two-day tutorial on the problems of arms control and international security that the National Academy of Sciences held for its membership in the spring of 1984. To assist the participants in preparing for the tutorial, the committee prepared a background paper on the major agreements and proposals directed at the control of nuclear arms. This paper has now been expanded and substantially reworked to form this book.

We have not attempted in this volume to reach conclusions or make recommendations on specific arms control proposals or issues. Rather, we have endeavored to present the reader with an overview of the historical development of present U.S. and Soviet positions on specific arms control proposals and to identify the underlying issues on which opinions are so divided. In presenting issues, we have chosen the approach of stating opposing points of view in order to illuminate the nature of the debate. In doing this, we have tried to avoid both extreme arguments that would unfairly discredit a position and compromise positions that would obscure the underlying issues. There are many variants to all of these arguments, and it is most unlikely that a spokesperson for any particular position would use or even support all of the arguments presented for that position. The Committee on International Security and Arms Control and its individual members obviously do not agree with all of the conflicting opinions set forth in this document, but they do believe that these opinions present a balanced view of the scope of the debate.

On behalf of the committee I would like to express our special appreciation to our fellow committeeman Spurgeon M. Keeny, Jr., Scholar-in-Residence, National Academy of Sciences, whose dedicated efforts made this volume possible. We also join in thanking Lori Esposito, who provided our staff support, for her invaluable contribution in researching and preparing drafts of many sections of this volume. We would also like to thank Charles van Doren for his assistance on the chapter on the non-proliferation of nuclear weapons, Lynn Rusten and Steve Olson for their editorial assistance, and Barbara Wollison for her secretarial support in preparing the many drafts that led finally to this volume.

In the preparation of this book we have all learned a great deal about the background and issues underlying the current debate on nuclear arms control. We hope that others will also find this book useful in their own efforts to understand the debate and to develop positions on the role of arms control in reducing the threat of nuclear war.

Marvin L. Goldberger
Chairman
Committee on International
Security and Arms Control

NUCLEAR ARMS CONTROL

BACKGROUND AND ISSUES

1 Overview

There is worldwide agreement that a general nuclear war would be a totally unprecedented human catastrophe. Civilization as we know it would be destroyed. The leaders of the United States and the Soviet Union have consistently proclaimed that a central objective of their national policies is to prevent such a disaster. Despite this common goal, the means of achieving it have been the subject of bitter controversy between the two superpowers. Within the United States, the lack of consensus is reflected in the heated debate on all aspects of the U.S.-Soviet relationship.

To many, the nuclear arms race has become a symbol of the danger of nuclear war. Failure to contain the arms race has resulted in widespread concern and even despair about the prospects of avoiding nuclear war. While not in itself the only, or even the central, cause of tensions between the superpowers, the nuclear arms race certainly increases the risk of nuclear war and the extent of the disaster that would result.

The dangers of the arms race can be limited by mutual agreements or independent actions by each nation. Arms control agreements are covenants between potential adversaries defining the boundaries between what is forbidden and what is allowed in military activities. Such a covenant is by its nature a complex and controversial undertaking. The contributions that this approach has made or can make to reduce the risk of nuclear war can only be judged by examining actual arms control agreements or proposals. Only in the specifics of these covenants do the problems and the opportunities of arms control come to life.

1

The national debate on nuclear arms control, which has grown steadily in intensity over the the past few years, has focused primarily on a half-dozen international agreements and proposals. These are the unratified SALT II Treaty and the START proposals, which are directed at strategic offensive systems; the SALT I Anti-Ballistic Missile (ABM) Treaty, which has been placed in jeopardy by renewed U.S. emphasis on strategic defense systems; a comprehensive freeze on nuclear weapons and delivery systems; the Intermediate-Range Nuclear Force (INF) negotiations which are directed at intermediate-range nuclear forces in Europe; a comprehensive ban on nuclear testing; and limitations on anti-satellite systems. To this list should be added a number of specific agreements and proposals directed at another aspect of the nuclear arms race, the proliferation of nuclear weapons to additional countries.

The chapters following this overview present the background and status of these specific agreements and proposals and examine the underlying issues as seen by supporters and critics. This overview chapter examines a number of underlying issues, on which opinions differ widely, that recur in the debate on each approach. These are:

- the desirability of arms control as a process;
- the basic objectives of arms control;
- the approaches to arms control agreements;
- the status of the U.S.-Soviet strategic relationship;
- the interaction of other nuclear states with the U.S.-Soviet strategic relationship;
- the requirements for verification;
- the record of compliance with existing agreements;
- "linkage" of arms control to other political or military objectives;
- the approach to the negotiating process; and
- domestic acceptance of arms control agreements.

ARMS CONTROL AS A PROCESS

Most observers would agree that the dangerous confrontation between the United States and the Soviet Union arises from the deep political differences that divide the two countries and generate mutual distrust. How best to moderate or eliminate these political differences has been the subject of recurring domestic debate. Some favor efforts to reduce tensions by developing areas of mutual interest that would encourage long-term improvement in political relations despite continuing differences on major issues. Others favor increasing the pressure on the Soviet Union in an attempt to force a long-term change in Soviet political attitudes by confronting it on fundamental issues and by seeking to isolate it politically and economically.

In the absence of real progress in resolving their political differences, the United States and the Soviet Union have to date sought to prevent nuclear war principally by developing more and improved nuclear weapon systems. These systems have been designed to deter the other side from using its own arsenal of nuclear weapons or from starting a conventional war on a scale or in circumstances that could escalate into general nuclear war. Thus, nuclear deterrence, where each side simultaneously threatens to destroy the other, has ironically become the principal shield against the outbreak of nuclear war. A strong case can be made that nuclear deterrence has prevented the outbreak of war directly involving the superpowers since World War II. Nevertheless, many people have long seen the very process of building up larger and more sophisticated strategic forces as substantially increasing both the likelihood and the consequences of nuclear war.

After World War II, nuclear disarmament leading to the elimination of nuclear weapons was widely advocated as the only way to prevent nuclear war. This approach, which was extended to include essentially all armaments as part of a program of "general and complete disarmament," was not only the slogan of pacifists but also of U.S. and Soviet leaders, who proclaimed it as an attainable goal at a time when both sides were rapidly building up their nuclear forces. By the mid 1960s complete nuclear disarmament, though still advocated by some as a realizable near-term goal, had lost much of its credibility. There were many reasons for this disillusionment, including the military doctrines of the nuclear powers, the high level of nuclear arsenals, the deep political divisions that existed not only between the United States and the Soviet Union but among other countries in many regions of the world, and the military significance of even small numbers of secretly retained nuclear weapons in an otherwise disarmed world.

For the past 20 years, arms control, as distinct from disarmament, has provided an alternative to sole reliance on unilateral military preparedness as a barrier to nuclear war. Arms control defines the effort to "manage" the nuclear confrontation by mutual agreement in ways designed to lessen the likelihood of nuclear war. The arms control process seeks to constrain the size and nature of nuclear arms and their delivery systems to stabilize the strategic relationship.

Despite wide differences as to the objectives and approaches of arms control, the basic concept is now generally accepted and considered an integral part of U.S. security policy. Domestic support for the concept, however, is not universal. Some critics assert that any agreement will inevitably constrain the United States in the exploitation of its technological advantages to improve its nuclear capabilities, which are the real deterrent to war. Other critics oppose the entire concept of arms control, contending that the net effect of any agreement, however favorable the terms, would be to lull the United States into complacency, with necessary and permitted military efforts being neglected while the Soviet Union continues its military buildup. In various forms, this underlying opposition to arms control reemerges in arguments against specific proposals. At the other extreme, arms control has been criticized as undercutting the real objective of nuclear disarmament. By accepting and institutionalizing nuclear arms and by adopting a graduated approach to nuclear reductions, according to this perspective, arms control impedes efforts to reduce drastically or eliminate nuclear arms. Between these extremes exists a broad spectrum of opinion on the desirability and urgency of various forms of control on nuclear weapons.

THE OBJECTIVES OF ARMS CONTROL

The underlying objective of arms control is to increase the stability of the military relationship of the nuclear powers, thus reducing the risk of nuclear war. The objective of stability can be divided into two separate, and sometimes conflicting, concepts, "arms race stability" and "crisis stability." Arms race stability is achieved by stopping or moderating the competition in nuclear arms. This competition increases the risk of war by introducing more threatening weapons and by making more nuclear weapons available for expanded roles and missions. Crisis stability, on the other hand, is achieved by eliminating the incentive for either side to launch a preemptive counterforce attack in an effort to obtain military advantage by significantly blunting the other side's capacity to retaliate. The danger of such a counterforce attack would clearly be greatest at the time of a major political crisis or military confrontation, when escalation to nuclear war might be judged a real possibility.

Agreements that establish mutual constraints on the size and quality of nuclear arsenals or ban certain activities completely contribute to arms race stability. By limiting existing forces and establishing a clear framework within which future forces are constrained, such agreements make the future U.S.-Soviet military relationship more predictable. This reduces the pressure on both sides to pursue developments and deployments based on worst case assessments of the other side's unconstrained future capabilities. Advocates of the importance of arms race stability argue that the elimination of such worst case assessments substantially reduces international tensions and the risk of war.

In an international context, efforts to prevent the further proliferation of nuclear weapons also contribute to arms race stability. If additional states acquire nuclear weapons, it clearly puts pressure on their adversaries to obtain their own nuclear weapon capability. The resulting nuclear arms competition in unstable areas of the world would dangerously increase the unpredictability of not only regional political-military relations but also of U.S.-Soviet relations as well.

Crisis stability, or the reduction of the risk of nuclear war in a crisis, can be increased by measures that assure the survival and effectiveness of retaliatory strategic forces in the face of a preemptive counterforce attack. Both the deployment of more survivable retaliatory systems and the elimination of highly vulnerable strategic systems that are tempting targets contribute to crisis stability. This objective can also be supported by constraining strategic offensive forces that threaten the survivability of retaliatory forces and by constraining strategic defensive forces that threaten to prevent retaliatory forces from reaching targets. A high level of crisis stability does not eliminate the possibility of military engagements escalating into nuclear war, but it does reduce pressure to preempt if nuclear war appears imminent by reducing the perceived need to use vulnerable weapons before they are destroyed.

U.S. military policy has long sought to improve crisis stability by unilaterally maintaining diversified, survivable strategic forces capable of penetrating Soviet defenses. The triad of land-based missiles, sea-based missiles, and bombers has been designed to assure retaliation after a Soviet preemptive counterforce attack that might severely degrade one or even two legs of the triad. Although much more dependent on potentially vulnerable land-based strategic missiles, the Soviet Union has also moved in the past 20 years to develop more survivable strategic forces by improving the sea-based component of its forces and by hardening the launchers of its land-based missiles.

Despite unilateral efforts to reduce the vulnerability of strategic forces, current technical developments could in principle increase the future vulnerability of both sides' strategic forces and decrease their ability to penetrate defenses to reach targets. These developments decrease crisis stability. Agreements that increase the survivability of retaliatory strategic forces, either by encouraging deployment of less vulnerable systems or by constraining the threat to these forces, would contribute to crisis stability above and beyond the unilateral measures that the United States and the Soviet Union might adopt.

Arms control should not be assessed exclusively on its immediate contributions to arms race and crisis stability. There are clearly secondary political and social objectives that some would argue may prove equally important in the long run. The process of negotiating mutually acceptable agreements dealing with matters that affect both sides' very survival, and then living within this self-imposed regime, can build understanding and confidence between the superpowers. The resulting atmosphere of constructive cooperation in reducing the risk of war can also significantly reduce international political tensions. By increasing the predictability in the military relationship, arms control not only reduces the pressure to plan on a worst case basis but reduces political concerns about future intentions. Thus, a successful arms control regime could be a major factor in gradually reducing the political tension and distrust that have intensified the arms race between the United States and the Soviet Union.

Another benefit often associated with arms control is that it would reduce military spending, freeing resources for the civilian economy. However, since less than 20 percent of the military budget is allocated to strategic nuclear forces, even far-reaching nuclear arms control agreements would affect only a small part of the military budget. There would also be strong pressures to invest savings from nuclear arms control in the modernization of conventional arms. In the longer term, however, if a successful arms control regime significantly reduced political tensions, the United States, the Soviet Union, and the world at large could profit enormously from the transfer of even a small fraction of military expenditures, now approaching $1 trillion annually, to constructive nonmilitary purposes.

APPROACHES TO ARMS CONTROL

Arms control agreements have sought to enhance arms race and crisis stability by one or more of the following general approaches: limits, freezes, restructuring, reductions, bans, and special stabilizing measures. Each of these ways to constrain nuclear weapons and their delivery systems is discussed below.

Limits

Limits on various types of weapons or other measures of nuclear military power are one approach to arms control agreements. By establishing limits at current or reduced levels of some agreed measure of military power (such as number of launchers or deployed missiles or total throw-weight), a quantitative ceiling can be placed on the arms race. Crisis stability can also be improved if modernization permitted under the ceilings results in more survivable systems. Limits that favor survivable systems and constrain threatening first-strike counterforce systems also favor crisis stability.

The SALT II Treaty sought to contribute to both arms race and crisis stability by establishing equal aggregate limits for both sides on their total number of strategic missile launchers and heavy bombers. A series of separate equal sublimits further constrained these systems. The sublimits together with certain qualitative limits constrained the number of multiple independently targetable reentry vehicles (MIRVs) and air-launched cruise missiles. The treaty also allowed the introduction of only one new type of intercontinental ballistic missile (ICBM), which could be deployed in a survivable mode, such as the proposed mobile MX with multiple protective shelters.

The U.S. proposal at the START negotiations seeks to contribute to arms race and crisis stability by establishing separate equal ceilings and subceilings on deployed strategic missiles, missile warheads, missile throw-weight, and strategic bombers. The U.S. proposal would cap the quantitative strategic arms race at substantially lower levels. It also seeks to encourage more survivable deployments and to reduce the counterforce threat of the large accurate force of Soviet ICBMs. The Soviet position at the START negotiations would essentially deepen the SALT II constraints. Both the SALT II and START approaches would allow continued testing, production, and deployment of new nuclear weapon systems within clearly defined limits.

The SALT I ABM Treaty is designed to prevent the deployment of nationwide ballistic missile defense systems. It limits the United States and the Soviet Union to a single ABM site apiece with no more than 100 fixed launchers and 100 interceptors. It also establishes specific limits on the radars associated with the system. Some system modernization is permitted within these quantitative limits, but specific provisions constrain the nature of the modernization so that it cannot be used to defeat the basic purpose of the treaty. For example, the two sides are not permitted to develop, test, or deploy launchers capable of launching more than one interceptor at a time or capable of rapid reload. Similarly, the treaty prohibits the development, testing, or deployment of ABM systems or components that are sea-based, air-based, space-based, or mobile land-based. By preventing the deployment of nationwide ballistic missile defense systems, it is argued that the ABM Treaty contributes substantially to both arms race stability and crisis stability because such a system, whatever its capabilities, would be perceived as an attempt to develop a shield to negate the deterrent effect of strategic retaliatory forces.

Freezes

A freeze would stop all new activity in the area covered. This is quite distinct from limits, such as those in SALT II and START, that would permit modernization within the agreed limitations unless specifically prohibited. A freeze would permit the continued use of existing inventories of the system in question. A comprehensive nuclear freeze would prohibit the further testing, production, and deployment of all nuclear weapons and delivery systems.

A comprehensive nuclear freeze would clearly meet the objective of arms race stability. There would be no further growth of the threat on either side, and the action/reaction cycle of the arms race would be broken. At the same time, a comprehensive freeze would have a mixed impact on crisis stability, because it would freeze the status quo with its problems as well as its advantages. On the one hand, such a freeze would prevent either side from introducing new, more destabilizing systems. On the other hand, it would prevent both sides from restructuring their strategic forces with new (or old) systems that would be less vulnerable or threatening.

A number of arms control proposals have used the freeze approach in a limited manner. For example, SALT II froze Soviet heavy missiles (no further deployments of any type and no testing of new types) and Soviet mobile SS-16 missiles (no further testing, production, or deployment). The Comprehensive Test Ban, on which final agreement was never reached, sought to freeze the development, production, and deployment of new types of nuclear weapons by prohibiting all nuclear testing, since testing was judged necessary to develop advanced new weapons.

Restructuring

Restructuring U.S. and Soviet nuclear forces could substantially improve crisis stability. Restructuring could eliminate threatening systems with a counterforce capability, such as highly MIRVed high-accuracy missiles, and vulnerable high-value systems, particularly fixed launchers for highly MIRVed missiles, which would be logical targets for a preemptive counterforce attack. In some cases, such as the Soviet SS-18s and SS-19s and the U.S. MX, the same system can be destabilizing in both respects.

Agreements can permit or even encourage desirable restructuring without actually requiring it. For instance, SALT II allows fixed, land-based missile launchers to be replaced within the agreed limits by sea-based missiles or bombers, but it does not permit the construction of additional new fixed land-based missile launchers. The U.S. START proposal goes further and essentially requires the Soviet Union to give up a significant fraction of its deployed land-based missiles. In seeking crisis stability, restructuring proposals can be inconsistent with arms race stability if they encourage or require major new nuclear arms programs to replace existing nuclear systems, such as the replacement of fixed land-based ICBMs with larger numbers of small, mobile, single-warhead missiles.

Reductions

Limits and restructuring can be at current or reduced levels, and a freeze can also lead to reductions. Even if modernization is permitted or encouraged, substantial reductions in measures of overall military nuclear power tend to enhance arms race stability. Selective reductions of threatening systems can obviously improve crisis stability, but strictly proportional reductions in all systems, threatening and retaliatory alike, would have little effect on this objective of arms control.

Substantial reductions could also reduce the risk of war by limiting the nuclear options available to both sides. This would make it less likely that nuclear weapons would be introduced in an escalating military situation. Substantial reductions could also reduce the possibility of accidental nuclear war, to the extent that this possibility depends statistically on the number of nuclear weapons deployed. At the same time, very large reductions could reduce crisis stability by making a preemptive attack more credible. Such an attack might be seen as reducing the opposing retaliatory force to a point where defenses could limit damage from retaliation to acceptable levels. With extremely large reductions, there would also be the problem that relatively small numbers of delivery systems that might be unaccounted for by the verification system could be judged a significant factor in assessing crisis stability.

Bans

The complete prohibition of an entire class of nuclear weapon systems, including the elimination of existing stockpiles, and the halt of future development, production, and deployment could contribute dramatically to both arms race and crisis stability if the activities banned constitute a significant present or future threat. To date, complete bans have generally been proposed for systems that do not yet exist. This avoids the difficult verification problems of dealing with undeclared stockpiles and standby production facilities. Such bans may still be very important if they close off a dangerous path of development before either side has a vested interest in it. The Outer Space Treaty, for example, banned the deployment of "weapons of mass destruction" in outer space at a time when there were no programs or plans to pursue such activities. As noted above, the SALT I ABM Treaty banned the development, testing, and deployment of ABM systems or components that are sea-based, air-based, space-based, or mobile land-based at a time when neither side had any known plans for such developments.

The proposed ban on anti-satellite weapons illustrates the problem of dealing with complete bans once systems have been tested and deployed. The Soviet Union has had a limited anti-satellite system in operation for a number of years, which raises the question of how to eliminate possible clandestine stocks that could be redeployed on short notice. In this and other cases, the significance of such a marginal capability must be balanced against the value of an overall ban on a dangerously destabilizing activity. A comprehensive test ban on all nuclear explosions, as noted above, amounts to a freeze on future nuclear weapon developments rather than a ban on nuclear weapons. It would stop the development and deployment of new types of nuclear weapons, but it would not affect existing stockpiles or the ability to produce and deploy additional weapons of existing designs.

Special Stabilizing Measures

A large range of measures that are not related to the size and nature of U.S. and Soviet nuclear forces could increase crisis stability and reduce the risk of nuclear war by accident or miscalculation. One class of actions would seek to reduce the risk of nuclear war, particularly by accident or miscalculation, by helping to assure that both sides are operating with a correct understanding of the threatening situation and the other side's intentions. Improved communication between governments might facilitate this objective, as might direct communication through permanent groups established to exchange information and resolve problems on a continuing basis.

Another class of stabilizing measures would seek a partial disengagement of nuclear forces so that unauthorized or precipitous use of nuclear weapons in a limited, conventional military engagement would be less likely. For example, the two sides might agree to keep their ballistic missile submarines at distances from the other side beyond the range of the submarines' missiles, a concept whose significance obviously depends on the range of the missiles. Another example would be the withdrawal of forward-based tactical nuclear weapons some distance behind the line dividing NATO and the Warsaw Pact, thus reducing the pressure to use them in the earliest phase of a conventional battle if it appeared that they would be overrun by an enemy advance. In both instances the question of the effect of such procedures on the credibility of the deterrent has to be faced.

Another class of stabilizing measures would seek to reduce the vulnerability of retaliatory systems and to build more time into the decision-making process. For example, limits on how close submarines can approach the coasts of the other side could prevent attacks with drastically reduced warning time on critical command and control facilities or strategic air bases. Another example would be to establish agreed sanctuary zones for ballistic missile submarines where neither side would engage in threatening antisubmarine warfare activities such as trailing the adversaries' submarines. These and other special measures could stand alone or be incorporated in any of the general approaches outlined above.

THE U.S.-SOVIET STRATEGIC RELATIONSHIP

Any arms control proposal must be assessed in terms of its impact on the present and future U.S.-Soviet military relationship. Thus, the status of the present and projected military forces of the United States, the Soviet Union, NATO, and the Warsaw Pact has been a critical factor in the arms control debate over the past 25 years. One can question whether "superiority" has any real military significance when the United States and the Soviet Union each have some 10,000 strategic warheads. Nevertheless, when either side has had, or has been perceived to have, a militarily or politically exploitable level of superiority, progress in achieving nuclear arms control agreements has proven difficult. The role of perceptions of military capabilities must not be underestimated; perceptions can play a much greater political role than the real military significance of apparently unbalanced forces. If either side seeks an agreement that establishes or permits a position of generally perceived superiority, there is little chance that it will be successfully negotiated and brought into force.

The problem of assessing the strategic balance between the United States and the Soviet Union is greatly complicated by the major asymmetries in the strategic nuclear forces and other military forces of the two countries. These asymmetries reflect many underlying differences between the two countries. These differences include such factors as attitudes acquired over centuries of radically different historical experience, military doctrine, political ideology, the economic base, relations with allies, and potential adversaries. In general, the asymmetries in strategic forces can be associated with a variety of geographic, military, technical, and bureaucratic differences between the two countries.

Geographically, the United States and the Soviet Union have radically different access to the sea and proximity to potential adversaries. Militarily, the United States has a much stronger tradition of naval and air power. The Soviet Union has a much stronger historical emphasis on massive land armies and defense of the motherland. In these circumstances, it is not surprising that the United States moved early to a triad of air, sea, and land forces while the Soviet Union emphasized land-based missiles and air defenses.

Technically, the United States, with its substantial advantages in technology, moved sooner to smaller, more sophisticated systems while the Soviet Union emphasized larger systems of less sophisticated hardware. For example, the United States at an early stage developed small solid-propellant missiles. This reflected the early availability in the United States of light thermonuclear warheads, miniaturized electronics, improved reentry technology, and advanced solid-fuel technology. The Soviet Union initially emphasized large liquid-fuel missiles that did not depend on these developments but could deliver large payloads.

Bureaucratic politics, reflecting the difference between an open and a closed society, have undoubtedly also played a major role in the development of the two countries' forces. Both countries experience strong institutional pressures to extend and expand ongoing programs. But in the United States there has been a relatively open and intense debate about the structure and procurement of strategic forces and their budgetary implications. The top Soviet military leadership has held these decisions very closely, and vested interests within the Soviet Union, such as major design bureaus and the military services, have been able to maintain the momentum of the Soviet buildup with less evident interference from other competitors for scarce resources.

With all of these basic asymmetries, it is not surprising that the structure and capabilities of U.S. and Soviet nuclear forces differ, notwithstanding a general pattern of action and reaction between the two sides and persistent Soviet efforts to match U.S. technological accomplishments. It is also not surprising that within the United States there has been a continuing controversy about the status of the strategic balance. The inherent complexity of the problem, particularly when coupled with the strong political emotions surrounding it in both countries, has been a major factor in the difficulty in negotiating mutually acceptable arms control agreements.

A brief historical review recalls the changing nature and perceptions of this strategic relationship and the reaction to it on both sides. In the late 1940s the United States had an absolute monopoly of nuclear forces. Throughout the 1950s it had a sufficient advantage in the number and quality of its strategic forces to be judged as having strategic superiority. At that time the United States declared a policy of massive retaliation to deter a perceived Soviet conventional military threat against an unprepared Western Europe. Yet even in this period of apparently overwhelming U.S. advantage, some sophisticated U.S. analysts believed as early as the mid-1950s that serious crisis instability was developing because of the perceived extreme vulnerability of U.S. strategic forces to surprise attacks by Soviet forces.

In the 1960s the United States developed a triad of air-, sea-, and land-based strategic forces. The objective was to assure that sufficient forces would survive in all circumstances to deter any Soviet preemptive attack. The Soviet Union, particularly after the Cuban missile crisis, moved to develop a comparable, survivable strategic force. It placed much greater relative emphasis on land-based missiles, which with the missile accuracies then available were thought to be survivable in hardened silos. Despite its continued quantitative and qualitative advantage during this period, the United States decided that meaningful superiority could not be maintained against a determined Soviet adversary. It concluded that the real measure of strategic forces was the ability to achieve "assured destruction" of the enemy by inflicting unacceptable damage on the other side after absorbing the worst credible preemptive attack. The Soviet Union had a similar capability that the United States could not defend against, and it was apparent the Soviet Union would not give up this capability. Thus emerged the recognition in this country, and presumably in the Soviet Union as well, that the underlying strategic reality was one of mutual assured destruction, which some soon identified and acclaimed as MAD.

During this period, consideration was given to supplementing the U.S. strategic posture with a major defense component, initially a nationwide civil defense program and then a ballistic missile defense system. On the basis of this experience, a widely based technical consensus emerged that it would not be possible for such systems to overcome the Soviet ability to inflict assured destruction on U.S. society. Nevertheless, the United States began developing multiple independently targetable reentry vehicles, in large part to assure that U.S. missile forces could overwhelm any ABM system that the Soviet Union might deploy.

In the 1970s the concept of assured destruction was formalized by the SALT I ABM Treaty, which prohibited efforts to achieve a nationwide ballistic missile defense system. At the same time, the concept of deterrence was extended to encompass other nuclear options. Reacting to NATO concerns about the credibility of the U.S. nuclear deterrent for a conventional Soviet attack on Europe, the Nixon and Ford administrations emphasized the importance of counterforce capabilities that would allow more limited nuclear options. It had always been technically possible to launch only a fraction of the total U.S. strategic force. This approach was reaffirmed in the latter part of the Carter Administration in its so-called countervailing strategy. This strategy emphasized the importance of flexible options and survivable command and control to assure deterrence against a wide range of threats. In a sense, this definition of extended deterrence simply made public policies and plans that had already existed for a long time.

The approach to strategy and arms control during the 1970s was also characterized in both the United States and the Soviet Union by acceptance of the fact that a rough "parity" existed between the two sides when all of the asymmetries in their strategic forces were taken into account, and that meaningful superiority was not attainable. After calling for superiority in his campaign, President Nixon called for a policy of "sufficiency." This essentially meant maintaining the level of forces necessary to assure deterrence, with no advantage accruing to the Soviet Union if it undertook a preemptive first strike, and maintaining "essential equivalence" in perceived forces so there would be no appearance of inferiority. The Nixon, Ford, and Carter administrations based their approaches to strategic procurement and arms control on this approach to the strategic balance. The SALT I Interim Agreement sought to cap the massive buildup in Soviet strategic forces that had begun in the mid-1960s. The SALT II Treaty formalized the status of parity between the two sides by establishing equal aggregate ceilings on strategic delivery systems and a series of equal subceilings on various components of the strategic forces.

In spite of these major arms control agreements, the 1970s saw a major buildup in potentially destabilizing U.S. and Soviet strategic forces. In the first half of the decade the United States deployed MIRVed warheads on a large portion of its land-based and sea-based strategic missiles; in the second half of the decade the United States developed long-range high-accuracy cruise missiles that essentially permitted the "MIRVing" of the strategic bomber force with air-launched cruise missiles. This upgrading of the existing strategic bomber force assured that its weapons could penetrate the extensive Soviet air defenses.

For its part, the Soviet Union in the first half of the 1970s completed the large-scale strategic buildup started in the mid-1960s. In the second half of the decade it moved rapidly to introduce MIRVs on its land- and sea-based strategic missiles. The impressively high accuracy of the first generation of Soviet land-based MIRVs, which almost matched that of comparable U.S. missiles, implied a high kill probability against U.S. fixed missile launchers. These developments concentrated an even larger fraction of the effective Soviet strategic power on land-based missiles, which were particularly destabilizing because of the threat they posed to U.S. fixed targets and because of their potential vulnerability to U.S. attack. The question remains how much larger and more threatening these forces might have grown without the restraints of SALT I and SALT II. Certainly, both sides were capable of building considerably larger and more threatening forces during this period.

In the 1980s the Reagan Administration has taken the position that the Soviet Union has achieved an unacceptable "margin of superiority." This advantage must be met by a major buildup of U.S. strategic forces, according to the administration, and must be taken into account in any future strategic arms agreements. Initially the administration emphasized a "window of vulnerability" caused by the threat to U.S. land-based missiles posed by Soviet land-based missiles. The U.S. position in the START negotiations has been based on these assertions of Soviet strategic superiority and U.S. vulnerability caused by Soviet land-based missiles. The Soviet Union has insisted that a strategic parity continues to exist between the two sides. Even as it began to build up its strategic forces, the United States further broadened the concept of extended deterrence to emphasize the capability to conduct nuclear war fighting missions, including the capability to wage protracted general nuclear war.

On March 23, 1983, President Reagan foreshadowed a fundamental change in the strategic doctrine that had prevailed in the previous two decades. He deplored the concept of deterrence based on assured destruction and called for a major scientific effort to develop an effective nationwide ballistic missile defense that could ultimately eliminate the strategic role of nuclear weapons. This approach, which has been widely challenged on technical, military, and arms control grounds, would reorient the longstanding U.S. "offense-dominated" nuclear strategy to a "defense-dominated" strategy. It remains to be seen how this radical new approach will in fact affect the United States' strategic doctrine and posture and the U.S. position on past and future arms control commitments and proposals.

OTHER NUCLEAR POWERS

The already complex problem of strategic asymmetries between the United States and the Soviet Union is further complicated by the existence and different perceptions of British, French, and Chinese nuclear forces. From the U.S. perspective these forces have consistently been considered separate from the U.S.-Soviet strategic balance. The United States argues that British and French forces were designed to provide an independent minimum deterrent against Soviet nuclear attack on those countries, that French nuclear forces have no commitment to NATO, and that the People's Republic of China, with no military association with the United States, has no role in the U.S.-Soviet strategic balance.

From its perspective the Soviet Union must deal with the prospect of facing the British, French, and Chinese nuclear forces in a general nuclear war with the United States, and must size its own strategic forces against the combined threat. At present the strategic forces of the other nuclear powers are very small compared with the vast U.S. and Soviet strategic arsenals. In a general war these forces would be incrementally insignificant. Alone, however, they would be capable of inflicting tremendous damage against urban targets. The United Kingdom now has four ballistic missile submarines with 64 missiles. France has five ballistic missile submarines with 80 missiles as well as a small number of land-based ballistic missiles capable of reaching the Soviet Union. It is estimated that the Chinese may have a few hundred nuclear weapons capable of striking Soviet targets. By the mid-1990s, however, the situation could be very different. If the United Kingdom goes ahead with plans to equip its submarines with Trident II missiles and France completes plans to MIRV its missile force, their combined strategic nuclear forces could approach 2,000 warheads.

British and French nuclear forces were a central issue in the SALT I negotiations. They were formally excluded from the agreement, but they appear to be compensated for by the higher levels of ballistic missile submarines and submarine-launched ballistic missiles permitted the Soviet Union. In SALT II the problem was resolved at Vladivostok when President Leonid Brezhnev dropped the Soviet position that forward-based systems threatening the Soviet Union, including British and French forces, must be included in the agreement in return for President Ford's dropping the U.S. demand that the number of Soviet heavy missiles be substantially reduced. In the INF negotiations this issue has become the fundamental point of difference between the two sides. The Soviet Union has insisted that, even if the United States reduced its deployment of Pershing IIs and ground-launched cruise missiles to zero, the Soviet Union would not reduce its SS-20 warheads within range of Europe below the level of the combined British and French strategic nuclear forces.

VERIFICATION

Given the status of U.S.-Soviet relations, an arms control agreement that affects vital U.S. security interests will not be acceptable unless it is possible to determine with some level of confidence that the Soviet Union is in fact honoring it. Since the systems to monitor compliance can never guarantee absolute verification, the question becomes, How much verification does a particular agreement require? The "rule of reason" standard that legislation governing the arms control process has established is that an agreement must be subject to "adequate" verification to be acceptable. While this standard is vague, it underscores both the requirement for verification and the fact that verification cannot, and need not, be absolute. In the case of the SALT agreements, the government interpreted adequate verification to be the ability to determine with high confidence compliance with treaty provisions to the extent necessary to safeguard national security and to detect significant violations in time to permit an appropriate response. In meeting this standard, one would also expect to detect with varying levels of confidence a much broader spectrum of less significant violations. The standard of adequate verification lends itself to widely differing interpretations of the verification required for a particular agreement.

The Reagan Administration has taken the position that SALT II did not meet its standard of verification and has implied that much more severe verification standards would be applied to its START proposals. It has been suggested that the U.S. START proposals would call for extensive use of cooperative measures and on-site inspection to improve verification capabilities. How fundamental a change may be involved and its actual effect on verification can only be judged by examining the specific START verification proposals, which have not yet been revealed by the administration.

Verification has always been associated with disarmament and arms control proposals. For a long time it seemed to provide an almost impenetrable barrier to progress in the field. President Eisenhower's proposal in the mid-1950s for open skies by means of aircraft reconnaissance and other highly intrusive inspection proposals had little prospect at that time of being accepted by a closed Soviet society and would probably not have been well received by many in the United States. In the 1960s reconnaissance satellites created a technological revolution in the possibilities of verification without highly intrusive measures. The SALT agreements established satellite monitoring as internationally accepted means of verification. Satellite monitoring systems—which together with any technical systems, such as radars and radio antennas, located outside the country under surveillance are now designated as National Technical Means (NTM)—have become the key to the verification of arms control proposals.

There are, however, proposals that clearly cannot be adequately verified by National Technical Means alone. In some cases the capabilities of National Technical Means can be extended by agreements providing for cooperative measures that facilitate the monitoring process. In certain cases, intrusive on-site inspection may still be the only technique available to provide adequate confidence, or any information at all, on certain activities. One example is the safeguarding of operating nuclear power facilities against diversions of materials to nuclear weapon production. On-site inspection is not a panacea for all verification problems. In some cases, such as the verification of the MIRV ceilings in SALT II, even very intrusive on-site inspection by itself would have been much less effective than National Technical Means used in conjunction with the counting rules for launchers. In this and other extremely intrusive cases, it is by no means clear that on-site inspection would have been acceptable to the United States in the very unlikely event that the Soviet Union agreed to it.

RECORD OF COMPLIANCE

Closely coupled with the verification issue is the record of compliance with existing agreements and the question of what can and should be done to enforce agreements. In presenting SALT II to the Senate, the Carter Administration contended that the record of compliance with the SALT I agreements had been good and presented extensive information to document its case. While a number of questionable Soviet activities had been detected and presented to the joint Standing Consultative Commission, which SALT I had created for this purpose, the U.S. government concluded that every case had been satisfactorily resolved. Some critics of SALT II challenged these conclusions.

In response to a Senate request, President Reagan submitted a report to Congress on January 23, 1984, stating that "The United States Government has determined that the Soviet Union is violating the Geneva Protocol on Chemical Weapons, the Biological Weapons Convention, the Helsinki Final Act, and two provisions of SALT II: telemetry encryption and a rule concerning ICBM modernization. In addition, we have determined that the Soviet Union has almost certainly violated the ABM Treaty, probably violated the SALT II limit on new types, probably violated the SS-16 deployment prohibition of SALT II, and is likely to have violated the nuclear testing yield limit of the Threshold Test Ban Treaty." The Soviet Union denied all of the charges and made its own charges of U.S. violations of existing agreements. Domestic critics have questioned some of the U.S. charges and have noted that the charges were considerably weakened by the assertion of the U.S. government that SALT II established only "political" and not "legal" obligations since the United States did not intend to ratify the treaty.

Whatever the correct explanation of these activities, the President's report to the Senate that the Soviet Union has violated formal agreements will make it much more difficult to cite past experience as evidence that the Soviet Union can be expected to comply with the provisions of future agreements. The U.S. charges also raise the question of how to deal with apparent violations or possible violations. In the past these problems were dealt with privately, with both sides seeking an explanation or mutually acceptable solution to preserve the integrity of the agreement. The question now is, Will a policy drawing public attention to the charges put pressure on the Soviet Union to comply with agreements, or will it weaken domestic confidence in arms control and Soviet willingness to participate constructively in a continuing discussion of compliance problems?

POLITICAL OR MILITARY "LINKAGE"

A fundamental issue in the development and negotiation of arms control agreements is the extent to which they should be "linked" to other political or military considerations. Perhaps the issue can be more realistically stated as the extent to which arms control negotiations can be isolated from other political and military activities. In the past the United States has sought in principle to avoid linkage in arms control negotiations with the Soviet Union. The argument has been that arms control agreements should stand on their own merits, even in times of heightened or reduced tensions. They should not be used as rewards for good behavior or withheld for bad behavior. In practice, some degree of linkage is probably inevitable in the United States, given the high political visability of arms control negotiations.

A striking example of the delinkage or isolation of the arms control process from potentially disruptive events occurred when the United States mined Haiphong Harbor and bombed Hanoi the day after Prime Minister Alexei Kosygin's visit to that city at a critical juncture in the SALT I negotiations. To the surprise of the U.S. delegation, the Soviet delegation did not walk out of the negotiations or even protest the action. In contrast, after the Soviet invasion of Afghanistan at the end of

1979, President Carter asked the Senate to discontinue the SALT II ratification process. This action reflected both the President's desire to link ratification of the treaty to acceptable Soviet behavior in an unrelated area and his recognition of the linkage of the upcoming Senate vote on ratification to the adverse public reaction to Soviet behavior in Afghanistan.

The recently terminated INF negotiation is the most obvious example of direct linkage of the arms control process to external objectives. Whatever the negotiating intentions of the two sides, it became clear as the negotiation was conducted increasingly in the open that both sides' immediate interest was the political impact of the negotiation on the impending U.S. deployment of Pershing IIs and ground-launched cruise missiles in NATO.

THE NEGOTIATING PROCESS

It is not particularly difficult to design nuclear arms control proposals that a broad political spectrum of American citizens would judge to be unambiguously advantageous to U.S. security interests. It is a much more difficult task to design and negotiate arms control agreements that the political and military establishments of both the United States and the Soviet Union see as mutually advantageous. A powerful sovereign state will clearly not be persuaded or coerced to accept an agreement that it judges to be contrary to its overall security interests. Although underlying motives and concerns may differ, there must be a significant area of common interest to make the negotiation of an arms control agreement possible.

A central issue in arms control negotiations is the extent to which the basic proposals should be "negotiable." A negotiation is a bargaining process between or among potential adversaries, and opening positions are seldom final positions. While a negotiable proposal may be formulated from the advocate's perspective, it is directed at perceived mutual objectives within a framework that can lead to a formulation acceptable to both parties. This can still be a very slow and difficult process. The SALT II negotiations extended over seven years, even though the basic objectives and framework were agreed on relatively early in the process.

An alternative approach that was openly favored at the outset of the Reagan Administration was to design proposals to optimize perceived legitimate U.S. interests without regard to their negotiability from the Soviet perspective. The administration argued that if the U.S. positions were right the Soviet Union might be persuaded; if the Soviet Union could not be persuaded, the United States would be in the strongest bargaining position in dealing with Soviet counterproposals. The concept of negotiability was rejected as a criterion in judging the acceptability of arms control proposals.

Whatever form proposals take, there are obvious advantages in coming to the negotiating table in a strong position. In this case, strength is measured not just in military forces but also in domestic and international political support. At the same time, experience indicates that the prospects for agreement are poor if either side has, or is perceived to have, real superiority in the military area under negotiation. The Reagan Administration initially emphasized that the United States would have to undertake a major arms buildup to be able to negotiate with the Soviet Union from a position of strength. To what extent the initiation, as opposed to the actual deployment, of new programs is judged by the administration to have met this criterion is not clear. The notion that one must "arm to parley" is not new, and critics would characterize it as simply the next phase of the arms race presented in a more palatable form.

Another controversial issue in the approach to the negotiating process is the value of "bargaining chips." Military assets that can be traded as bargaining chips against present or future components of an adversary's forces can in theory play a useful role in a negotiation. To be really effective negotiating tools, such bargaining chips must have real military significance. Militarily significant systems are prone to take on a life of their own, however, and to be judged so valuable as to be nonnegotiable. In SALT I, some may have considered the new MIRV technology as a powerful bargaining chip, but it soon became clear that it was not negotiable by either side. In SALT II and START, the MX program was frequently supported as a critical bargaining chip, but it has never really been used for this purpose. On the other hand, it has been argued that U.S. plans in the late 1960s and early 1970s to deploy ballistic missile defense systems contributed to the successful negotiation of the ABM Treaty in 1972.

At the other extreme, there is the question of whether unilateral restraint can contribute to the successful negotiation of an agreement. At the beginning of the trilateral negotiation for a comprehensive nuclear test ban in 1958, President Eisenhower's declaration of a one-year moratorium on nuclear testing undoubtedly lent an air of seriousness and urgency to the first real effort at negotiated arms control. Critics would note, however, that Eisenhower unilaterally terminated the moratorium at the end of 1959 and that a comprehensive test ban was not achieved. Currently, the Soviet Union has declared a moratorium on the testing and deployment of anti-satellite weapon systems. Congress has also passed legislation, over administration objections, mandating a moratorium on the full testing of anti-satellite systems while efforts are pursued to reopen the U.S.-Soviet negotiations on such systems. The question remains whether such actions accelerate the negotiating process and point it in a favorable direction or weaken negotiating leverage to obtain the best agreement.

DOMESTIC POLITICAL ACCEPTABILITY

The extent of public and political support for arms control agreements is a serious issue underlying the arms control process. Public support for arms control has been closely linked with the varying fortunes of the U.S.-Soviet political relationship. These changing public attitudes are coupled with the unique requirement of the U.S. Constitution for a favorable vote of two thirds of the Senate to "advise and consent" on treaty ratification.

The remarkable changes in public and congressional attitudes are illustrated by the different reactions to SALT I and SALT II. In 1972 the U.S. Senate approved both the SALT I ABM Treaty and the SALT I Interim Agreement by separate votes of 88 to 2. In 1979, before the Soviet invasion of Afghanistan, it was judged a close call whether SALT II would receive the necessary two-thirds vote in the Senate. After the Afghanistan invasion, all observers agreed that it would be impossible to obtain Senate approval of the treaty at that time. This was the political reality even though SALT II did not raise basic doctrinal issues as did the ABM Treaty and appeared to answer many of the criticisms of the earlier SALT I Interim Agreement.

Despite the intense public interest in arms control today, a national consensus on this issue clearly does not exist. For its part, the Congress has increasingly shown its independence on issues of national security and foreign policy, which have previously been largely delegated to the executive branch. Consequently, arms control proposals must be judged not only on their negotiability with the Soviet Union or other nations but also on their acceptability within the complex political process of the United States.

SPECIFIC PROPOSALS

This chapter has outlined the broad range of general issues underlying the nuclear arms debate today. The debate itself can only be understood and assessed in terms of specific proposals. The following chapters present the background and issues relating to each of the principal current agreements and proposals directed at nuclear arms control. In each case the issues are addressed from the point of view of the various protagonists in the debate. It is hoped that this approach will help readers arrive at their own conclusions on these matters of such critical importance to the United States and the world at large.

2 Strategic Offensive Nuclear Arms Control

This chapter discusses the current arms control agreements and proposals directed specifically at the control and limitation of strategic offensive nuclear weapon systems. These are the SALT I Interim Agreement on Strategic Offensive Arms, the SALT II Treaty, and the current START negotiations. These agreements and negotiations have sought to limit the central strategic systems, usually defined as land-based intercontinental ballistic missiles, submarine-launched ballistic missiles, and long-range heavy bombers with their armaments. Other arms control agreements and proposals—including restrictions on defensive systems, proposals for a freeze on all nuclear systems, and the proposal to limit intermediate nuclear forces—also relate directly or indirectly to the objective of controlling strategic offensive nuclear arms. These related but separate issues are addressed in subsequent chapters.

PART I

THE STRATEGIC ARMS LIMITATION TALKS (SALT)

INTRODUCTION

The Strategic Arms Limitation Talks between the United States and the Soviet Union began in November 1969 under the Nixon Administration. These negotiations were directed at limiting the major buildup in strategic offensive systems and the emerging competition in ballistic missile defensive systems. On May 26, 1972, Presidents Nixon and

Brezhnev signed a five-year Interim Agreement that took the first step toward limiting strategic offensive arms by placing ceilings on land-based and submarine-based offensive nuclear forces. At the same time, they signed the SALT I ABM Treaty, a treaty of unlimited duration that drastically limited the future deployment of anti-ballistic missile (ABM) systems. A few months later the Senate and the House of Representatives approved the Interim Agreement and the Senate advised ratification of the ABM Treaty by overwhelming majorities.

In November 1972 the United States and the Soviet Union began the second phase of the Strategic Arms Limitation Talks (SALT II). The goal was a comprehensive treaty limiting strategic offensive nuclear systems of the two sides. The negotiations, which lasted for almost seven years under Presidents Nixon, Ford, and Carter, produced the SALT II Treaty, which was signed on June 18, 1979, in Vienna by Presidents Carter and Brezhnev. The treaty provided for equal quantitative and qualitative limits on central strategic systems, including intercontinental ballistic missile (ICBM) launchers, submarine-launched ballistic missile (SLBM) launchers, and strategic long-range bombers together with their armaments. It also began a process of reductions.

The ratification process for the SALT II Treaty was suspended after the Soviet invasion of Afghanistan at the end of 1979. No further action toward ratification has been taken on the treaty, which was to expire at the end of 1985. The Reagan Administration has stated that, since the United States has no intention of ratifying it, the SALT II Treaty has no legal status, but that the United States would not undercut the treaty prior to its expiration at the end of 1985 as long as the Soviet Union acted likewise. The Soviet Union has stated that it is acting in compliance with the treaty.

BACKGROUND

The Origins

As discussed in Chapter 1, a number of major developments during the 1960s in the technology, doctrine, and perceptions of strategic armaments set the stage for the initiation and successful pursuit of strategic arms control negotiations between the United States and the Soviet Union. The United States completed the deployment of a powerful triad of air-, sea-, and land-based strategic forces designed to be capable of surviving any attack and successfully delivering an assured devastating retaliatory strike. The Soviet Union, particularly after the Cuban missile crisis, undertook a massive buildup of its strategic forces, with particular emphasis on land-based missiles. By the mid-1960s a consensus was growing within the United States that meaningful nuclear superiority was no longer possible. Rather, U.S. security was seen to depend on the development of a stable U.S.-Soviet strategic relationship based on mutual deterrence and the acceptance of strategic parity.

During this period there was also a growing U.S. consensus that an effective nationwide ballistic missile defense, on which a great deal of research and development effort had been and was being expended, was technically unachievable. Moreover, attempts to deploy such systems were seen as inevitably leading to further major expansions in strategic offensive capabilities as both sides sought to assure their ability to penetrate potential defenses. After initially rejecting this negative assessment of ballistic missile defense, Soviet leaders by the late 1960s apparently accepted this coupling of offensive and defensive strategic arms as a driving factor in the nuclear arms race. Concurrently, the rapid development of satellite technology produced a variety of increasingly capable reconnaissance systems that not only greatly improved the quality of intelligence but opened up the possibility of verifying arms control measures that had previously not appeared to be verifiable without very extensive and intrusive inspection. In the light of these developments, the Johnson Administration determined to explore the possibility of stabilizing the evolving strategic relationship by negotiating arms control agreements with the Soviet Union on offensive and defensive strategic systems.

In January 1967, President Lyndon Johnson announced that the Soviet Union had begun deploying a ballistic missile defense around Moscow and declared that the United States was prepared to initiate discussions with the Soviet Union on the limitation of ABM deployments. To facilitate these negotiations, the President stated that the United States would delay its own ABM defense. While Moscow agreed in principle to discuss "means of limiting the arms race in offensive and defensive missiles," a Soviet commitment to talk was delayed for a year. In the absence of a Soviet response, the United States announced the decision to deploy a light ABM defense against an anticipated modest Chinese missile threat, to provide some protection for the U.S. Minuteman ICBM force, and to protect against the possibility of accidental missile launches. To counteract what was perceived as the potentially destabilizing effect of the Soviet Union's anticipated nationwide ABM system, the United States was also vigorously pursuing the technology of multiple independently targetable reentry vehicles (MIRVs) to assure that any potential Soviet missile defenses could be overwhelmed. Finally, the Johnson Administration undertook a high-level policy review to develop arms control proposals designed to limit both defensive and offensive strategic systems and to engage the interest of the Soviet Union in negotiations.

On July 1, 1968, at the signing of the Nuclear Non-Proliferation Treaty, President Johnson announced that the United States and the Soviet Union had agreed to start strategic arms negotiations. About a month later, on August 19, 1968, the Soviet Union informed the White House that it was prepared to begin negotiations on September 30. But the following day the Soviet Union invaded Czechoslovakia, and the United States postponed the talks. President Johnson's interest in the problem continued throughout the final days of his term in office, but as a lame duck president he was unable to initiate the postponed talks, despite intense private efforts.

SALT I Negotiations

Although as a presidential candidate Richard Nixon had proposed that the United States should regain strategic nuclear superiority, the new administration soon adopted a doctrine of "sufficiency." This doctrine essentially continued the policy of deterrence based on the capability to retaliate and inflict unacceptable damage in all circumstances. It also called for a strategic posture that would be perceived politically as providing "essential equivalence" with Soviet forces. In this context, President Nixon responded favorably to renewed Soviet overtures to start strategic arms talks. Along with a desire to improve U.S.-Soviet relations, the Nixon Administration recognized the potential value of arms control in restraining the rapid, ongoing Soviet construction of ICBM launchers and ballistic missile submarines and in stabilizing the strategic balance between the superpowers. After nine months of intensive preparation, the Strategic Arms Limitation Talks began in Helsinki on November 17, 1969. The Nixon Administration considered arms control a central element of the array of issues between the two superpowers, including the resolution of the Vietnam conflict.

When the SALT negotiations began, U.S. and Soviet offensive strategic forces differed in many respects. For historical, geographic, bureaucratic, and technical reasons, the strategic forces of the two countries had developed in substantially different ways. The United States, with its strong tradition of air and naval power, had developed a triad of air, land, and sea forces that increased confidence in a survivable deterrent. For a variety of technical reasons, including the early development of light thermonuclear warheads, miniaturization of electronics, improved reentry technology, and the development of solid missile fuel technology, the trend in U.S. strategic weaponry was toward smaller missiles. By 1967 the United States had completed the deployment of its second generation of strategic missiles. The U.S. strategic force included 1,054 land-based ICBM launchers, 656 SLBM launchers on Polaris submarines, and almost 600 heavy bombers (B-52s). The United States then shifted its emphasis from construction of more missiles and missile launchers to the development of MIRVs for use on missiles in existing launchers. This was to assure penetration of a future Soviet ABM system and to increase target coverage. With an advantage in MIRV technology, the United States looked forward to developing a lead in the number of missile warheads while retaining its major lead in the number and quality of strategic bombers.

For its part the Soviet Union, with a large land mass having poor access to the sea and with relatively little experience in strategic bombing, emphasized the development of land-based ballistic missiles. The large size of these missiles was initially dictated by the less advanced state of Soviet technology and by the Soviets' approach to military hardware. After the Cuban missile crisis in 1962, the Soviet Union was determined not to find itself again in an inferior strategic position and started a rapid buildup of its strategic forces. By 1969 the Soviet Union had overtaken the United States in the number of land-based ICBMs. It was also rapidly increasing the number of its submarine-based launchers, although it was still far behind the United States in submarine technology. The Soviet Union was also several years behind in the development of MIRV technology and missile accuracy, and it was uncertain how rapidly the Soviets would advance in these areas. When the Soviet Union subsequently developed accurate MIRVs, the large throw-weight of its big land-based missiles with the potential to carry many warheads presented a special threat.

From the outset the two sides were separated by a number of fundamental differences in their perspectives about the negotiations. Perhaps the most serious difference was the definition of the systems to be covered by the agreement. The Soviet Union sought to define as "strategic" any U.S. or Soviet weapon system capable of reaching the territory of the other side. This would have included U.S. forward-based systems, chiefly medium-range bombers based in Europe or on aircraft carriers, and it would have excluded Soviet intermediate-range missiles and aircraft that were aimed at Western Europe and could not reach the United States. The United States held that the weapons to be negotiated in SALT were those that had an intercontinental range, and therefore that its forward-based forces should not be included since they countered Soviet medium-range missiles and aircraft aimed at U.S. allies.

After initial attempts to reach a comprehensive agreement failed, the Soviets sought to restrict negotiations to anti-ballistic missile systems, proposing that limitations on offensive systems be deferred. The United States argued that to limit ABM systems but allow the unrestricted growth of offensive weapons would be incompatible with the basic objectives of SALT. A long deadlock was finally broken when an understanding was reached to concentrate on a permanent treaty to limit ABM systems but at the same time to work out interim limitations on offensive systems that would be incorporated into a comprehensive treaty in future negotiations. The Interim Agreement on Strategic Offensive Arms (Appendix A) was signed by Presidents Nixon and Brezhnev on May 26, 1972, in Vienna at the same time as the ABM Treaty (see Chapter 4).

The Interim Agreement, which was to remain in force for five years, until 1977, was intended as a holding action. The agreement essentially froze at existing levels the number of strategic ballistic missile launchers, operational or under construction, on each side. It did permit construction of additional SLBM launchers up to an agreed level for each party, provided that an equal number of older ICBM or SLBM launchers were destroyed. Within these limitations, modernization and replacement of missiles were permitted. But to prevent further increases in the number of the very large Soviet ICBMs (originally SS-9s, now replaced by SS-18s), launchers for light or older ICBMs could not be converted into launchers for modern heavy ICBMs. The Interim Agreement also formalized the principle of verification by National Technical Means (NTM). These means included all sources of technical intelligence in space or outside the boundaries of the country being monitored. Limitations were stated in terms of "launchers," which could be verified by existing intelligence collection systems, rather than in terms of total missiles, which could not be directly verified by National Technical Means alone. Among the systems and characteristics not limited by the Interim Agreement were strategic bombers, forward-based systems, mobile ICBMs, MIRVs, and missile accuracy. The different numerical limits in the Interim Agreement were considered to be balanced by those forces and by other advantages not limited in the accord.

The U.S. Congress voted overwhelmingly for the Joint Resolution approving the Interim Agreement. The Senate endorsed the Interim Agreement 88 to 2, the same vote by which it advised ratification of the ABM Treaty. Yet despite the almost unanimous vote for the Interim Agreement, some senators expressed concern about the unequal ceilings in the agreement and about the buildup in the throw-weight of the Soviet missile force, as exemplified by the heavy SS-9 missile. As a result, the resolution approving the Interim Agreement included an amendment sponsored by Senator Henry Jackson that established an ambiguous criterion of "equality" for the comprehensive treaty that was to follow the Interim Agreement. Specifically, it placed the Congress on record as requesting "the President to seek a future treaty that *inter alia* would not limit the United States to levels of intercontinental strategic forces inferior to the limits provided for the Soviet Union."

The SALT II Negotiations

In accordance with the Interim Agreement, the SALT II negotiations began in November 1972, only one month after the Interim Agreement had been approved. The principal U.S. objectives were to establish equal ceilings for the two sides on central strategic nuclear delivery vehicles, to restrain qualitative developments that could threaten future stability, and to begin reducing the number of delivery vehicles. In response to some domestic criticism about the ambiguities and lack of detail in the SALT I agreement, the United States sought to ensure that the provisions of the SALT II Treaty would be sufficiently detailed to minimize potential loopholes or misunderstandings.

Considerable progress in developing a formal treaty was achieved in the next two years. However, the positions of the sides still differed widely on a number of fundamental issues. The most important differences concerned limits on Soviet heavy missiles, for which there were no U.S. counterparts; on U.S. and NATO forward-based systems, for which there were no Soviet counterparts; and on MIRVs. These differences were resolved in principle at a meeting in Vladivostok between Presidents Ford and Brezhnev in November 1974. At Vladivostok it was agreed that the strategic offensive arms treaty, which was to be of ten years' duration, would contain the following elements: equal aggregate limits of 2,400 on strategic nuclear delivery systems (ICBM launchers, SLBM launchers, and heavy bombers); equal aggregate limits of 1,320 on MIRVed systems; a continuation of the ban on construction of new land-based ICBM launchers (which implied a ban on additional Soviet heavy ICBMs); limits on the deployment of new types of strategic offensive arms; incorporation of the important elements of the Interim Agreement on verification; and inclusion of mobile ICBMs and air-launched strategic missiles within the overall ceiling. Essentially, the United States had withdrawn its demand for reductions in Soviet heavy missiles in exchange for a Soviet withdrawal of its demand for inclusion or compensation for U.S. and NATO forward-based systems.

When negotiations resumed in Geneva in early 1975, it soon became clear that the two sides still disagreed on two major issues that had not been resolved at Vladivostok. These were whether cruise missiles, which the United States planned to use in large numbers as armaments on its B-52 heavy bombers, were to be counted individually in the overall aggregate, and whether the new Soviet Backfire bomber should be considered a heavy bomber and counted in the 2,400 aggregate. These issues remained unresolved throughout the remainder of the Ford Administration.

The new Carter Administration placed renewed emphasis on SALT. In March 1977 it presented a comprehensive proposal to the Soviets that was a significant departure from the draft of the SALT Treaty previously negotiated. This proposal added significant reductions and qualitative constraints to the ceilings agreed upon at Vladivostok. It called for a reduction of the overall aggregate from 2,400 to 1,800, a sublimit of 550 on MIRVed ICBMs, and a reduction of Soviet heavy ICBMs from 308 to 150. It also called for limits on ICBM flight tests, no new land-based missiles, and no mobile ICBMs. At the same time, the United States presented an alternative proposal for a SALT II agreement similar to the framework agreed to at Vladivostok, with the Backfire and cruise missile issues deferred until SALT III. Initially, the Soviet Union angrily rejected both proposals as inconsistent with its understanding of the Vladivostok Accord.

In subsequent negotiations the sides developed an agreement that accommodated both the Soviet desire to retain the Vladivostok framework and the U.S. desire for more comprehensive and detailed limits in SALT II. This agreement (Appendix B), which was signed by Presidents Carter and Brezhnev in Vienna on June 18, 1979, consisted of three parts: a treaty that would be in force through 1985; a protocol of three years' duration that dealt temporarily with certain unresolved issues to be considered further in SALT III; and a joint statement of principles that set guidelines for the SALT III negotiations. Separate statements associated with the SALT II Treaty placed quantitative and qualitative limits on the Soviet Backfire bomber. The treaty established a framework of equal ceilings and subceilings and qualitative constraints within which the strategic systems could evolve and future reductions could be undertaken.

The Senate ratification debate on the SALT II Treaty continued for several months. Critics challenged not only the treaty's basic provisions but a broad range of foreign and defense policies and their interrelationship with arms control. Senatorial attention was also deflected by concern over the unrelated Iranian hostage crisis and by charges that a Soviet combat unit had been stationed in Cuba. Before a vote could be taken, the debate ended with the Soviet invasion of Afghanistan late in

December 1979. Since it was apparent that a favorable vote could not be obtained under the circumstances, President Carter asked the Senate on January 3, 1980, to postpone action on the treaty. He announced, however, that the United States would abide by the treaty as long as the Soviet Union did.

Describing the SALT II Treaty as "fatally flawed," presidential candidate Ronald Reagan said he would withdraw it from the Senate if elected. Subsequently, the Reagan Administration has taken the position that SALT II will not be ratified and has no legal status under international law, but that the United States will not undercut the treaty at least through 1985 as long as the Soviet Union does likewise. The Soviet Union has simply stated that it is in compliance with SALT II.

Despite the Reagan Administration's refusal to ratify the SALT II Treaty and a growing problem with compliance, the status of the treaty remained an active issue in the summer of 1984. Democratic candidate Walter Mondale consistently supported the treaty and strongly criticized the Reagan Administration for failing to ratify it. The 1984 Democratic platform pledges "to update and resubmit the SALT II Treaty to the Senate for its advice and consent."

Another important issue has been the Soviet record of compliance with the SALT I and the SALT II accords. Previous administrations satisfactorily resolved earlier compliance problems between the two countries. However, in the fall of 1983, in response to a Senate request, President Reagan sent a report to Congress on the record of Soviet compliance with existing arms control agreements. The classified report, which the President presented to Congress in late January 1984, charged the Soviet Union with seven violations or probable violations of arms control agreements. Three of these related to the unratified SALT II agreement. The Soviet Union denied the charges and leveled a series of countercharges against the United States. Although the President stated that the report did not mean that the United States should give up its search for arms control agreements, administration officials added that the outstanding arms control issues raised in the report had to be resolved for the process to succeed.

THE PROVISIONS OF SALT I AND SALT II

The SALT I Interim Agreement

The SALT I Interim Agreement of 1972 (Appendix A), an agreement of five years' duration, was designed to complement the SALT I ABM

Treaty by limiting competition in strategic offensive arms while providing time for further negotiations. The agreement established a ceiling on the aggregate number of ICBM and SLBM launchers operational or under construction. The number of ICBM launchers was frozen at those then operational or under construction. SLBM launchers could be increased beyond those operational or under construction up to an agreed level for each party, but only if a corresponding number of older ICBM or SLBM launchers were dismantled or destroyed. At the date of the signing, the United States had 1,054 operational land-based ICBMs and none under construction. The Soviet Union had 1,618 land-based ICBMs operational and under construction.

Under the terms of the agreement, the United States was permitted to reach a ceiling of 710 SLBM launchers on 44 submarines. At the time it had 656 SLBM launchers on 41 submarines, to which it could add by replacing 54 older ICBM launchers. The Soviet Union had an initial ceiling of 740 SLBM launchers on modern nuclear-powered submarines. This could be increased to 950 launchers by replacing older ICBM launchers on a one-for-one basis. Launchers for light or older ICBMs could not be converted into launchers for modern heavy ICBMs, and the dimensions of launch silos could not be significantly increased. Mobile ICBMs were not covered, although the U.S. negotiators unilaterally stated that the deployment of such missiles would be considered contrary to the objectives of the treaty. Heavy bombers were not constrained at all by the treaty. At the time the United States had some 600 heavy bombers while the Soviet Union had only around 150 significantly less capable bombers.

The SALT II Treaty

The SALT II Treaty of 1979 (Appendix B) is composed of three parts: (1) a treaty providing for equal aggregate limits and sublimits on strategic nuclear delivery vehicles until December 31, 1985; (2) a protocol providing for limits on cruise missile and mobile ICBMs until December 31, 1981; and (3) a joint statement of principles to serve as guidelines for future negotiations. The SALT II Treaty is a detailed technical contract that establishes precise definitions and provisions in an effort to close potential loopholes.

Specifically, the SALT II Treaty provides for:

• Equal aggregate limits on the number of ICBM and SLBM launchers and heavy bombers—initially 2,400, with a reduction to 2,250 by the end of 1981.

• Equal aggregate limits of 1,320 on the total number of MIRVed ballistic missile launchers and heavy bombers equipped for launching cruise missiles with ranges over 600 km.

• Equal limits of 1,200 on the total number of MIRVed ballistic missile launchers and 820 on MIRVed land-based ICBM launchers.

• A freeze on the number of heavy ICBM launchers and on new heavy ICBMs.

• Ceilings on the throw-weight and launch-weight of light ICBMs.

• A ban on the testing and deployment of new types of ICBMs, except for one new type being permitted on each side.

• A freeze on the number of reentry vehicles (RVs) on current types of ICBMs, a limit of 10 RVs on the one new type of ICBM, and a limit of 14 RVs on new SLBMs.

• A limit of 28 on the average number of air-launched cruise missiles (ALCMs) with ranges over 600 km deployed on heavy bombers carrying ALCMs, and a limit of 20 ALCMs on current bombers.

• A ban on the testing and deployment of ALCMs with ranges over 600 km on aircraft other than those counted as heavy bombers.

• A ban on heavy mobile ICBMs, heavy SLBMs, and heavy air-to-surface ballistic missiles (ASBMs).

• A ban on certain types of strategic offensive systems not yet employed by either side, such as ballistic missiles with ranges over 600 km on surface ships.

• Advance notification of certain ICBM test launches.

In addition, the treaty included the following provisions designed to facilitate its verification by National Technical Means (NTM):

• A ban on interference with the NTM used to verify the agreement.

• A ban on all deliberate concealment measures that impede verification by NTM of the provisions of the agreement.

• A specific ban on the encryption of telemetry (test data relayed by radio) when such encryption would impede verification of provisions of the agreement.

• Agreed counting rules to facilitate verification by using launchers, which are easily identifiable and distinguishable into classes, as the measure of aggregate missile and MIRVed missile capabilities.

• Cooperative measures to distinguish aircraft with different missions by requiring observable differences related to the missions, referred to as FRODs (functionally related observable differences).

• A periodically updated data base to assist in measuring compliance with the various limits and sublimits.

• Use of the U.S.-Soviet Standing Consultative Commission (SCC) established in SALT I to consider compliance questions and other problems under the treaty and to develop necessary procedures to implement the agreement.

The SALT II Protocol dealt with certain issues on which the parties were unable to agree for the entire term of the treaty. It established the following temporary limitations through 1981:

- A ban on the flight testing of ICBMs from mobile launchers and on the deployment of mobile ICBM launchers.
- A ban on the testing and deployment of long-range air-to-surface ballistic missiles.
- A ban on the deployment of ground-launched and sea-launched cruise missiles having ranges greater than 600 km.

The SALT II Joint Statement of Principles provided guidance for subsequent negotiations on the limitation of strategic arms. In the statement the sides agreed to pursue further reductions and further qualitative limitations on strategic systems and to work to resolve the issues covered by the protocol. Each side was explicitly permitted to bring up any other pertinent topic it wished to discuss.

On the controversial Backfire bomber issue, Presidents Carter and Brezhnev exchanged documents and statements during the Vienna Summit that were considered part of the SALT II negotiating record. President Brezhnev handed President Carter a written statement that the Backfire was a medium bomber and that the Soviet Union would not upgrade it to an intercontinental bomber or increase its production. He further confirmed that the Soviet Union would not produce more than 30 Backfire bombers per year. In response, President Carter stated that the United States entered into the SALT II agreement on the basis of the commitments contained in the Soviet statement and that it considered these commitments essential to the obligations assumed under the treaty. President Carter also asserted for the record that the United States had the right to an aircraft comparable with the Backfire bomber.

THE MAIN ISSUES SURROUNDING SALT II

The Strategic Relationship

SALT II Supporters

The main premise underlying the SALT process was that an overall "parity" or "essential equivalence" existed between the strategic forces of the United States and the Soviet Union. This strategic balance, which was acceptable to the security interests of the United States, permitted the development of an arms control framework based on the existing forces of the two countries despite large asymmetries in the detailed structure of these forces. Proponents of SALT II argued that there was essential equivalence between the forces when one took into consideration a combination of static measures (numbers of warheads, numbers of delivery vehicles, throw-weight, equivalent megatonnage, etc.) and dynamic measures of real military capability. Despite a growing vulnerability of fixed land-based forces, both sides had a survivable and reliable deterrent that could be maintained with reasonable prudence. In this context, arms control limitations could be formulated in terms of equal ceilings of quite different delivery vehicles while still maintaining essential equivalence.

Supporters of SALT II emphasized that essential equivalence did not require U.S. and Soviet forces to be symmetric in detail. For example, while the Soviets had more ballistic missiles with larger payloads and more megatonnage, the United States had more strategic warheads, greater accuracy, and better submarine and bomber forces. Supporters argued that, despite a major modernization of the Soviet force, essential equivalence had been maintained throughout the 1970s by a U.S. strategic modernization program that included a vigorous MIRV program for Minuteman III, Poseidon, and Trident I missiles; improved Minuteman accuracy and yield; increased hardening of Minuteman silos; the Trident I missile and the Trident submarine program; the air-launched cruise missile program; and the use of advanced avionics to upgrade the B-52 force. In assessing the strategic balance, SALT II supporters also emphasized that one should take into account other factors, such as geographic asymmetries and the location, capabilities, and reliability of allies, all of which tended to favor the United States.

Under SALT II and for the foreseeable future, according to SALT II supporters, the United States would maintain essential equivalence if it proceeded with certain modernization programs allowed under SALT II. These permitted programs included the deployment of the MX missile in a survivable basing mode, such as the multiple shelter racetrack system, the development and deployment of the Trident II missile, and the development and deployment of an advanced bomber. The U.S. programs for developing and deploying cruise missiles in air-, sea-, and ground-launched modes would also not be impeded. The modernization permitted under SALT II ensured that U.S. bombers and submarines would continue to be far more capable than the corresponding Soviet forces. This capability would give the United States a range of devastating retaliatory responses, including selective attacks on military and command and control targets, even in the unlikely event of a total loss of Minuteman silos.

In short, SALT II supporters argued that the agreement would not impede any planned U.S. modernization but would break the momentum of the Soviet buildup, which would otherwise require a further U.S. response. In the future, as in 1979, Soviet advantages in some areas would be offset by U.S. advantages in others, and the overall flexibility, power, and survivability of the U.S. forces would ensure that deterrence and equivalence would be maintained. This strategic balance would be maintained at lower levels and less cost within the limits prescribed by the SALT II Treaty.

SALT II Critics

Some critics of SALT II challenged the underlying premise of overall strategic parity. They argued that Soviet strategic forces were in fact superior to those of the United States and that a treaty based on the false premise of strategic parity was inequitable and would prevent the United States from regaining equality. A more extreme position held that not only was the existing strategic balance unfavorable to the United States but that the security of the United States required strategic superiority to deter Soviet aggression.

Another line of criticism held that, although essential equivalence may have existed in 1979, the SALT process would lull the United States into a false sense of security. It would permit the Soviet Union to pursue its ongoing strategic buildup within the limits of the treaty while the United States failed to do the same. These analysts stated that even though SALT II did not prohibit any of the planned U.S. programs, the greater momentum of the Soviet programs, which had been maintained during the entire SALT process while the United States had reduced its efforts, would cause the United States to fall behind the Soviet Union strategically. Beginning in the early to mid-1980s, the United States would find itself relying on an ICBM force that would be useful only if launched on warning, a bomber force increasingly vulnerable to SLBM attack and with a declining capability to penetrate Soviet air defense to targets, and an SLBM force that would become an increasingly valuable target to potential Soviet antisubmarine warfare (ASW) breakthroughs. Consequently, the United States would find itself strategically inferior to the Soviet Union by the expiration of the treaty in 1985, if not sooner.

Critics projected that the Soviet Union would have as many warheads as the United States by 1985, undercutting this important U.S. advantage. Also, due to the greater throw-weight of its missiles and the rapidly improving accuracy of its warheads, the Soviet strategic force would have twice the area of destructive capability, five times the hard target kill capability of ICBMs and SLBMs combined, three times the megatonnage, and twice the throw-weight of the comparable U.S. strategic forces.

These analysts argued that the large Soviet advantage in ICBM capabilities gave, or soon would give, the Soviet Union the capability to conduct, or threaten to conduct, a successful counterforce attack against U.S. land-based missiles. At the same time, the United States would have no comparable capability against Soviet land-based missiles. Critics also questioned the capability of B-52 bombers with cruise missiles to offset the Soviet counterforce potential. These factors, plus the Soviet Union's greater air defense and civil defense program and its harder and more diverse command and control facilities, were asserted to give the Soviets meaningful strategic superiority as early as 1982 unless the United States took urgent and prompt steps to reverse the trend.

The Rationale for SALT II: Preserving Essential Equivalence

SALT II Supporters

SALT II approached the problem of preserving essential equivalence between asymmetrical strategic forces in several ways. It sought to place equal ceilings and subceilings on the central strategic systems (ICBMs, SLBMs, and heavy bombers) and on the warheads carried by these systems, to complement these numerical limits with selective qualitative constraints, and to begin the process of reductions within this framework.

SALT II supporters claimed that the main benefit of SALT II's numerical constraints would be to help assure essential equivalence by preventing either side from gaining a numerical advantage that could be exploited militarily or politically. SALT II's overall ceiling of 2,400 central systems capped the race for advantage in numbers of missile launchers and heavy bombers. Its 1,200 subceiling on the total number of launchers for MIRVed ICBM and SLBM missiles and its 820 subceiling on launchers for MIRVed ICBM missiles put a cap on the overall number of launchers for MIRVed missiles, which were considered the most destabilizing element of the arms race. These subceilings also kept the numbers of Soviet launchers for MIRVed missiles well below what they might have been without the agreement.

Supporters of SALT II also argued that these aggregate limits and sublimits enhanced longer-term stability by providing a framework for future incremental reductions while providing sufficient flexibility for each side to maintain or deploy forces that it judged to be survivable. The provisions of SALT II began this process of reductions by lowering the initial overall ceiling of 2,400 to 2,250 by the end of 1981. These ceilings would have the practical effect of reducing the Soviet strategic force by some 300 delivery vehicles. Although these would presumably be the least effective components of the Soviet force, they represent a tremendous amount of destructive capability. At the same time, the United States would have the option of increasing its strategic forces by some 150 delivery vehicles, if this was deemed necessary. Moreover, the sides were committed to negotiate substantial reductions in the number of strategic offensive arms in the next stage of SALT. By approaching arms control and reductions as a process, according to SALT II supporters, each side could adjust its forces to lower levels in a practical manner suited to its requirements for security. The provisions allowing modernization to continue within agreed constraints would enhance stability by improving the survivability of the forces remaining at reduced levels.

SALT II supporters also argued that qualitative constraints in the agreement helped assure stability and provide predictability in the planning of both sides' strategic forces. The number of warheads on existing types of ICBMs and SLBMs was frozen, and ceilings were established on the number of warheads that could be placed on new SLBMs and the one new type of ICBM permitted. This meant that the Soviet Union could not exploit the full potential of its advantage in ICBM throw-weight for MIRVed missiles. The treaty banned new types of ICBMs with the exception of one new type of light ICBM for each side. This one new type of ICBM and new SLBMs could not have larger throw-weights than the largest current light ICBM, the Soviet SS-19. Constraints written into this provision required that improvements to existing types of ICBMs be limited to such verifiable characteristics as numbers of warheads and 5 percent changes in throw-weight, launch-weight, length, and diameter. Thus, SALT II, through its qualitative restraints, sought to begin the process of controlling those characteristics that could be destabilizing while allowing limited modernization to continue in areas that could not be adequately verified.

SALT II supporters argued that the numerical ceilings and the qualitative restraints of SALT II were mutually reinforcing. Taken together they limited the ability of both sides to increase their military potential significantly. In effect, SALT II capped most of the major indexes of central strategic power. Equal aggregate ceilings capped the first index of strategic power—the number of missile launchers and heavy bombers. Subceilings on the number of MIRVed missile launchers, "fractionation" limits on the numbers of warheads that could be put on a given missile, limits on the number of heavy bombers carrying air-launched cruise missiles, and limits on the number of cruise missiles per heavy bomber put an upper limit on the second index—the total number of warheads. The ban on additional heavy ICBMs and the upper limits on the size of both heavy and light ICBMs capped the third index—total throw-weight. Together these limits made the planning of strategic forces and the problem of land-based missile vulnerability much more manageable. For instance, in the case of the Minuteman missile's vulnerability the SALT II fractionation and launcher limits not only effectively capped Soviet throw-weight but facilitated the development of survivable deployment plans for the MX missile by limiting the number of warheads the Soviet Union could target against the system. Specifically, the Carter Administration proposed a system of multiple protective shelters in which a single MX launcher moved on a closed road system, or racetrack, containing 23 hardened shelters, each of which would have to be considered a target in a preemptive Soviet attack.

Supporters of SALT II argued that the framework of equal aggregate ceilings in the context of essential equivalence gave the two sides the flexibility to resolve certain extremely difficult problems related to the asymmetric structures of their forces. Specifically, this framework provided a basis for dealing with the critical asymmetries in Soviet heavy missiles, U.S. forward-based systems in Europe, U.S. cruise missiles, the Soviet Backfire bomber, and British and French strategic forces. For example, at Vladivostok, when the equal aggregate approach was accepted, the United States dropped its insistence that the Soviet Union substantially reduce the number of its heavy missiles, for which there was no comparable U.S. system, and the Soviet Union withdrew its demand that U.S. forward-based systems (aircraft in Europe and on carriers) capable of striking the Soviet Union be included in the aggregate. This represented the most significant example of the trade-off of asymmetric capabilities that had proved a major barrier to progress in the negotiations. Subsequently, the United States did not press for equal rights for heavy ICBMs because it had no plans for such a system and because it did not wish to pay a price for this unwanted option, which might have included a Soviet demand to replace the SS-18 with a more advanced heavy missile.

As another example of this negotiating flexibility, supporters cited the case of air-launched cruise missiles. The Soviets maintained that it had been agreed at Vladivostok to count these missiles on a one-for-one basis as part of the overall aggregate of missile launchers and heavy bombers. The United States countered that there had been a misunderstanding and refused to accept this approach. The problem was resolved by considering heavy bombers with air-launched cruise missiles as analogous to MIRVed missiles. A special sublimit was negotiated including both of these categories of delivery systems, and separate limits were placed on the average number of cruise missiles that could be carried by heavy bombers.

Finally, SALT II supporters pointed to the manner in which the Backfire bomber issue, a particularly difficult barrier to agreement, was resolved within the context of essential equivalence. Although this bomber probably had the capability to reach the United States on one-way missions with special flight profiles, the Backfire, which was in fact being deployed for theater and naval missions, was not an intercontinental system in the same sense as the other central strategic systems covered under the SALT ceilings. Consequently, while the Backfire might contribute marginally to the strategic balance, it was difficult to argue that it should be included on a one-for-one basis. Moreover, the Backfire was in part compensated for by the FB-111, a U.S. bomber smaller than the B-52 but capable of hitting Soviet targets from U.S. bases. In the context of essential equivalence, the two sides agreed that it would be adequate to obtain a commitment that the Backfire would not be upgraded technically or operationally so as to have a more truly intercontinental capability and that its rate of production would not be increased. Although the Soviets refused to include the Backfire in the treaty unless the issue of forward-based systems was reopened, they did agree to make such a commitment on the Backfire in a separate statement signed by President Brezhnev. This statement together with an oral statement by President Brezhnev on the specific rate of production of the Backfire were considered integral parts of the treaty by the United States. For their part, the Soviets did not insist on counting or receiving compensation for French and British strategic nuclear forces. Rather, they settled for a joint commitment in Article XII "not to circumvent the provisions of this Treaty, through any other state or states or in any other manner." These and other compromises, which SALT II supporters considered acceptable to U.S. security interests, were possible because the SALT II approach allowed enough flexibility to deal realistically with asymmetric strategic forces.

In addition, SALT II supporters pointed out that the SALT II Treaty contained a variety of specific, detailed qualitative constraints that contributed to stability. For example, SALT II banned ICBM rapid-reload capability, which, if technically feasible, could greatly increase the military value of existing ICBM launchers. To reinforce this provision, SALT II banned the storage of excess missiles in the vicinity of launch sites.

In general, SALT II supporters argued that the SALT approach contributed significantly both to crisis stability—by assuring both sides a survivable deterrent—and to arms control stability—by placing predictable limits on future force postures and by providing a workable framework for future reductions in force levels.

SALT II Critics

SALT II was criticized from two fundamentally different perspectives. One group of critics argued that the SALT II limits were inequitable because they did not bring the Soviet advantages in destructive capability into balance. The other group of critics argued that, although the SALT limits were equal, SALT was not an acceptable approach to arms control because it did not provide for significant reductions and in fact "institutionalized" the arms race.

The primary criticism against the SALT II Treaty came from those who argued that the numerical ceilings did not really provide equality. These critics argued that the quantitative ceilings and subceilings provided only the appearance of equality. Because these provisions did not establish equal limits on the destructive characteristics of missiles, such as throw-weight, they allowed the Soviet Union to maintain a large lead in the destructive capability of its land-based missiles. A central reason cited for this inequity in destructive power was the Soviets' retention of heavy ICBMs. It was asserted that the destructive capability of these systems alone exceeded the destructive capability of all U.S. strategic missiles. Because of these inequities in destructive power, the agreement was inherently unequal. Thus it would lock the United States into a position of inferiority.

These analysts argued that the vulnerability of U.S. ICBMs to Soviet attack, which heavily contributed to the United States' inferior strategic position by limiting its retaliatory response, could not be corrected during the term of the treaty. Despite the freeze on the number of warheads on existing MIRVed missiles and the equal limit of ten warheads on the one new type of ICBM, the Soviet Union would still be able to deploy enough warheads during that period to destroy the U.S. ICBM force. Moreover, the fractionation limits on MIRVs would not assure the survivability of the future MX deployment even in a 23-shelter racetrack mode because the treaty would expire before the MX could be deployed. The critics also argued that the freeze on warheads for existing systems was inherently inequitable because the yield of Soviet warheads was at least double that of U.S. warheads and because the Soviets were allowed ten warheads on their heavy missiles (the SS-18s) and six warheads on their largest light missile (the SS-19) while the United States was allowed only three warheads on its Minuteman III. In effect, said the critics, the SALT II Treaty allowed the Soviet Union a three to one numerical and a five to one throw-weight advantage in ICBM warheads, which are the most destabilizing strategic weapons.

Critics also asserted that the quantitative and qualitative limits did not effectively cap the strategic arms race or effectively begin the process of reductions in a way that would assure essential equivalence. Instead of forcing a reduction, SALT II would permit a large increase in Soviet capabilities. It was argued that the ceilings in the treaty were so high that they would not provide a useful cap on the arms race. Similarly, the capabilities of the Soviet systems were so great that the reductions agreed upon would have little or no military significance. The lower ceiling of 2,250, which the Soviet Union would have to reach by 1982, would in effect involve only scrapping obsolete systems. In 1982 the United States would still have only 2,050 operational systems, which some critics claimed would be less capable than the Soviet force. After this reduction in Soviet forces, the United States would still have to build up its forces to achieve equality with the Soviet Union.

The critics argued that the restriction to one new type of ICBM did not effectively or equally cap the quantitative and qualitative arms race. It was claimed that the Soviet Union could still develop and deploy more than one type of ICBM simply by operating within or close to the envelope of characteristics defining a missile type. Thus the largely unconstrained potential of Soviet destructive capability would put the Soviet Union strategically ahead of the United States in the early 1980s and could be further exploited once the treaty expired.

The treaty was also inequitable, according to SALT II critics, because it did not effectively limit Soviet Backfire bombers. Critics claimed that the Backfire bomber was a strategic bomber because it could strike the United States on unrefueled one-way missions. Consequently, by not including the Backfire bomber in the aggregate, the United States was allowing the Soviet Union to increase by as much as one third the already large destructive power that it could deliver against the United States. By not having to count Backfires in the aggregate total, the Soviet Union would also not have to eliminate an equal number of ICBMs and SLBMs to reach the 2,250 total.

A further criticism was that the provisions of SALT II made a rapid breakout from the treaty a serious threat. For example, the provisions designed to limit the Soviets' ability to reload their ICBM silos could not be counted on to be effective for more than a number of hours. Also, any bomber with hard points on its wings could rapidly be converted to carry cruise missiles, and any cruise missile operable from a plane could rapidly be adapted for launch from ground- and sea-based launchers.

Some critics emphasized that the most serious danger of the SALT II Treaty was that it would lull the American public into a false sense of security while in fact locking the United States into a position of inferiority. To present negotiable proposals, the United States sacrificed its goals of limiting significant military characteristics such as throw-weight, according to these critics. It reached an agreement that limited the wrong strategic characteristics and allowed the Soviet Union not only to retain but also to enhance its strategic destructive advantage during the term of the treaty.

An entirely different group of critics argued that SALT did not stop the arms race but rather institutionalized the qualitative arms race. They contended that arms control should stop the arms race, comprehensively constrain modernization, and significantly reduce the number of weapons systems. In this view, the SALT II approach to reductions was too slow, its qualitative restraints too limited. In particular, these analysts denied the need for one new type of ICBM, which in the United States was to be the MX missile with ten highly accurate warheads deployed in a very expensive and controversial multiple-shelter race-track mode. In short, these critics charged that SALT II may have provided essential equivalence but only at unacceptably high levels.

Verification

The SALT II Treaty was structured to facilitate "adequate" verification by the existing technical intelligence systems of the United States and the Soviet Union. These intelligence systems, designated as National Technical Means (NTM), include reconnaissance satellites (with photographic, infrared, radar, and other sensors) and ground-based technical systems (such as radars and radio antennas) located outside the borders of the country under surveillance. The standard of "adequate" verification, as was set forth in the Arms Control and Disarmament Act and enunciated by President Nixon in his instructions to the first session of SALT in 1969, had been the stated objective of verification throughout the SALT process. Adequate verification has generally been interpreted as meaning a level of verification which would assure with high confidence that compliance could be determined to the extent necessary to safeguard national security and that violations could be detected early enough to permit an appropriate response.

SALT II Supporters

Supporters of SALT II asserted that the agreement, which had been carefully designed to take maximum advantage of existing intelligence monitoring capabilities, was "adequately" verifiable by any reasonable criteria. They pointed out that this conclusion was supported by the intelligence community and the Joint Chiefs of Staff on the basis of extremely detailed studies that considered not only normal Soviet practices but possible Soviet efforts to conceal their activities. It was emphasized that these conclusions were drawn from the total collective experience of the intelligence community in monitoring the growth of Soviet strategic forces since the end of World War II. During this period the U.S. intelligence community had developed a very detailed understanding of Soviet strategic forces, including the number and capabilities of deployed missiles and bombers, the location of production and test facilities, the structure of the command system, training and operational procedures, and missile testing practices. On the basis of this in-depth knowledge, it was possible not only to establish confidently the baseline of the forces in being at the time of the agreement but also to judge retrospectively the effectiveness and timeliness of the intelligence system in monitoring the specific systems limited in each of SALT II's provisions.

Supporters of SALT II emphasized that the intelligence community had also had several years' experience monitoring the provisions of SALT I. These provisions dealt with the same systems and many of the same problems as SALT II.

Supporters of SALT II also emphasized that information relating to the verification of specific provisions usually came from several independent systems. This provided both cross-checks and redundancy in the system. It was argued, for example, that despite charges to the contrary even the loss of important collection facilities in northern Iran had not significantly reduced the verification capabilities for the Interim Agreement and SALT II. Furthermore, the Soviet Union did not know the full extent or capabilities of the intelligence resources of the United States and its allies, and it would have to operate very cautiously in any attempts to violate or circumvent the treaty. It was also pointed out that programmed major improvements in the intelligence collection system would substantially increase verification capabilities in ways that the Soviets could not project with confidence. Finally, the

Soviet Union could never be sure that information from espionage operations or defectors might not expose concealed activities that might have initially escaped detection by NTM.

Supporters of SALT II argued that the provisions of the treaty protected and enhanced the existing capabilities of national intelligence. By acknowledging that National Technical Means would be the method of verification, the treaty incorporated national technical intelligence into the body of acceptable international activities. This was a matter of major significance in the case of space-based reconnaissance systems, which are extremely vulnerable to attack. Moreover, the specific provision in the treaty banning interference with NTM used to verify the agreement would protect essentially all U.S. technical intelligence collection systems, since they all contributed to the verification process. In addition, the ban on deliberate concealment measures that would impede verification of treaty provisions substantially enhanced the capabilities of national intelligence, since without this agreement there would be no legal constraints on such concealment. In this regard, it was emphasized that the explicit ban on the encryption of test telemetry related to the verification of specific provisions of the treaty assured the availability of important information that would otherwise almost certainly be denied. Based on extensive experience, it was noted that the authenticity of unencrypted telemetry could be established with confidence.

Supporters of SALT II pointed out that the counting rules used to establish the aggregate ceilings and subceilings in the agreement had been defined so that they depended on information attainable by NTM, namely, numbers of missile launchers, missile test data, and numbers of aircraft. Moreover, while each side would rely on its own verification capabilities, the treaty provided for a data exchange to provide an agreed basis for purposes of compliance. Finally, the authority of the Standing Consultative Commission, which was originally established in SALT I, would be extended to consider compliance issues and other problems relating to the verification of SALT II. SALT II supporters pointed out that the SCC had proven in practice to be a very effective mechanism in dealing with a range of compliance problems with SALT I. As a result of detailed private discussions at the SCC, all compliance problems had been resolved to the satisfaction of the U.S. government by the time the ratification of SALT II was being considered by the Senate. The identification of compliance problems demonstrated the capabilities of the verification system, and their referral to the SCC demonstrated the willingness of the United States to release sensitive intelligence information and address potentially confrontational issues.

In view of the experience of the intelligence community and the verification provisions in the treaty, supporters of SALT II argued that the major provisions of SALT II could be adequately verified. The verification of these major provisions is discussed below.

The Aggregate Limit on ICBM and SLBM Launchers and Heavy Bombers. The aggregate limit on ICBM and SLBM launchers and heavy bombers can be accurately verified with high confidence since (1) ICBM launchers (silos) are inherently easy to identify, are built in large complexes for reasons of security and control, take over a year to build, and require extensive support facilities; (2) SLBM launchers are located in fixed numbers on ballistic missile submarines, which are currently constructed over a period of years at a single location and are subsequently outfitted in the open; and (3) heavy bombers are large, distinctive aircraft that have been produced at only a few well known plants and deployed at a limited number of bases. The easily identifiable missile launchers are a suitable measure of actual missile capabilities. The reload of SLBM launchers would be impossible in wartime, and the reload of ICBM launchers (silos), which would take many hours or days, would not be practical with missile fields under attack in wartime. Nevertheless, to minimize this latter possibility, the treaty specifically banned the storage of reload missiles or facilities suitable for that purpose at missile sites.

The U.S. insistence on permitting the introduction of mobile ICBM launchers after 1981 complicates the verification process. But the intelligence community concluded that it would still be possible to make reasonably reliable estimates of the numbers of these mobile launchers because they would be very large, unique vehicles that would probably require identifiable support and command facilities. The success in monitoring the deployment of the SS-20 mobile system was cited as evidence of this capability. Moreover, as a special precaution, the agreement banned the testing, production, and deployment of the SS-16, which was essentially the SS-20 with a third stage, to avoid the rapid upgrade of the known SS-20 force to intercontinental capability. To establish standards for cooperative measures in the verification of mobile systems, the United States revealed its plans to deploy the mobile MX in a multiple protective shelter mode. This deployment incorporated a number of major features to assure the Soviet Union that not more than one missile was associated with the 23 shelters at each of the easily identifiable MX sites.

The Subceilings on MIRVed Launchers. Supporters argued that the subceiling of 1,200 on the number of launchers for ICBMs and SLBMs equipped with MIRVs and the subceiling of 820 on the number of launchers for ICBMs equipped with MIRVs could be verified with high confidence because of the powerful counting rules that make use of the fact that launchers for different types of missiles are easily distinguished. The counting rules provide that all missiles of a type that has ever been tested with MIRVs (which is verifiable by NTM with high confidence) will be considered as MIRVed missiles and that any launcher of a type that has ever contained or launched such a MIRVed missile will be considered a launcher of MIRVed missiles. This conservative counting rule, which depends strictly on observable characteristics, includes within the MIRVed launcher subceilings all launchers that have a capability of launching MIRVed missiles even if the missile in the launcher has only a single warhead (as was believed to be the case in some instances). This approach was considered to be far more effective than on-site inspection. Inspections would require very intrusive procedures to ascertain whether a missile was in fact MIRVed, and a MIRVed warhead could be temporarily replaced with a single warhead during inspections.

Constraints on Qualitative Modernization of Missiles. Supporters of SALT II argued that the constraints on changes in launch-weight and throw-weight and the freeze on the number of RVs on missiles could also be verified with adequate confidence. Any significant changes would involve extensive testing, which would be monitored by a number of independent NTM, including the collection of telemetry. It would be extremely difficult to increase the number of warheads on a particular type of MIRVed ICBM without detection because testing would be necessary and the number of warheads released in a test can be monitored confidently by a number of independent techniques. Moreover, detailed provisions in the treaty limit certain testing activities that do not involve the release of additional warheads but might be directed at circumventing this important limitation. Launch-weight and throw-weight can be measured quite accurately from test data, and changes in these characteristics would be particularly obvious for existing missiles whose characteristics are well known from the scores of development and training tests conducted over the years. While the 5 percent limit on changes in these parameters in existing missiles admittedly presses verification capabilities, the limit was set as low as possible to provide a basis for challenging any detected changes and to minimize the incentive to introduce a new ICBM as a permitted modernization of an existing type.

One New Type of ICBM. Supporters argued that the restriction on both sides to develop only one new type of ICBM with no more than ten warheads could be verified with high confidence. Any new missile would require 20 to 30 tests over a period of a couple of years before it could be deployed. During this test period a variety of independent techniques, including telemetry, could establish with confidence whether the launch-weight, throw-weight, and other parameters were outside the envelope of permitted modernization of existing missiles and whether the missile had a MIRV capability in excess of ten warheads. If the new missile was within the envelope of any old missile and thus did not qualify as a new type, it could not have more warheads than the old type. It would therefore not have significantly greater capability other than possibly accuracy.

In summary, supporters of SALT II argued that it was possible to verify with high confidence whether or not the Soviet Union was complying with the provisions of the treaty. They also held that any violation large enough to threaten the security of the United States would be discovered in time to permit an appropriate reaction.

SALT II Critics

Many critics of SALT II asserted that the treaty did not, in fact, meet the criterion of adequate verification. Some went further and rejected the concept of adequate verification, saying it was not sufficiently stringent to meet the demands of national security. Some critics questioned the record of the U.S. intelligence community in the area of strategic weapons monitoring over the past 20 years, pointing out alleged underestimates of the size and capabilities of Soviet strategic missile forces. There were suggestions that the intelligence community exaggerated the confidence that could be placed in its current assessments, and that these assessments might in fact substantially underestimate the present threat. Other critics challenged the inherent capability of certain intelligence systems to provide information that would permit timely assessments with the accuracy required to meet treaty provisions, particularly those relating to qualitative constraints. In this connection, the loss of Iranian collection facilities was claimed to have seriously degraded verification capabilities, thus illustrating the fragility and unreliability of many intelligence resources.

The strongest criticisms, however, focused on the ability of the intelligence community to operate with the indicated level of confidence if the Soviet Union deliberately undertook to violate or circumvent the treaty by clandestine procedures. It was argued that the intelligence community exaggerated its ability to deal successfully with concealment problems that it had not previously confronted. The concern about clandestine activities applied not only to potential direct violations of the treaty but to activities that might be designed to permit a rapid breakout from the treaty, either by abrogation or upon its expiration at the end of 1985.

Some critics expressed particular concern about the fact that the treaty did not limit Soviet missile production and that the cumulative stockpile of various strategic missiles was not known with confidence. They argued that the focus on ICBM silos and SLBM launchers obscured the real threat, the number of missiles. These missiles might be used for reload, be deployed clandestinely, or be held in reserve for rapid deployment after the expiration of the treaty.

Some of these critics minimized the significance of the treaty provisions designed to enhance verification. They argued that the ban on interference with NTM prevented an activity from which the Soviet Union was currently deterred in any event, and that the ban on concealing of activities from NTM was largely meaningless since really effective concealment would be difficult to detect. In this connection, the value of the partial ban on encryption was dismissed on the grounds that one could not be certain that critical data, or even the real data, were not in the encrypted portion of the telemetry. The value of the SCC was questioned by some who suspected it had become a Soviet device to probe sensitive U.S. intelligence capabilities. Others believed the SCC was being used to shield the Congress and the American people from serious compliance problems.

Critics of SALT II challenged the adequacy of verification of the treaty's major provisions along the following lines:

The Aggregate Limit on ICBM and SLBM Launchers and Heavy Bombers. Critics argued that the enumeration of ICBM silos and SLBM launchers did not adequately verify the real threat, the number of Soviet missiles. Since the production of missiles was neither limited nor adequately verifiable by NTM, large numbers of additional missiles could have been or might be produced. These missiles might be deployed in ordinary industrial-type buildings near the production facilities. They could also be stored in the general vicinity of existing launchers for rapid reload, despite treaty provisions to the contrary. In any event, these missiles could be available for deployment in the event the treaty was abrogated or expired. The ability to verify the number of missiles deployed in a mobile mode was also challenged, since these missiles could be kept under cover and exercised individually. It was argued that the verifiability of the Soviet mobile SS-20 system should not be taken as a precedent for future deployment practices. Moreover, critics held that there was no reason to believe that the Soviet Union would follow the U.S. lead in incorporating extensive cooperative measures to facilitate verification, such as those originally proposed for the deployment of the MX. Concern was expressed that despite the special limitations on the SS-16, the SS-20 could be upgraded to intercontinental range without U.S. knowledge. The ability to count aircraft was not challenged, but questions were raised about the ability to verify that some Bison and Bear aircraft could be excluded from the aggregate total on the grounds that they were committed to and outfitted for other missions. The Backfire presented particularly controversial verification problems, since many critics thought it should be included in the aggregate as a heavy bomber. They challenged both the assessment of its capabilities and the ability of NTM to detect the upgrading of those capabilities.

The Subceiling on MIRVed Launchers. As in the case of the aggregate limits, critics argued that the subceiling on MIRVed launchers did not really permit adequate verification of the number of MIRVed missiles that had been produced. Moreover, if excess missiles were produced for clandestine deployment, reload, or deployment after a breakout from the treaty, they would very likely be MIRVed.

Constraints on Qualitative Modernization of Missiles. Critics argued that qualitative constraints of the treaty could not be verified with adequate confidence or precision. In particular, they asserted that it was not possible to measure the launch-weight or throw-weight of a missile to within 5 percent, the limit on permitted changes to an existing missile. This already questionable capability, the critics continued, had been further degraded by the loss of critical ground-based NTM facilities in Iran. Critics also challenged the ability to verify with high confidence the ban on increased numbers of warheads on existing MIRVed missiles. Since a MIRV dispensing system can be tested without actually releasing its full complement of warheads, verifying the maximum number of warheads the missile can carry depends on detailed analyses of telemetry, which can be encrypted or otherwise concealed.

One New Type of ICBM. Critics argued that the limitation to one new type of ICBM could not be verified with high confidence and would provide a potential loophole for more extensive ICBM developments. The provision in fact permits new ICBMs provided they fit within the envelope defining permitted modernization. Citing the limited ability to verify some of these parameters, critics argued that it would not be possible to prove conclusively that a new missile involving considerably greater changes was not in fact a permitted modernization. Moreover, they emphasized that the Soviet Union was known to have several new missiles under preliminary development that might be tested and deployed on this basis.

In summary, critics of SALT II argued that it was not possible to verify with confidence that the Soviet Union was not violating the provisions of SALT II to an extent that might threaten the security of the United States.

Compliance

The Soviet record of compliance with the SALT I agreements was a central issue in the SALT II ratification hearings. Since that time, the debate has continued and grown to include questions about Soviet compliance with the unratified SALT II agreement. On January 23, 1984, in response to a congressional request, President Reagan sent Congress a classified report with an unclassified summary dealing with seven compliance issues. It charged the Soviet Union with violations and probable violations of five provisions of the SALT agreements. On January 29, 1984, the Soviet Union released a diplomatic note that charged the United States with numerous violations of the SALT agreements.

Compliance with SALT I from 1972 to 1979

During the SALT II ratification hearings the Carter Administration took the position that the overall record of Soviet compliance with SALT I had been good and presented the Senate with full documentation to support this conclusion. Early in SALT I the decision had been made to raise certain compliance issues in the Standing Consultative Commission even though they involved sensitive intelligence information. Prior to the SALT II hearings the United States had taken eight potential problems to the SCC for clarification. After extensive discussion in the SCC the government concluded in each case either that there was in fact no problem in the light of additional information, that an ambiguity in the agreement had been clarified to mutual satisfaction, or that the questionable activity had ceased.

Supporters of SALT II argued that an objective examination of specific compliance cases demonstrated not only the power of the verification system but also the effectiveness of the SCC. The following three cases, which were potentially the most serious compliance problems, were used to illustrate their point.

In 1973 the United States observed the initiation of construction of what appeared to be new silos at a number of missile fields. New silos were clearly prohibited by the SALT I Interim Agreement. When this suspicious activity was raised at the SCC, the Soviet representatives explained that the silolike structures were to house launch control facilities, as would become apparent. On the basis of subsequent information about the structures, the United States agreed that this was the case.

In 1973 and 1974, technical intelligence indicated that a radar associated with the Soviet SA-5 air defense system had apparently tracked a Soviet ballistic missile during a test flight. To prevent a permitted air defense system (such as the SA-5) from being upgraded to have a marginal ABM capability, the SALT I ABM Treaty prohibited the testing of such a system or any of its components in an "ABM mode." In a unilateral statement accompanying the treaty, the United States had interpreted this term to include the testing of such a system's radar against a ballistic missile reentry vehicle. In the SCC the Soviet representatives denied that the radar was being tested in an "ABM mode" and noted that the use of radars for instrumentation and range safety was not prohibited. Whatever the true nature of the activity, the practice ceased. Subsequently, more detailed interpretations of this complex technical provision were worked out in the SCC to the mutual satisfaction of both sides.

In 1975, when the SS-19 deployment began, the United States brought the matter before the SCC since it underscored a troublesome ambiguity in the Interim Agreement, although it was not a violation of the agreement. The agreement prohibited the conversion of launchers for light ICBMs to heavy ICBMs but failed to define the dividing line between the missiles. In the negotiations the U.S. delegation had unilaterally stated that it would consider any missile with a volume substantially greater than that of the largest Soviet light missile (the SS-11) to be a heavy missile. The Soviet delegation had rejected this definition and had informally told a member of the U.S. delegation that when deployed the SS-19 would have a volume "less than midway between the volume of the SS-11 and the SS-9." While this appeared to be the case, the United States wanted to make clear its concern about an erosion of the distinction between light and heavy missiles. Subsequently, the SALT II negotiators agreed on a clear demarcation for missile launch-weight and throw-weight between light and heavy ICBMs.

SALT II supporters emphasized that these three cases, as well as all of the other cases of less potential significance for security, were resolved to the satisfaction of the United States. It was also pointed out that a number of additional compliance cases that had been reported in the press had not been referred to the SCC, since on careful examination they proved to be incorrect. For example, the press reported that the Soviet Union had been "blinding" U.S. reconnaissance satellites (apparently a consequence of a large gas fire in the Soviet Union) and that the Soviet Union had tested and deployed a mobile ABM.

The Soviet Union also raised a number of compliance questions in the SCC about U.S. practices under SALT I prior to the SALT II ratification hearings. The most serious issue related to the use of prefabricated environmental shelters over Minuteman silos during construction to modernize and increase the hardness of the silos. Starting in 1973, the Soviet SCC representatives objected to the practices as being a form of prohibited concealment, since SALT I placed specific limits on the extent to which launchers could be modified and since it was necessary to distinguish between Minuteman II and Minuteman III silos. Although the United States took steps to reduce the size of these shelters, they were not removed. The Soviets continued to press this issue until 1979, when the shelters were removed in connection with the signing of SALT II, which specifically banned their use.

Some critics of SALT II acknowledged the list of compliance cases presented by the Carter Administration during the ratification hearings but argued that the potential significance of the cases had been underestimated. For example, they asserted that the silolike hardened command and control modules were in fact suitable for dual use as missile silos; that the SA-5 radar may have been tested sufficiently before the testing was stopped to permit the entire widely deployed SA-5 system to be upgraded to a significant terminal ballistic missile defense system; and that the decision to accept deployment of the SS-19 as consistent with the Interim Agreement greatly increased the Soviet counterforce threat since the SS-19, with its six warheads and the highest accuracy of any Soviet missile, was being deployed in large numbers (360). Other critics took a more extreme view, suggesting that additional compliance problems—such as the problem of rapidly transportable, if not mobile, ABM systems—were being ignored. In fact, some critics suggested that the United States was ignoring a clear and continuing pattern of violations and circumventions. A few critics went so far as to claim that the Soviet Union had carefully designed the entire SALT process to permit a program of violations that on abrogation or expiration of the treaty would give the Soviet Union a decisive strategic advantage.

Compliance with SALT I and SALT II from 1980 to 1984

In the period since the SALT II ratification debate, a steady flow of alleged Soviet violations of various arms control agreements has been reported in the press. At first these were largely restatements of previous charges connected with SALT I and other earlier treaties. But recently several potentially significant violations of the unratified SALT II Treaty have been widely reported. On January 23, 1984, in response to a congressional request, President Reagan submitted a classified report to Congress on "Soviet Non-Compliance with Arms Control Agreements," which reviewed seven major compliance issues. The President's transmittal message states: "The United States Government has determined that the Soviet Union is violating the Geneva Protocol on Chemical Weapons, the Biological Weapons Convention, the Helsinki Final Act, and two provisions of SALT II: telemetry encryption and a rule concerning ICBM modernization. In addition, we have determined that the Soviet Union has almost certainly violated the ABM Treaty, probably violated the SALT II limit on new types, probably violated the SS-16 deployment prohibition of SALT II, and is likely to have violated the nuclear testing yield limit of the Threshold Test Ban Treaty."

The report drew a careful distinction between the ratified SALT I ABM Treaty, which was a "legal obligation," and the unratified SALT II Treaty, which was a "political commitment." It emphasized that, because the U.S. government had formally announced in 1981 that it would not ratify SALT II, the legal obligation under international law not to take actions that would "defeat the object and purpose" of a signed but unratified agreement did not apply. The report noted, however, that the United States has observed a "political commitment" to refrain from actions that would "undercut" SALT II as long as the Soviet Union does likewise.

With regard to the ratified SALT I ABM Treaty, the report found that a new large phased-array radar under construction near Krasnoyarsk in central Siberia "almost certainly constitutes a violation of legal obligations under the ABM Treaty of 1972 in that in its associated siting, orientation and capability, it is prohibited by the Treaty."

With regard to the unratified SALT II Treaty, the report addressed three problem areas: (1) encryption, (2) the new Soviet SS-X-25 missile, and (3) the SS-16.

In the case of encryption, the report found that "the Soviet encryption practices constitute a violation of a legal obligation prior to 1981 and a violation of their political commitment subsequent to 1981. The nature and extent of encryption of telemetry on new ballistic missiles is an example of deliberate impeding of verification of compliance in violation of this Soviet political commitment."

In the case of the SS-X-25 missile, the report found that "while the evidence is somewhat ambiguous, the SS-X-25 is a probable violation of the Soviets' political commitment to observe the SALT II provision limiting each party to one new type of ICBM. Furthermore, even if we were to accept the Soviet argument that the SS-X-25 is not a prohibited new type of ICBM, based on the one test for which data are available, it would be a violation of their political commitment to observe the SALT II provision which prohibits (for existing types of single reentry vehicle ICBMs) the testing of such an ICBM with a reentry vehicle whose weight is less than 50 percent of the throw-weight of that ICBM."

In the case of the possible deployment of banned SS-16 ICBMs at Plesetsk, the report found that "while the evidence is somewhat ambiguous and we cannot reach a definitive conclusion, the available evidence indicates that the activities at Plesetsk are a probable violation of their legal obligation not to defeat the object and purpose of SALT II prior to 1981 during the period when the Treaty was pending ratification, and a probable violation of a political commitment subsequent to 1981."

The report also found that the Soviet Union had violated its legal obligations under the Biological Weapons Convention of 1972 and the Geneva Protocol of 1925 and its political commitments under the Helsinki Final Act concerning the notification of military exercises. Finally, in the case of the Threshold Test Ban Treaty (TTBT), which limits underground nuclear tests to 150 kt, the report found that, "while the available evidence is ambiguous, in view of ambiguities in the pattern of Soviet testing and in view of verification uncertainties, and we have been unable to reach a definitive conclusion, this evidence indicates that Soviet nuclear testing activities for a number of tests constitute a likely violation of legal obligations under the TTBT." Under international law the United States and the Soviet Union have a legal obligation to the unratified TTBT until one of them declares that it does not intend to ratify it, as the United States did with the SALT II Treaty.

The Soviet Union in effect denied all of these charges. It identified the large radar near Krasnoyarsk as a space track radar, which would be permitted under the ABM Treaty. Encryption practices were held not to impede the verification of the treaty's provisions, and the United States had refused on security grounds to answer Soviet questions as to the specific information denied. The SS-X-25 was identified as a permitted modernization of the SS-13, and the charge that its reentry vehicle weighed less than half of its throw-weight was denied. The charge that the SS-16 had been deployed at Plesetsk was also denied. As to the other charges, the Soviet Union denied that there had been any tests over 150 kt and dismissed the charges arising from the Biological Weapons Convention and the Geneva Protocol as political propaganda.

The Soviet Union also responded by publicly releasing a diplomatic note on January 29, 1984, that charged the United States with a long list of alleged violations of SALT I and SALT II as well as other arms control agreements. The Soviet note charged that by deploying Pershing II missiles and long-range cruise missiles in Western Europe, the United States had directly violated the "non-circumvention" provision in SALT II, since from the Soviet point of view these missiles were strategic in character. With regard to the ABM Treaty, the note charged the United States with violating specific provisions by developing both a mobile and a space-based ABM radar system; by developing multiple warheads for ABM intercepters; by building and upgrading large phased-array radars on its coasts (Pave Paws) that, despite their early warning function, could cover large parts of the United States and serve as battle management radars for an ABM system; and by incorporating ABM capabilities in the intelligence radar on Shemya Island. The note also reopened earlier questions about the shelters placed over Minuteman silos during construction work. It charged that the Minuteman II silos had been modernized to be compatible with MIRVed Minuteman III missiles and suggested that such missiles may in fact be deployed there now.

In the absence of more detailed information on the U.S. charges about Soviet violations, initial domestic criticism of the President's action by supporters of SALT II focused on the undesirable consequences of formally and publicly charging the Soviet Union with treaty violations, particularly when some of the evidence was admittedly "ambiguous." It was argued that this action would make it extremely difficult to conduct constructive discussions or work out mutually acceptable solutions to these problems in the SCC. Moreover, concern was expressed over the long delay that had occurred before the SALT II issues had been raised in the SCC. Finally, it was emphasized that the force of the U.S. position on SALT II compliance issues, which had been weakened by the failure of the United States to ratify the treaty, was essentially destroyed by the formal statement, underscored in the President's report, that the United States had no intention of ratifying the treaty.

PART II

THE STRATEGIC ARMS REDUCTION TALKS (START)

The Strategic Arms Reduction Talks between the United States and the Soviet Union opened in Geneva, Switzerland, on June 29, 1982. The U.S. negotiating position, which has gone through several revisions, rejects the SALT approach to equal aggregates. Instead, it seeks major reductions, particularly in ICBMs, to establish equal destructive power of U.S. and Soviet missile forces. The Soviet Union continues to support the SALT approach in the START negotiations and seeks modest reductions within the SALT II framework. Despite various revisions in both sides' proposals, there had been little significant progress in narrowing the fundamental differences between the two positions by the end of the fifth round of START. At that point the Soviet negotiators refused to set a date to resume the negotiations, contending that the U.S. deployment of intermediate-range missiles in Europe had created a new strategic situation that had to be reexamined.

BACKGROUND

The Origins

During the 1980 presidential campaign, candidate Ronald Reagan opposed the unratified SALT II Treaty and promised, if elected, to withdraw the "fatally flawed" treaty from the Senate. He argued that the treaty did not limit throw-weight, the true measure of destructive power, and did not close the "window of vulnerability" caused by accurate Soviet ICBM warheads aimed at U.S. ICBMs. After several months in office the new administration announced that while it reviewed arms control policy, the United States would not undercut the provisions of the SALT II Treaty as long as the Soviet Union did likewise.

The new administration did not initially announce its own approach to strategic arms control, although it did state that a prerequisite for genuine future arms control was to redress the strategic imbalance and restore a margin of safety with the Soviet Union. When the President announced his military program, he called for a 10 percent increase in the military budget over each of the next five years "to restore our defensive forces and to close that window of vulnerability that was opened in recent years with the superiority of Soviet forces." The administration emphasized that it would approach arms control as only a single element in a full range of political, economic, and military efforts. The administration also stressed the need for more effective verification in its new approach to arms control, citing the alleged failure of the Soviet Union to comply with existing agreements.

As domestic and NATO pressure for arms control increased, the President announced in November 1981 that strategic arms talks would possibly begin the following year. He stated that these negotiations, which would be called Strategic Arms Reduction Talks, or START, would have the goal of substantially reducing strategic nuclear arms. Although the negotiations on intermediate-range nuclear forces (INF) began in late November 1981 under strong political pressure from the NATO allies, the START negotiations did not actually begin for another eight months.

The issue of "linkage" of arms control negotiations with the overall U.S.-Soviet relationship, which had been a recurring problem in SALT, arose at the beginning of 1982 in connection with the Polish crisis. This played a role in postponing initiation of the START negotiations. However, by March 1982 the administration came under increasing domestic pressure to initiate negotiations, with nuclear freeze resolutions being introduced in the House of Representatives and the Senate. Shortly afterward the administration, which opposed a nuclear freeze on the grounds that it would leave the United States in a position of strategic inferiority, publicly set forth its preferred approach to nuclear arms control. On March 31, 1982, in his first prime time news conference, the President invited the Soviet Union to join with the United States in negotiations to reduce nuclear weapons substantially. The President also endorsed the Jackson-Warner freeze resolution, which called for reductions to equal levels prior to a freeze. The President contended that since the Soviet Union had "a definite margin of superiority," an immediate freeze would put the United States in a dangerous and disadvantageous position.

Initial START Proposals

President Reagan outlined the elements of the START proposal on May 9, 1982, in an address at Eureka College. In the first phase of the proposal, the United States and the Soviet Union would reduce their arsenals of nuclear warheads on land- and sea-based ballistic missiles from the current levels of around 8,000 to 5,000, with no more than half, or 2,500, of those warheads on land-based missiles. The first phase would also include a limit of 850 on "deployed ballistic missiles," the unit of measure introduced to replace launchers, the SALT II measure of ballistic missiles. In the second phase of the proposal, both nations would accept an equal ceiling on the throw-weight of all nuclear missiles.

The President said that the U.S. goal was to enhance deterrence and achieve stability through significant reductions in "the most destabilizing nuclear systems—ballistic missiles, and especially intercontinental ballistic missiles—while maintaining a nuclear capability sufficient to deter conflict, underwrite our national security and meet our commitment to our allies and friends." Strategic long-range bombers were not included in the President's outline of the START proposals, but under questioning, administration officials said that the United States would be prepared to deal with bombers and cruise missiles throughout both phases of the arms control talks with the Soviet Union.

In declaring a readiness to negotiate an accord with the United States on May 18, 1982, Soviet President Brezhnev stated that the proposed U.S. approach would require a unilateral reduction in the Soviet arsenal. He proposed instead that the accord should either ban or severely restrict the development of all new types of strategic armaments. Brezhnev also called for a nuclear freeze "as soon as the talks begin." When the United States and the Soviet Union simultaneously announced their agreement to begin the START negotiations, President Reagan again pledged to "refrain from actions which would undercut" the unratified SALT II Treaty so long as the Soviets showed the same restraint.

The START negotiations began in Geneva on June 29, 1982. In response to the U.S. proposals, the Soviet Union presented a proposal that included an interim freeze on strategic arms, limits based on the SALT II framework (involving a 20 percent reduction of the SALT II ceilings on the aggregate of central strategic systems from 2,250 to 1,800), and unspecified reductions in the various SALT II subceilings. In presenting this proposal, the Soviet Union emphasized that parity presently existed between both sides' strategic systems. Over the next year the two sides slowly elaborated the details of their proposals, but little progress was made in bridging the gap between the two radically different approaches.

The Scowcroft Commission and Build-Down

As the negotiations proceeded in Geneva, the congressional debate on the nuclear freeze and the MX missile intensified. When it became apparent that the latest MX basing mode, known as dense pack, was unacceptable to Congress, President Reagan established the Special Commission on Strategic Forces in January 1984, under the chairmanship of General Brent Scowcroft, to review the U.S. strategic modernization program, particularly the future of the land-based ICBM. At the same time, Senator William Cohen (R-Maine) began his efforts to mobilize members of Congress who were opposed to the freeze approach, concerned about the apparent nonnegotiability of the U.S. START position, and interested in an arms control formula that would accommodate modernization of U.S. strategic forces.

On February 3, 1983, Senator Cohen and Senator Sam Nunn (D-Ga.) introduced the guaranteed "build-down" resolution, which at this stage called on each side to eliminate two older nuclear warheads for each new warhead added to its force. Senator Cohen explained that the resolution provided for reductions while embodying the principle that weapons modernization could be stabilizing and also provided for reductions. Although the administration did not publicly endorse the build-down approach, President Reagan privately supported the idea in a conversation with Senator Cohen. Nuclear freeze advocates, on the other hand, criticized the build-down as a political device to make it easier for members of Congress to vote for the MX. Nonetheless, 43 senators agreed to cosponsor the build-down resolution within seven weeks of its introduction in Congress.

On April 6, 1983, the Scowcroft Commission gave the President its report. The report proposed a threefold approach to the modernization of the ICBM force: deploying 100 MX missiles in existing Minuteman silos to satisfy the immediate needs of the ICBM force; initiating engineering design of a small single-warhead ICBM (Midgetman) to reduce the value of individual targets and to permit flexibility in basing for better long-term survivability; and seeking arms control agreements designed to enhance strategic stability by counting warheads rather than deployed missiles. The report called for a higher missile limit than the 850 in START, while maintaining the ceiling on warheads, to encourage both sides to move to smaller, single-warhead missiles. The report also minimized the ICBM "window of vulnerability" problem, noting that the different components of U.S. strategic forces should be assessed collectively and not in isolation.

Despite the Scowcroft report's suggested modification in the START position and its deemphasis of the window of vulnerability, President Reagan endorsed the report and called on the Congress for prompt approval of the MX. Several key moderate Democratic members of Congress championed the report and sought a bargain with the administration whereby they would support the MX program if the administration would adopt a more forthcoming approach at START and fund the Midgetman program. After a contentious debate following the release of the report, the House of Representatives passed a much amended nuclear freeze resolution by a vote of 278 to 149, with both supporters and opponents of the freeze claiming victory. With the freeze resolution on its way to the Republican-controlled Senate and some moderate congressmen conditioning their MX vote on a more flexible approach to arms control, President Reagan publicly committed himself to incorporate a new build-down approach into the U.S. START position. Shortly afterward, in mid-May 1983, the House of Representatives and the Senate released research and development funds for the MX missile.

The Revised U.S. START Proposal

President Reagan followed the action on the MX with the announcement that he had given the U.S. negotiators at START "new flexibility" in an effort to obtain an agreement. The U.S. position at START was reportedly modified to increase the original ceiling of 850 deployed ballistic missiles to 1,250. On the issue of throw-weight, the President said, "We believe, as does the Scowcroft Commission, that stability can be increased by limitations on the destructive capability and potential of ballistic missiles. As a consequence, we will continue to propose such constraints which indirectly get to the throw-weight problem while making clear to the Soviets our readiness to deal directly with the corresponding destructive capability if they prefer." He explained that throw-weight could be addressed "indirectly" by counting missiles and warheads. He also stated that the administration was giving "high priority" attention to how the concept proposed by Congress of a "guaranteed build-down" of U.S. and Soviet strategic nuclear weapons could be implemented within the context of the modified START proposal.

In mid-July 1983 the United States presented a draft treaty that incorporated the new position. The draft treaty reportedly included the equal missile warhead ceilings of 5,000, with no more than 2,500 land-based; the newly increased level of 1,250 deployed ballistic missiles; a separate bomber ceiling of 400, which included the Soviet Backfire bomber; a limit of 20 air-launched cruise missiles per bomber; and alternative approaches to limiting throw-weight. The three approaches to throw-weight limitations were (1) indirect limitations by subceilings on heavy and medium missiles, (2) a direct ceiling on aggregate missile throw-weight (with the United States reportedly insisting on a level far below the Soviet 5.6 million kilograms and approaching the U.S. level of

1.8 million kilograms), or (3) an alternative approach the Soviet Union might suggest for reducing its superiority in throw-weight.

At the same START session the Soviet Union presented further details of its proposal. For the first time it explicitly indicated that not only the aggregate ceiling but the SALT II subceilings as well would be reduced by approximately 20 percent. The Soviet Union also eliminated provisions that would have banned the new U.S. Trident II D-5 missile, limited long-range cruise missiles on aircraft to a range of 600 km, stopped the deployment of new Trident submarines at four to six, and limited the missiles on each submarine to 16. The Soviet proposal did maintain the provision for limiting the range of sea- and ground-launched cruise missiles to 600 km.

The Soviet Union reacted publicly to the proposed modifications in the U.S. proposal by saying that they did not change the fundamental inequity in the U.S. position and offered no more promise of an agreement than did the previous position. The United States observed that the elaboration of the Soviet position showed that there had been some movement in the talks, but acknowledged that the two positions remained far apart.

Renewed congressional skepticism about the administration's commitment to arms control and the plan to put the MX in vulnerable silos became apparent when in July the House endorsed the MX missile by a vote of only 220 to 207. Although the Senate then authorized the first group of MX missiles by a vote of 58 to 41, Senate build-down supporters who had voted for the MX were annoyed by the administration's slow pace in incorporating the build-down into START. With another MX vote scheduled for the fall, a small coalition of senators and representatives decided to use the leverage of the upcoming vote to obtain further revisions in the U.S. START position. Despite the deep chill in U.S.-Soviet relations caused by the Soviet downing of a Korean airliner, the congressmen persisted in their pressure on the President.

The build-down resolution suffered a setback in late September when the Senate Foreign Relations Committee failed to muster a majority for either the nuclear freeze or the build-down and sent both resolutions to the Senate floor stating it agreed with neither. Nevertheless, strong congressional support continued for the build-down concept, which was broadened in congressional negotiations with the administration to include bombers by introducing a measure of "destructive capability" that related such factors as missile throw-weight, aircraft takeoff weight, and MIRV and ALCM capabilities. The administration did not accept the proposed definition of destructive capability, but it agreed that some version of this approach might form the basis from which to proceed with a guaranteed build-down.

The U.S. START Proposal Incorporating Build-Down

On October 4, 1983, President Reagan announced that the United States would incorporate the build-down concept into the basic U.S. negotiating position. The build-down concept unveiled by the President was much more detailed than the build-down originally proposed by Senator Cohen. The new U.S. position included a proposal that the reduction to 5,000 missile warheads be carried out in whichever of the two following ways produced the greatest annual reduction in warheads: a link between warhead reductions and modernization that would use variable ratios to identify how many existing nuclear warheads must be withdrawn as new warheads of various types are deployed, or a guaranteed annual reduction of 5 percent in the total number of missile warheads. Specifically, the build-down provision reportedly called for the removal of two old warheads for each new MIRVed land-based missile warhead, three old warheads for every two new submarine-based missile warheads, and one old warhead for each new single-warhead land-based missile. In addition, the President stated that the U.S. delegation would be prepared to discuss the build-down of bombers and additional limitations on the air-launched cruise missiles carried by bombers, and to negotiate trade-offs that would take into account Soviet advantages in missiles and U.S. advantages in bombers in ways that would give each side maximum flexibility while maintaining movements toward greater stability. At the same time, the administration made clear that it was keeping intact the main features of the basic U.S. START proposal, including the reduction of missile warheads to 5,000, the limit on deployed ballistic missiles of 1,250, the need to reduce the throw-weight discrepancy between the two sides, and a ceiling of 400 on bombers.

Congressional supporters of the build-down hailed the President's action as a positive move in the arms control process. They stated that it demonstrated the willingness of the United States to make trade-offs between the U.S. lead in bombers and the Soviet lead in missiles. Congressional opponents of the build-down, particularly those who supported the comprehensive nuclear freeze, questioned the President's initiative. They emphasized that it would still allow dangerous, destabilizing first-strike systems to be produced and deployed and that it did not necessarily give the Soviet Union more flexibility in structuring its reductions, since the variable ratios discriminated against the land-based missiles that constitute 70 percent of the Soviet force. On October 31, 1983, the Senate rejected a legislative amendment supporting the comprehensive nuclear freeze and avoided a direct test on the build-down by voting in a parliamentary maneuver to postpone further debate on the approach.

The Soviet reaction to the new initiative was swift. Within a day of the President's offer, the Soviet news agency Tass dismissed the new U.S. proposal as a public relations ploy aimed at securing congressional approval of the MX missile and the planned deployment of medium-range nuclear arms in Europe. The same day as the President's offer, the Soviet Union had proposed at the United Nations a comprehensive nuclear freeze resolution, which was described as not inconsistent with their proposals in Geneva. Several weeks later, in a more detailed editorial on the new build-down initiative, *Pravda* called it entirely one-sided because it aimed chiefly at reducing the number and destructive power of ICBMs. With 70 percent of the Soviet force in ICBMs and only 20 percent of the U.S. force in these systems, *Pravda* stated that the plan was aimed at weakening the Soviet Union while allowing the United States to go ahead with all of its planned deployments for its strategic arsenal.

In Geneva the Soviet delegation reportedly showed no interest in the build-down proposals, arguing that the proposal still focused in a discriminatory manner on slashing Soviet ICBMs. At the end of Round V of START, which followed the Soviet walkout from the INF negotiations, the Soviet delegation did not set a resumption date for the talks, saying that the deployment of Pershing II and cruise missiles in Europe had changed "the overall strategic situation," which had to be reexamined. In response, President Reagan stated that the move was "more encouraging than a walkout" and that he hoped Soviet negotiators would return in 1984.

U.S. AND SOVIET START PROPOSALS

Complete descriptions of the U.S. and Soviet negotiating proposals at START have not been made public, but the main elements of the revised U.S. and Soviet START positions at the end of Round V in December 1983 have been announced by the U.S. government or reported authoritatively in the press.

The U.S. START Proposal as of December 1983

The revised U.S. START proposal at the end of Round V included the following elements:

- Reductions to equal levels of 5,000 for both sides in the aggregate number of warheads on land- and sea-based ballistic missiles.
- Equal limits on deployed land- and sea-based ballistic missiles of 1,250 (originally 850).
- Equal ceilings on aggregate missile throw-weight by one of the following approaches: (1) indirectly, by a sublimit of 2,500 on warheads on deployed land-based missiles and sublimits on the number of deployed land-based medium and heavy ballistic missiles (originally the United States proposed a sublimit of 210 on these missiles, of which no more that 110 could be heavy ballistic missiles); (2) by unspecified equal ceilings on overall missile throw-weight that would be substantially below the present Soviet level of 5.6 million kilograms (originally this approach was to be the second phase of the START negotiations); or (3) by an alternative approach to be suggested by the Soviet Union to reduce its superiority in throw-weight.
- The proposed reductions to a ceiling of 5,000 missile warheads would be accomplished in annual increments by whichever of the following two procedures produced the greater annual reduction: (1) a guaranteed annual reduction of 5 percent in the number of missile warheads or (2) build-down in missile warheads by reductions linked to any modernization programs by variable ratios defining the number of existing strategic missile warheads that must be withdrawn as new strategic missile warheads are introduced. Reportedly, to encourage modernization toward more stable systems, the build-down would require the removal of two old warheads for each new MIRVed land-based missile warhead, three old warheads for every two new submarine-based missile warheads, and one old warhead for each new single-warhead land-based missile.
- An equal ceiling for both sides of 400 strategic bombers (to include the Soviet Backfire bomber), with a limit of 20 cruise missiles per bomber.
- A willingness by the U.S. delegation to (1) address the build-down of bombers, (2) discuss additional limitations on the air-launched cruise missiles carried by bombers, and (3) negotiate trade-offs that would take into account Soviet advantages in missiles and U.S. advantages in bombers in ways that would give each side maximum flexibility while maintaining movements toward greater stability.
- Unspecified verification measures involving more comprehensive and intrusive measures than in previous agreements to ensure compliance. No encryption of flight test data must be permitted.
- A series of confidence-building measures.

The Soviet START Proposal as of December 1983

The revised Soviet START proposal at the end of Round V included the following elements:

- An interim freeze of unspecified coverage on strategic nuclear arms while the negotiations are in progress.
- A limit of 1,800 on the aggregate number of ICBM launchers, SLBM launchers, and heavy bombers (reduced by 20 percent from the SALT II limit of 2,250).
- A limit of 1,200 on MIRVed missile launchers plus bombers equipped with air-launched cruise missiles (reduced from the SALT II limit of 1,320).
- A limit of 1,080 on MIRVed missile launchers (reduced from the SALT II limit of 1,200).
- A limit of 680 land-based MIRVed ICBM launchers (reduced from the SALT II limit of 820).
- Unspecified equal aggregate limits on missile warheads and bomber weapons.
- A number of modernization constraints, including a ban on the deployment of ground- and sea-launched cruise missiles with a range greater than 600 km.
- Corresponding verification provisions.

The Soviet Union also dropped earlier provisions that would have banned the Trident II missile and long-range cruise missiles on aircraft, limited the U.S. deployment of new Trident submarines to four or six, and reduced the number of missiles on future Trident submarines from 24 to 16.

THE MAIN ISSUES SURROUNDING START

The Strategic Relationship

START Supporters' Assessment of the Strategic Relationship

Underlying the Reagan Administration's approach to START is the premise that the United States is strategically inferior to the Soviet Union. Consequently, the United States must first redress the strategic balance with military programs that will build up U.S. strategic forces and provide a necessary margin of safety. Any arms control agreement must therefore either await the restoration of the strategic balance by a

U.S. military buildup or achieve the strategic balance at lower levels with a substantial restructuring of forces.

In the Reagan Administration's view, the strategic forces of the two sides were roughly in balance when the SALT I agreements were signed in 1972. The Soviet Union achieved this balance because, in the late 1960s and early 1970s, the United States decided not to contest Soviet efforts to attain equality in strategic forces. According to START supporters, this equality was lost during the 1970s, when the United States exercised unilateral restraint in its strategic programs. Once the MIRV programs for the Poseidon SLBM and Minuteman III ICBM were completed in the first half of the 1970s, the United States canceled or stretched out a number of new strategic programs. The B-1A bomber program was canceled, the cruise missile program was cut back, construction of the Ohio-class Trident ballistic missile submarines was delayed, and the development of the MX was stretched out.

In contrast to this U.S. restraint, according to the administration, the Soviets since 1972 have introduced three new MIRVed ICBM types (the SS-17, SS-18, and SS-19), which markedly increased the Soviet throw-weight advantage; four new SLBMs (the SS-N-8, SS-N-7, SS-N-18, and, in development, the SS-NX-20); three types of Delta-class ballistic missile submarines; the new large Typhoon ballistic missile submarine; the Backfire bomber; and in development, the Blackjack bomber. The administration holds that during this period, by any measure, the Soviet Union achieved strategic superiority over the United States.

With regard to the general military balance, President Reagan has stated that "in virtually every measure of military power the Soviet Union enjoys a decided advantage." He has emphasized Soviet advantages in total numbers of intercontinental missiles and bombers and the fact that the Soviet Union has deployed over a third more land-based missiles than has the United States, with the number of U.S. ICBMs essentially frozen since 1965. The President has stated that the Soviet Union has put 60 new ballistic missile submarines to sea in the last 15 years, whereas until last year the United States had not commissioned any in the same period. With regard to strategic bombers, the President has noted that the Soviet Union has built over 200 modern Backfire bombers—and is building 30 more a year—whereas the United States has deployed no new strategic bombers for 20 years. Finally, the President has emphasized that the Soviet Union invests 12 to 14 percent of its gross national product in military spending, which is approximately twice the U.S. percentage.

In short, the United States, according to the administration, finds itself in a position of dangerous strategic inferiority that must be overcome by unilateral rearmament or by a new approach to arms control.

START Critics' Assessment of the Strategic Relationship

Domestic critics of the Reagan Administration's assessment of the strategic relationship between the United States and the Soviet Union argue that the United States is not strategically inferior to the Soviet Union and that essential equivalence continues to exist today. Although there has been a major Soviet strategic buildup during the last decade, the modernization of U.S. strategic forces has prevented any significant shift in the overall strategic balance. Consequently, the United States should continue to approach arms control in the context of essential equivalence.

Critics disagree with the administration's assessment that strategic parity was lost during the 1970s. They point out that during this period the United States continued to modernize its strategic forces within the constraints of SALT I and maintained parity in the strategic balance between the superpowers. The U.S. modernization program in the 1970s involved all three legs of the strategic triad. In this period the United States deployed more than a thousand MIRVed missiles, thereby increasing the total number of U.S. missile warheads nearly fourfold. The United States also substantially increased the capability of the B-52 force by deploying short-range attack missiles (SRAMs), by incorporating improved avionics, and then by initiating the air-launched cruise missile program. In addition, development was under way on the Stealth bomber that uses advanced technology to penetrate air defenses. The C-4 Trident missile was developed and retrofitted into the Poseidon submarines, and the first new Trident submarine went on patrol in late 1981. The United States upgraded the Minuteman missile force with Mark 12A warheads, which increased accuracy and yield. The survivability of the land-based force was also increased by hardening Minuteman silos. Finally, as permitted in SALT II, the United States was developing the MX missile, which was originally intended to be deployed in a survivable basing mode.

Critics state that when measuring the forces of the two superpowers, it is important to bear in mind the asymmetry of their arsenals. These asymmetries reveal Soviet advantages in some areas and U.S. advantages in others. For example, the Soviet Union today has more ballistic missiles with larger payloads and more megatonnage. But to offset this advantage the United States has more warheads with greater accuracy and major advantages in the operating effectiveness of its submarine and bomber forces. Critics also argue that it is misleading to compare numbers without looking at missions, geography, and the forces of allies.

These critics also challenge the President's statements on the strategic balance as being misleading. Concerning the statement that the United States has not increased its number of ICBMs since 1965, critics note that the Soviet Union as well as the United States froze the number of land-based missile launchers in the SALT I Interim Agreement of 1972. Since then the Soviet Union has in fact decreased the number of fixed land-based ICBM launchers by some 200 in exchange for an equal number of additional SLBM launchers as allowed under SALT I. Concerning the buildup in new ballistic missile-firing submarines over the last 15 years, critics point out that the U.S. submarine force has been substantially upgraded during that time by the deployment of MIRVed Poseidon missiles. American submarines now carry many more ballistic missile warheads per submarine than do Soviet submarines. Moreover, it is generally agreed that U.S. ballistic missile submarines are decidedly superior to their Soviet counterparts in overall performance, since U.S. submarines spend more time at sea and operate much more quietly, which reduces the possibility of detection.

Critics of the President's assessment also note that the U.S. and Soviet development cycles for these systems are out of phase. New U.S. submarines and missiles, whose development cycle began ten years ago after the last new submarines had been completed, are now just beginning to be deployed. In response to the President's statements about the buildup of Soviet Backfire bombers, the critics assert that, despite its age, the B-52 is a far better long-range bomber than either the Backfire, which has questionable strategic capability, or the standard Soviet long-range bombers, Bears and Bisons, which have not been modernized to nearly the same extent as the B-52. In the 1970s, for cost-benefit reasons, the United States decided that instead of procuring a new bomber it would upgrade the B-52 bombers, first by developing short-range attack missiles and then by developing highly accurate long-range cruise missiles to ensure the ability to penetrate Soviet defenses. Finally, the estimate that the Soviet Union spends 12 to 14 percent of its gross national product on arms compared with the U.S. figure of 6 to 7 percent is misleading, since the U.S. gross national product is almost double that of the Soviet Union and the method of calculation tends to inflate the Soviet military budget. Moreover, recent U.S. intelligence analyses indicate that the growth rate in Soviet military spending since 1976 has been only about 2 percent per year—about the same as the growth rate of the Soviet gross national product—with no increase in the procurement sector of the military budget. This is far less than estimates in the late 1970s and early 1980s.

In short, these critics believe that essential equivalence continues to exist today and that the United States does not need a major strategic arms buildup or major asymmetric reductions to enter a mutually advantageous strategic arms control agreement.

The Soviet View of the Strategic Relationship

For its part, the Soviet Union insists in its public statements that an approximate military balance or parity exists now and is being maintained between the Soviet Union and the United States. It also stresses that this approximate parity is sufficient for its defense needs and that it does not seek strategic superiority. In official statements and documents the Soviet government has emphasized that by the mid-1970s an approximate balance or equilibrium had been struck in the quantity and quality of strategic nuclear arms between the two nuclear superpowers. The Soviet government asserts that since the signing of SALT II it has done nothing in the field of strategic armaments to disturb this equilibrium.

With regard to U.S. assertions that the Soviet Union has achieved strategic superiority, Soviet spokesmen argue that U.S. assessments are misleading because they compare selected components from the overall mass of strategic weaponry. These assessments focus only on land-based missiles, say the Soviets, ignoring U.S. ballistic missile submarines and heavy bombers, where the United States has a major advantage. According to Soviet statements, the United States also has a greater number of nuclear warheads.

Soviet statements also specifically reject the U.S. government's assessment of the window of vulnerability and its assertion that the United States froze its forces in the 1970s. Soviet officials argue that growth of U.S. strategic forces has been uninterrupted. They point out that three new weapon systems were produced in the United States in large quantities during the 1970s. Five hundred and fifty Minuteman III intercontinental ballistic missiles became operational, each with three MIRVed warheads. Some 496 Poseidon C-3 missiles, each with 10 to 14 warheads, were placed on 31 nuclear submarines. The accuracy of these systems was more than double that of the previous systems. The SRAM and ALCM missile systems were introduced in the armaments of the upgraded U.S. strategic bomber force. Finally, by the end of the 1970s the U.S. Navy began to retrofit Trident I missiles, which have greater range, throw-weight, and accuracy, in existing Poseidon submarines and new Trident submarines, and by the early 1980s air, land, and sea versions of long-range cruise missiles were being deployed.

Soviet officials emphasize that there is in fact no window of vulnerability, as the Scowcroft report finally acknowledged, and that the strategic forces of the United States and the Soviet Union continue to be in equilibrium. They contend that the U.S. government's present assessment of Soviet superiority is simply propaganda designed to gain domestic support for its new nuclear programs, which have the objective of gaining strategic superiority over the Soviet Union.

In connection with their postponement of further START negotiations, Soviet officials went further and stated that the U.S. deployment of Pershing II and ground-launched cruise missiles in Europe has altered the strategic balance and therefore requires a new assessment of their strategic arms control proposals.

The Rationale for START: Selective Deep Cuts to Restore Stability

START Supporters' Approach

The Reagan Administration has emphasized that the deep reductions proposed in START will lead to equal overall limits on missile throw-weight, which is the true measure of the "destructive capability" of strategic forces. It has argued that the present strategic relationship is destabilizing because of the large Soviet advantage in the throw-weight of its ICBMs, which are capable of carrying large numbers of high-yield, accurate warheads.

The key objective of START is to reduce radically the number of medium and large Soviet ICBMs, which account for a large percentage of the throw-weight of Soviet strategic forces. The Soviet medium and heavy ICBMs are the most threatening systems because they combine large numbers of warheads with high kill probabilities due to the high accuracy and yield of the warheads. Today these systems carry four to ten warheads; potentially they could carry as many as three times those numbers. These Soviet ICBMs not only threaten present hardened U.S. land-based retaliatory systems and command and control networks, but are also destabilizing because they are themselves vulnerable to attack, which creates pressure for a dangerous launch-on-warning doctrine. Moreover, an excess in throw-weight capability gives the Soviet Union a capability to add more warheads to existing missiles. Consequently, deep reductions of these systems are the best way to ensure the survivability of U.S. deterrent forces.

The administration argues that the major restructuring of strategic forces implicit in the U.S. START proposals will lead to a much more stable strategic relationship. Under the U.S. START proposals the Soviet Union would be forced to decrease its dependence on land-based ICBMs. The build-down provision would encourage future Soviet force modernization in the direction of submarine-based forces or small single-warhead ICBMs. This restructuring would stabilize the strategic balance, since submarine-based forces and small single-warhead ICBM systems are more survivable and therefore less likely to provoke a "use it or lose it" stance by either superpower.

The administration initially emphasized that a major advantage of its START proposals was that the Soviet Union would have to reduce the number of warheads on its ICBMs from 6,000 to fewer than 2,500. It suggested that this reduction in accurate Soviet ICBM warheads would help close the window of vulnerability by making it easier to solve the problem of U.S. ICBM vulnerability. Such a solution was important not only because a successful attack would reduce the U.S. capability for prompt retaliation but also because concern about vulnerability may lead to a destabilizing launch-on-warning policy. After a preemptive Soviet strike the U.S. retaliatory capability would be qualitatively impaired, because the ICBM force is the only part of the strategic triad that can quickly respond with a high-accuracy attack on the remaining Soviet strategic forces.

START supporters argued that the upward revision of the limit on deployed missiles from 850 to 1,250, as recommended by the Scowcroft Commission, would further help alleviate the vulnerability problem by providing more flexibility for the deployment of small single-warhead ICBMs. The proposed variable build-down ratios for reductions would also favor the move toward small single-warhead missiles, since there would be a one-for-one trade-off of warheads for new missiles of this type if either side decided to move in this direction. Supporters of the small missile argue that it would be cost effective and could be deployed in either a semihardened mobile mode or superhardened silo mode by 1990, and presumably sooner in Minuteman-type silos. However deployed, small single-warhead ICBMs would contribute to stability by increasing the survivability of both sides' land-based strategic forces. Increased survivability would result from both the reduced vulnerability and reduced target value of an ICBM force made up of low-value, single-warhead missiles.

The administration argues that the absence of constraints on modernization in its START proposal would allow both sides to develop their forces in more survivable modes. The United States would not be limited to a single new missile, as in SALT II, but could develop the MX, the Midgetman, and other land-based systems as well. In addition, the adoption of the variable build-down ratios would enhance the incentive to move toward stabilizing systems.

Proponents of the build-down argue that it would both permit stabilizing modernization and reduce warhead totals without requiring as drastic a restructuring of Soviet strategic forces as the original START proposal. The build-down approach coupled with the direct or indirect U.S. requirement for reductions in the aggregate missile throw-weight would move the Soviet Union away from its heavy dependence on destabilizing land-based systems. But proponents argue that the revised proposal should be more negotiable, since the President has indicated a willingness to negotiate unspecified trade-offs that would take into account U.S. advantages in bombers and Soviet advantages in missiles. Although the U.S. proposal retains a bomber limit of 400 (which would include the Soviet Backfire bomber), START proponents note that the administration has indicated it is willing to discuss proposals for the concurrent build-down of bombers and further restrictions on air-launched cruise missiles.

The administration has argued that the Soviet Union also stands to gain from the U.S. START proposal. The proposal would cap U.S. strategic forces and foster strategic stability, thereby reducing the risk of war.

Domestic Criticisms of START

The basic domestic criticism of the U.S. approach to START is that it cannot realistically be expected to provide the basis for an agreement. Instead of taking into account the asymmetry of the U.S. and Soviet strategic forces, START seeks to take unilateral advantage of the structural differences in these forces, according to this view. Critics maintain that the new build-down initiative, when taken in the context of the overall U.S. proposal, has not significantly altered this situation.

Critics point out that while the U.S. START approach would require the Soviet Union to undertake a radical restructuring of its forces, the United States could modernize its forces according to existing plans. Basically, the U.S. proposal calls for drastic reductions in Soviet land-based ICBM forces, which account for 70 percent of the Soviet Union's strategic assets. Specifically, the original ceiling of 2,500 on ICBM warheads (which is still retained as part of one of the approaches to an equal ceiling on throw-weight) would require a reduction of 60 percent in Soviet ICBM warheads. Moreover, the sublimit on medium and heavy missiles (which was originally set at 210, of which no more than 110 could be on heavy missiles) would require the dismantling of almost three quarters of Soviet modern, MIRVed ICBMs, whether or not there were replacements. Given the constraint of 1,250 deployed launchers, the Soviet Union would also have little incentive to replace these missiles with new small single-warhead missiles. Critics point out that the build-down initiatives do not alleviate these inequitable requirements. The ratios for building down, which favor submarine deployments or small single-warhead missile deployments, when combined with the explicit or implicit requirement for almost equal missile throw-weights, would not have any practical impact on this problem from the Soviet point of view.

The deep reductions in the U.S. START approach would have a much less drastic impact on U.S. strategic forces. Since only 20 percent of the U.S. strategic warheads are on land-based ICBMs, while two thirds of the U.S. strategic warheads are on submarines, much less restructuring of forces would be required. In fact, under the U.S. proposal the United States would be able to increase the number of warheads on its land-based ICBMs by 350. Moreover, the sublimit of 210 on medium and heavy missiles would permit the United States to deploy up to that number of MX missiles. Even under the new build-down approach to reductions, the United States would not have to restructure its forces as they are reduced to lower levels. Furthermore, the United States could continue to take advantage of those areas where it has a technological lead by continuing with its plans to deploy the MX, the Trident II, the B-1 and Stealth bombers, air-launched cruise missiles, and sea- and ground-launched cruise missiles.

Although the United States has proposed to negotiate trade-offs between missiles and bombers, some critics point out that there will be little room for such trade-offs, since the administration has established requirements for an equal missile throw-weight ceiling near the current U.S. level and an equal bomber ceiling. Moreover, critics point out that by separating strategic missiles and aircraft into two independent categories of 1,250 deployed missiles and 400 aircraft, the U.S. START proposal further complicates any trade-offs between areas of U.S. and Soviet advantage. The limit of 1,250 deployed missiles would require a major reduction in Soviet missiles, while the limit of 400 bombers would allow the United States to retain its entire active and planned bomber force. The modernized B-52 force, armed with short-range attack missiles and several thousand long-range cruise missiles, is a far more effective strategic force than the 150 Soviet Bison and Bear long-range bombers. Nevertheless, the Soviet Union appears to have a numerical advantage because the United States has included in the overall bomber total some 250 Backfire bombers, which the Soviet Union maintains perform theater and naval missions and do not have a strategic capability.

Critics also argue that the U.S. START proposal's call for equality in throw-weight, either directly or indirectly, makes it nonnegotiable. There has been a long history of controversy on how to quantify throw-weight, particularly as it relates to bombers, and on whether it is an effective measure of strategic capability. If throw-weight as defined by the United States were used as a measure of the strategic balance, the Soviets would have to cut their existing missile throw-weight by at least 60 percent to match the current U.S. capability. The Soviet Union has rejected this proposal. Some critics also argue that throw-weight is not an appropriate measure of strategic capabilities, since it does not reflect the current overall parity between the superpowers' strategic forces when all quantitative and qualitative factors are considered. For example, as the accuracy of warheads improves, throw-weight becomes less significant. Similarly, concern about the Soviet breakout potential, where the greater throw-weight of Soviet missiles would allow the deployment of more warheads on their missiles, is not an urgent problem, since the major undertaking of adding a substantial number of warheads to missiles would require testing that the United States would detect well in advance of deployment.

The START proposal has also been criticized because it does not include any qualitative restraints on the modernization of both sides' strategic forces. Even after including the build-down provisions, according to this argument, the U.S. proposal would do nothing to halt the qualitative arms race toward improved first-strike systems. Specifically, the agreement would allow the United States to continue to develop and deploy the MX and the Trident II missiles, cruise missiles, and B-1 and Stealth bombers, while equivalent improved systems could be developed and deployed on the Soviet side.

Some critics also point out that the U.S. START approach was originally advanced in part to deal with the vulnerability of U.S. land-based ICBMs. But reducing the Soviet land-based warheads from 6,000 to 2,500 would do little to reduce the vulnerability of the U.S. ICBM force, because it would also have to be reduced significantly to stay within the deployed missile limit. These analysts have emphasized that basing the MX in Minuteman silos will only heighten instability under the START reductions by creating vulnerable targets of particularly high value.

The build-down ratios proposed by the administration are designed to promote the development of small single-warhead Midgetman-type missiles. This represents a longer-term solution to the problem of ICBM vulnerability. It has in turn given rise to a variety of criticisms and questions relating to both the START provisions and the long-term posture of U.S. strategic forces. Some critics, while endorsing the general concept of Midgetman both as a less vulnerable land-based system and as a step toward the deMIRVing of strategic missiles, have raised questions as to whether enough Midgetman missiles could be deployed to constitute a credible independent force, within the ceiling of 1,250 deployed missiles, given other ICBM and SLBM forces that would presumably be retained. Technical questions have also been raised as to whether a mobile system could be hardened sufficiently to permit it to be confined to military reservations or whether it would have to move cross country or on public roads. The latter requirement could provoke domestic opposition in the United States that would not have to be faced in the Soviet Union. Critics have also argued that putting small missiles in hardened fixed silos or on relatively soft mobile launchers would do little to alleviate the vulnerability problem, because not enough missiles could be deployed under the U.S. START constraints to assure survivability against the number of accurate warheads that the Soviet Union could have under the proposed numerical ceiling on warheads.

Other critics argue that the U.S. approach to START arose largely from undue concern over the vulnerability of U.S. ICBMs, which the Scowcroft Commission has now put in better perspective. These critics note that land-based ICBMs are only 20 percent of the U.S. strategic force and that the different components of the U.S. strategic forces should be assessed collectively and not in isolation. The U.S. strategic forces are designed as an air, land, and sea triad so that any leg of the triad can deter attack. Even if all of the U.S. land-based ICBM force were destroyed, the remaining U.S. strategic capability in submarines and/or bombers would still be able to deliver a devastating retaliatory strike. Thus a Soviet preemptive counterforce attack on the U.S. land-based force, or the threat of such an attack, would serve no rational purpose. Some critics question the desirability of a Midgetman program stimulated by START. Without an arms control framework to limit the deployments, such a program could become a problem in itself. In the absence of an arms control agreement, this new system may be unconstrained and eventually even include a MIRVed payload. Unless questions of missile characteristics and verification can be managed within an arms control framework, according to these critics, the deployment of a large force of small mobile missiles on both sides could prove to be a major new factor in arms race instability due to the uncertainty in the number of missiles the other side might be deploying.

The Soviet Approach to START

The Soviet approach to the START negotiations has been to build on the SALT framework. Its proposals call for cuts of approximately 20 percent in the SALT II ceilings and subceilings. The Soviet government emphasizes that it took many years for the two sides to agree on a SALT framework that accounted for the different structures of the two sides' strategic forces and quantified the parity of forces that existed between the two sides. Soviet officials state that their proposal, which is based on the assessment that parity still exists, would substantially reduce the number of nuclear warheads to equal, agreed-upon ceilings. They also state that their proposal would severely limit the channels available for the continuation of the strategic arms race, and that the Soviet Union would be prepared to negotiate deeper reductions within this framework in the future.

According to the Soviet Union, its willingness to consider strict qualitative constraints has been demonstrated by its support for a nuclear freeze. Both Presidents Brezhnev and Andropov called for a freeze on strategic armaments while the START negotiations were in progress. Such a freeze was reportedly proposed without detail at the outset of the START negotiations. In October 1983 the Soviet Union introduced a resolution at the UN General Assembly calling for a comprehensive freeze on the testing, production, and deployment of nuclear weapons. In presenting these proposals the Soviet Union explicitly stated that they do not interfere with or contradict their START proposals for reductions in the SALT ceilings.

The Soviet Union argues that the U.S. approach to reductions in START selectively favors the United States and provides no qualitative constraints. Soviet officials point out that even with the build-down the U.S. proposal would require the Soviet Union to destroy a large fraction of its ICBM force while the United States proceeds unhindered with its plans to create new strategic weapon systems. Specifically, the U.S. proposal would allow the United States to deploy the MX, the Midgetman, Trident I and Trident II, and the B-1 and Stealth bombers. Consequently, far from building down U.S. long-range weaponry, the U.S. proposal would permit a massive buildup of U.S. forces, according to Soviet officials, and the U.S. warhead total, when cruise missiles are included, would rise significantly. The Soviet press has also noted that the Reagan Administration was trying to lock in an American advantage in heavy bombers by insisting that the Soviet Backfire bomber be included in the calculations.

The U.S. government has stated that the Soviet approach to START is not acceptable. According to U.S. officials, it would not correct the basic inequities between the strategic forces of the two countries and would lock the United States into a position of strategic inferiority. Although welcoming the Soviet acceptance of significant reductions from existing levels, the U.S. government has argued that the specific Soviet approach to reductions would not reduce the large relative Soviet advantage in MIRVed ICBMs, which are the most dangerous and destabilizing strategic weapons. U.S. officials contend that this problem is inherent in the Soviet approach to reductions, because the SALT II aggregate limits and sublimits are not directed at the proper measures of destructive capabilities. Finally, they emphasize that the Soviet approach is fundamentally flawed, because it is built on the incorrect premise that there is overall parity between the strategic forces of the United States and the Soviet Union.

Verification

The details of the verification provisions in the U.S. START proposals have not been publicly disclosed, and it is not clear how much, if anything, has been said about them in the negotiations. The administration has stated publicly that the United States would insist on going beyond the previous reliance on National Technical Means (see the section on verification in Part I of this chapter). President Reagan has stated that the United States cannot be sure the Soviet Union has complied with current arms control agreements because the verification provisions have been inadequate. Experience has shown, according to the administration, that agreements lacking adequate provisions for verification and compliance become a source of suspicion, tension, and distrust rather than reinforcing the prospects for peace.

Administration officials have stated that the verification provisions of a START agreement would have to include cooperative measures and on-site inspection to supplement National Technical Means. Among the cooperative measures that the United States reportedly would call for in START are a complete ban on the encryption of telemetry, an expanded exchange of data on nuclear forces, and notification of all ICBM and SLBM launches. A number of other measures have been discussed publicly, but it is not clear which, if any, have been included in the U.S. START proposal so far in the negotiations. These measures include prior notification of removal, dismantling, and destruction; on-site presence during removal, dismantling, and destruction; on-site presence at any facility intended for, or capable of, production or stockpiling of weapons or equipment banned or limited by the agreement; designation of the deployment areas for weapons and equipment limited by the agreement; and on-site presence to control and count the numbers of weapons and equipment that enter and leave designated deployment areas.

Critics argue that the administration appeared to be developing such high verification standards that concrete verification proposals, when they emerged, would be nonnegotiable. The calls for more on-site inspection have generated the most concern. Critics maintain that in many cases on-site inspections are actually less effective than National Technical Means, particularly when the NTM are supported by effective cooperative measures. Above all, critics warn that intrusive on-site inspection requirements can easily and unnecessarily become insuperable barriers to the successful negotiation of an agreement.

3 *The Nuclear Freeze*

INTRODUCTION

The rapid growth of public support for a comprehensive nuclear freeze has been a remarkable political phenomenon, reflecting a deep and widespread sense of frustration over the lack of progress in arms control negotiations to date. Although freeze proposals have taken somewhat different forms in various local, state, and congressional resolutions, they share the common objective of seeking a verifiable freeze by the United States and the Soviet Union on the testing, production, and deployment of nuclear weapons and their delivery systems. The freeze concept has increasingly become a political symbol of commitment to arms control and opposition to the Reagan Administration's approach to arms control. The Reagan Administration opposes a comprehensive freeze at current levels as being contrary to U.S. security interests.

BACKGROUND

The Origins

Over the years the United States and the Soviet Union have advanced a variety of nuclear freeze proposals as possible approaches to nuclear arms control. For example, a freeze or cutoff of fissionable material for nuclear weapons purposes, which was first suggested in the Eisenhower-Bulganin letters in the mid-1950s, surfaced as a concrete U.S. arms control proposal in the early 1960s. In 1964 the United States formally proposed to the Soviet Union a partial freeze on the number and characteristics of strategic nuclear offensive and defensive vehicles. The Soviet Union rejected the proposal, saying that it would freeze the Soviet Union into a position of strategic inferiority. In 1970, during the SALT I negotiations, the U.S. Senate passed a freeze resolution calling on the President to propose to the Soviet Union an immediate suspension by both countries of "the further development of all offensive and defensive nuclear strategic weapons systems." Many other arms control proposals have in fact been partial freezes. Among these are the SALT I and SALT II agreements and the proposed comprehensive ban on nuclear tests, which has been the subject of intermittent negotiations since the late 1950s.

Beginning in 1980, substantial grass-roots support has developed throughout the United States for the proposal of a comprehensive nuclear freeze on all nuclear weapons and their delivery vehicles. Within a few years the nuclear freeze had come to the forefront of the public debate on how best to control the nuclear arsenals of the superpowers. The origin of the current comprehensive freeze initiative is generally attributed to Randall Forsberg, founder and director of the Institute for Defense and Disarmament Studies in Massachusetts. In 1980 Forsberg prepared a public memorandum entitled "Call to Halt the Nuclear Arms Race," which challenged the United States and the Soviet Union to stop the nuclear arms race by adopting a comprehensive freeze on the testing, production, and deployment of nuclear weapons and their delivery systems. In the memorandum, Forsberg argued that ending the nuclear arms race with a comprehensive freeze was the crucial first step that the superpowers needed to take at this time, because the next generation of more dangerous counterforce nuclear weapons would disrupt the present balance of forces and increase the likelihood of nuclear war in a crisis.

The simple, straightforward language of the comprehensive nuclear freeze proposal attracted the attention of a private funder in Massachusetts who contributed the initial money in 1980 that set the freeze campaign in motion. The state senatorial districts in western Massachusetts were the first to pass the nuclear freeze referendum based on Forsberg's memorandum in November 1980. In March 1981 a national conference of peace groups met in Washington, D.C., where the groups decided to concentrate on promoting the freeze as a common strategy. The national freeze campaign that developed out of the 1981 Washington conference consisted of a loose coalition of grass-roots networks, including both existing antinuclear groups and new groups established by local citizens to promote the freeze proposal.

By early 1982 the freeze campaign was rapidly increasing its momentum. On ballots and in town meetings in Vermont and California, the freeze had achieved impressive successes. Within two years there were active freeze campaigns in every state and two thirds of the congressional districts in the nation. The rapid growth of popular support for the comprehensive nuclear freeze proposal can be attributed to a variety of factors: widespread anxiety about U.S.-Soviet political and military relations, the administration's early statements about fighting and surviving nuclear war, the administration's calls for a vastly increased defense budget, the administration's delay in initiating nuclear arms control negotiations with the Soviet Union and its failure to achieve progress once it did so, and the simple, direct language of the proposal.

The Congress and Freeze Resolutions

The nuclear freeze movement was raised to a national level on March 10, 1982, when identical nuclear freeze resolutions were introduced in the U.S. Senate and House of Representatives by Senators Edward Kennedy (D-Mass.) and Mark Hatfield (R-Oreg.) and Congressmen Edward Markey (D-Mass.) and Silvio Conte (R-N.Y.). The Kennedy-Hatfield freeze resolution stated that the United States and the Soviet Union should decide when and how to achieve a mutual verifiable freeze on the testing, production, and further deployment of nuclear warheads, missiles, and other delivery systems. The freeze would then be followed by negotiated reductions. The main premise of the resolution was that the strategic forces of the two superpowers were in a state of essential parity. This parity provided short-term stability in the strategic relationship that the freeze should urgently seek to preserve.

The introduction of the freeze resolutions in Congress sparked a heated policy debate. The Reagan Administration immediately rejected the Kennedy-Hatfield freeze approach. On March 31, 1982, President Reagan invited the Soviet Union to join the United States in substantially reducing nuclear weapons. But he specifically rejected an immediate freeze on the grounds that the Soviet Union's "definite margin of superiority" would make a freeze disadvantageous and dangerous to U.S. security and would militate against subsequent reductions. The President instead embraced an alternate resolution proposed by the late Senator Henry Jackson (D-Wash.) and Senator John Warner (R-Va.), which stated that the current nuclear imbalance was destabilizing and that a long-term, mutual, and verifiable freeze should occur after reductions brought the sides to an equal and sharply reduced level of forces.

Leading supporters of the SALT process were split on the freeze. Some endorsed the freeze, while others challenged the approach on the grounds that, despite its apparent simplicity, it would in fact take many years to negotiate and would deflect attention from more promising approaches to arms control, such as the signed but unratified SALT II Treaty. Moreover, it did not address the instabilities in the strategic balance or the need to reduce force levels. Other SALT supporters, while sharing doubts about the freeze, endorsed it as an effective political vehicle to apply pressure on the Reagan Administration to resume arms control negotiations. Still others supported the freeze while proposing various types of partial freezes that, they argued, might be more easily negotiated than a comprehensive freeze. As a result of this debate, support for the freeze approach began to take on a variety of meanings. Meanwhile, public support for the simply worded grass-roots freeze resolutions continued to grow. An AP/NBC news poll on April 6, 1982, reported that 74 percent of those polled supported a bilateral verifiable freeze, 18 percent opposed it, and 8 percent were not sure.

In May 1982 the nuclear freeze movement gained further international attention when Soviet President Leonid Brezhnev, in declaring the Soviet Union's readiness to negotiate an accord with the United States that would either ban or severely restrict the development of all new types of strategic armaments, called for a nuclear freeze "as soon as the talks begin." Brezhnev said that strategic armaments should be frozen quantitatively and that their modernization should be limited to the utmost. The Soviet press praised the U.S. proponents of the freeze and criticized the Reagan Administration for its militant policies and its rejection of the freeze. Secretary of Defense Caspar Weinberger responded to Brezhnev's statement on the freeze by saying that a nuclear freeze might tempt the Soviet Union to try nuclear blackmail or even a first strike against the United States because of the U.S. disadvantage in nuclear forces. On May 31, 1982, President Reagan announced that the Strategic Arms Reduction Talks (START) would open in June with a focus on substantial reductions. The President also pledged not to undercut SALT II.

As Congress continued to consider the freeze resolutions and the Democratic party began to assess the issue, the freeze movement emerged as a political symbol of commitment to arms control and opposition to the administration's approach to the problem. The Senate Foreign Relations Committee rejected a resolution that called for a freeze and commended the administration's START proposal. However, on August 5, 1982, after extended debate, the full House rejected by the remarkably close vote of 204 to 202 a nonbinding resolution that called for a comprehensive freeze followed by reductions. An alternative resolution, which called for a freeze after reductions to equal levels, won with the aid of extensive pressure by the White House.

Because the vote in the House was much closer than political analysts had predicted, the freeze movement was viewed as posing a significant challenge to the Reagan Administration's approach to arms control and its assessment of a strategic imbalance. In July 1982 the Democratic party endorsed the nuclear freeze at its national miniconvention, which set an agenda for both the 1982 congressional elections and the 1984 presidential election. Three months later, in November 1982, various forms of the simply worded nuclear freeze resolution were on 28 state or local ballots, winning in 25 of them. Freeze supporters claimed a net gain of 20 to 30 seats in the House as a result of the freeze movement's impact on congressional races.

Several weeks after the election the struggle between the administration's approach to arms control and the grass-roots freeze movement took on a new dimension when President Reagan stated that he believed that a number of "sincere" Americans who were supporting the freeze were being manipulated by foreign interests who wanted to weaken America, and that "foreign agents" had helped "instigate" the freeze movement. The leaders of the freeze campaign were outraged and reaffirmed that one of the their projected goals for the 98th Congress was to send a joint freeze resolution to the President. After weeks of delay and contentious deliberations in Congress, a much-amended nuclear freeze resolution finally passed the Democratically controlled House by a vote of 278 to 149. Key congressional freeze supporters, who were largely Democrats, claimed victory in the vote, maintaining that the resolution kept the wording that the freeze should come first, followed by reductions in weapons. Opponents of the freeze, who succeeded in adding major amendments to the resolution, also claimed victory, noting that the resolution required the freeze to end if reductions were not achieved in a specified period of time. The legislative strategy of the opponents, most of whom supported the administration's arms control policy, was to delay the freeze within the Democratically controlled House and complicate the final resolution with so many amendments that the impact of the simple freeze resolution would be lost. By the final vote the freeze resolution was no longer a simple, comprehensive proposal but a complex set of sometimes contradictory provisions.

President Reagan denounced the amended freeze resolution, proclaiming his confidence that if the resolution were debated in the Senate "the doubts and opposition to a simple freeze . . . will continue to grow." As the debate over the comprehensive freeze resolution began in the Senate, several moderate senators, unsatisfied with both the apparently one-sided nature of the government's START proposals and the deficiencies of the freeze, sought to mobilize a consensus around an arms control formula that would accommodate reductions in overall forces and modernization of U.S. forces while not requiring radical restructuring of Soviet forces. This initiative, called the build-down, initially required the retirement of two old warheads for every new one deployed (see Chapter 2).

In September 1983 the Senate Foreign Relations Committee failed to muster majorities for either a nuclear freeze resolution or a build-down resolution. Both resolutions were sent to the Senate floor with the word that the committee agreed with neither. On October 31, 1983, the full Senate in effect rejected the Kennedy-Hatfield freeze resolution by voting 58 to 40 to table a freeze amendment offered by Senator Kennedy to the debt ceiling bill. Proponents of the freeze were not expecting a victory in the Republican-controlled Senate, but they wanted to get all senators on record for or against the freeze before the start of the election year. The build-down amendment was then offered to provide senators who had voted against the freeze a chance to support the build-down. However, by the time of this vote a much more detailed version of the build-down had been incorporated into the U.S. START position, which complicated a straight up or down vote on the build-down. After a complex set of parliamentary maneuvers, the Senate voted 84 to 13 not to table the build-down amendment on the condition that the resolution's sponsors would pull it from the Senate floor. The withdrawal of the amendment prevented a direct test of Senate support for the build-down.

In the meantime the Soviet Union had formally submitted a freeze resolution at the United Nations on October 4, 1983. Soviet Foreign Minister Andrei Gromyko stated in his speech, which was read in absentia because his plane had not been permitted to land in New York, that the Soviet Union proposed to cease, under effective verification, the buildup of all components of nuclear arsenals, including all kinds of delivery vehicles and nuclear weapons; to renounce the deployment of new kinds and types of such arms; to establish a moratorium on all tests of nuclear weapons and new kinds and types of nuclear weapon delivery vehicles; and to stop the production of fissionable materials for the purpose of creating arms. Gromyko added that the freeze could initially apply to the Soviet Union and the United States on a bilateral basis, by way of example to other nuclear states. The Soviet proposal received little attention in the United States because of the tense atmosphere between the two superpowers after the Soviet downing of a South Korean airliner. On December 15, 1983, the UN General Assembly adopted the Soviet freeze resolution by a vote of 84 in favor, 19 opposed, including the United States, and 17 abstaining.

In December 1983, representatives of the nuclear freeze campaign, which now had organizations in two thirds of the congressional districts in the nation, held their fourth national conference. In reassessing the freeze movement's goals for election year 1984, the conference decided to pursue a more forceful legislative strategy that would promote the use of congressional power of the purse to enact parts of a freeze. This approach differed from that of the earlier freeze resolutions, which expressed the sense of Congress without carrying the force of law. The new tactic was to press Congress to implement a limited freeze by suspending funds for the testing of nuclear warheads and the testing and deployment of new ballistic missiles and anti-satellite weapons, provided the Soviet Union halted the same activities. Once the moratorium was enacted, negotiations between the United States and Soviet Union would immediately begin on a comprehensive freeze, including any elements of testing, production, and deployment of nuclear weapon systems not covered in the moratorium. This would be followed by negotiations to reduce the number of nuclear weapons systems of both countries. The freeze campaign's new emphasis split the freeze movement in Congress. Members could no longer support this freeze resolution and still vote for the MX missile, as some had with the more general freeze resolution of the first session. The result was to diminish the ranks of the freeze supporters in Congress.

By the spring of 1984 the freeze had lost some of its preeminence, having become one of several arms control initiatives to emerge from Congress. Nevertheless, in the summer of 1984 the freeze promised to reemerge as a significant issue in the presidential campaign. The Democratic platform stated that on January 20, 1985, as a first, practical step, "a Democratic President will initiate temporary, verifiable, and mutual moratoria, to be maintained for a fixed period during negotiations so long as the Soviets do the same, on the testing of underground nuclear weapons and anti-satellite weapons; on the testing and deployment of all weapons in space; on the testing and deployment of new strategic ballistic missiles now under development; and on the deployment of nuclear-armed sea-launched cruise missiles." The platform went on to state that "these steps should lead promptly to the negotiation of a comprehensive, mutual, and verifiable freeze on the testing, production, and deployment of all nuclear weapons." Democratic candidate Walter Mondale, who had long been a supporter of the freeze, announced his intention to pursue, if elected, the position set forth in the platform.

DESCRIPTION OF THE COMPREHENSIVE NUCLEAR FREEZE PROPOSAL

Despite public enthusiasm for the comprehensive freeze, there has not been an "authoritative" detailed statement of the provisions of the proposal. Since the original nuclear freeze resolutions appeared on state and local referendums in November 1980, there have been many general formulations of the proposal that differ in scope and detail. The leadership of the nuclear freeze campaign has consistently maintained, however, that their objective is to stop the arms race by an immediate, mutual, verifiable, comprehensive freeze that would prevent further testing, production, and deployment of nuclear weapons and delivery systems.

This position was initially described by Randall Forsberg in her 1980 paper "Call to Halt the Nuclear Arms Race," and was subsequently elaborated in an article by her in the November 1982 *Scientific American*. Specifically, the freeze she describes in these sources would stop the following activities: the production of fissionable material (uranium-235 and plutonium) for nuclear weapons; the testing of nuclear weapons; the fabrication and assembly of nuclear warheads; the testing, production, and deployment of all missiles designed to deliver nuclear warheads; and the testing, production, and deployment of any new types of aircraft or additional aircraft designed primarily to deliver nuclear weapons. The freeze would also prohibit modernization of nuclear weapons or delivery systems, but it would provide for the maintenance and replacement of existing systems until they are removed by an agreed process of reductions. Submarines are not included in the freeze and could be replaced on a one-for-one basis if they contained only existing missiles.

The freeze offers several approaches to the complex problem of dual-capable systems, such as tactical aircraft, that can deliver both nuclear and conventional weapons. First, dual-capable systems might be allowed under this formula, but only with a conventional capability. Alternatively, new dual-capable systems could be produced, but only on a one-for-one replacement basis. If these approaches proved too difficult to verify, these systems could be excluded from the freeze, and efforts at control would then be focused on the freeze of associated nuclear warheads. The comprehensive freeze would not restrict nonnuclear defensive systems beyond those restraints already included in the SALT I ABM Treaty.

The Kennedy-Hatfield resolution, which was introduced jointly in the Senate and the House of Representatives on March 10, 1982, but which has never been passed in its original form, is probably most widely identified as the legislative formulation of the freeze proposal. The text of this resolution in full is as follows:

Resolved by the Senate and the House of Representatives of the United States of America in Congress assembled, that

(1) as an immediate strategic arms control objective, the United States and the Soviet Union should
  (a) pursue a complete halt to the nuclear arms race;
  (b) decide when and how to achieve a mutual and verifiable freeze on the testing, production and further deployment of nuclear warheads, missiles and other delivery systems; and
  (c) give special attention to destabilizing weapons whose deployment would make such a freeze more difficult to achieve.

(2) Proceeding from the freeze, the United States and the Soviet Union should pursue major, mutual and verifiable reductions in nuclear warheads, missiles and other delivery systems, through annual percentages or equally effective means in a manner that enhances stability.

The nuclear freeze resolution that actually passed the House of Representatives on May 4, 1983, was so extensively amended that both supporters and opponents of the freeze claimed victory. The resolution states in part:

That consistent with the maintenance of essential equivalence in overall nuclear capabilities at present and in the future, the Strategic Arms Reduction Talks between the United States and the Soviet Union should have the following objectives:

(1) Pursuing the objective of negotiating an immediate, mutual and verifiable freeze, then pursuing the objective of negotiating immediate, mutual and verifiable reductions in nuclear weapons.

(2) Deciding when and how to achieve a mutual verifiable freeze on testing, production and further deployment of nuclear warheads, missiles and other delivery systems and systems which would threaten the viability of sea-based nuclear deterrent forces, and to include all air defense systems designed to stop nuclear bombers. Submarines are not delivery systems as used herein.

(3) Consistent with pursuing the objective of negotiating an immediate, mutual and verifiable freeze, giving special attention to destabilizing weapons, especially those which give either nation capabilities which confer upon it even the hypothetical advantages of a first strike.

(4) Providing for cooperative measures of verification, including provisions for

on-site inspection, as appropriate to complement National Technical Means of Verification, and to ensure compliance.

These provisions were followed by an extensive series of amendments that called for such diverse and far-reaching requirements as the following: incorporating the Intermediate Nuclear Force (INF) negotiations into START; maintaining in the negotiations the ability of the United States to preserve freedom; providing in the negotiations for the maintenance of a vigorous program of research, development, and safety-related improvements to assure that the U.S. nuclear deterrent forces would not be limited to levels inferior to those of the Soviet Union; providing for a stable international balance and the enhancement of the survivability of the U.S. nuclear deterrent forces; and assuring to the extent possible full compliance by all parties with preexisting international treaties.

THE MAIN ISSUES SURROUNDING THE COMPREHENSIVE NUCLEAR FREEZE*

The Strategic Relationship: Equivalence Versus Inferiority

Supporters of the Comprehensive Freeze

A central aspect of the freeze debate has been the assessment of the current strategic relationship between the superpowers. Supporters of the freeze emphasize that overall nuclear parity exists between the United States and the Soviet Union. They argue that a bilateral freeze would preserve this parity and prevent further destabilizing developments that would begin a dangerous new phase in the U.S.-Soviet nuclear arms race. They further assert that the development of new U.S. strategic systems with a preemptive counterforce capability—in particular the MX, the Trident II missile, and the Pershing II—would make such an arms race inevitable and result in a less stable strategic balance.

In asserting that the United States and the Soviet Union are today closer to nuclear parity than they have been at any time since World War II, freeze supporters compare the numbers of strategic ballistic missiles and heavy bombers and the numbers of nuclear warheads they carry. Whereas the Soviet Union has more strategic missiles and more and larger land-based intercontinental ballistic missiles (ICBMs) that

---

*This discussion assumes a comprehensive freeze along the lines outlined in the November 1982 *Scientific American* article by Randall Forsberg and generally supported by the leadership of the freeze movement.

carry more warheads, the United States has more warheads, owing to the large number of warheads on submarine-launched ballistic missiles (SLBMs) and strategic bombers. The United States also has many more intercontinental bombers, with much larger payloads, and a substantial lead in the new technology of small, long-range, high-accuracy cruise missiles. Finally, freeze supporters emphasize that more meaningful than comparisons of numbers of weapons is the fact that both countries have acquired enormous "overkill," that is, each has many times the number of weapons necessary to destroy the other's urban population and society.

Critics of the Comprehensive Freeze

Administration officials and many other critics of the freeze reject the assessment that parity exists between the superpowers. They assert instead that the nuclear forces of the United States are dangerously inferior to those of the Soviet Union and that a freeze of present force postures would lock the United States into this position. Such an unfavorable strategic balance not only places the United States in a poor political bargaining position but in a crisis could encourage the Soviet Union to launch a preemptive attack.

These critics assert that Soviet modernization efforts have outstripped the U.S. efforts, particularly in the development and deployment of intercontinental ballistic missiles, which now pose a serious threat to the U.S. land-based ICBM force. In the last ten years the Soviets have deployed three ICBMs (the SS-17, SS-18, and SS-19), the Typhoon and Delta submarines, new submarine-launched ballistic missiles, the Backfire bomber, and the SS-20 missiles capable of striking targets in NATO and the Far East. During this same period, according to this view, the United States exercised restraint, deploying no new ICBMs or intermediate-range missiles in Europe. Consequently, according to these critics, the freeze would prevent the United States from correcting the existing deficiencies in its nuclear forces caused by the sustained Soviet buildup. Specifically, it would stop ongoing U.S. programs (including those for the MX, the Midgetman, the B-1 and Stealth bombers, and the Trident II missile), extensive future deployments of cruise missiles on bombers and submarines, and the deployment of Pershing II and ground-launched cruise missiles in Europe.

Other critics argue that, although overall parity exists between the two sides, it would not improve crisis stability to lock the two sides into their present force structures, since some of the systems on both sides are inherently vulnerable.

Rationale: Freeze at Parity to Stop the Arms Race

Supporters of the Comprehensive Freeze

General Issues. Supporters of the comprehensive nuclear freeze argue that it would immediately stop the wasteful and dangerous nuclear arms race in both its quantitative and qualitative aspects. Once achieved, a freeze would create a strategic environment in which the nuclear powers will seriously pursue the reduction of nuclear arms with some hope of success. This would be in sharp contrast to other arms control approaches, which simply manage the arms race and accept continued modernization in the illusory hope of achieving greater stability. According to supporters of the comprehensive freeze, the nuclear arms race itself is a prime cause of instability in the relations between the two superpowers. Consequently, new or additional nuclear weapons and delivery systems will contribute to this dangerous instability.

Supporters of the freeze emphasize that it would prevent the introduction of new counterforce weapons that threaten the survivability of the other side's deterrent forces. These weapons increase the risk of a nuclear war by putting additional pressure on leaders to place their nuclear forces in a dangerous launch-on-warning status in peacetime and to launch their weapons first in a crisis. Specifically, the freeze would prevent new destabilizing U.S. advances in counterforce capability that the Soviet Union would inevitably seek to match. Preventing these developments on both sides would move the two countries away from counterforce and war-fighting strategies that increase the likelihood of war.

Comprehensive Coverage. Supporters of the freeze argue that by including *all* nuclear weapons and delivery systems, the freeze makes it possible for the first time to speak realistically of stopping the nuclear arms race between the superpowers. Under a limited freeze that permits modernization within limits or leaves categories of delivery vehicles unconstrained, the arms race would continue, or even accelerate, in permitted systems.

By banning the testing of nuclear weapons and new delivery systems, the comprehensive freeze would effectively stop their development, since significant advances could not be made without testing. This would essentially eliminate the qualitative nuclear arms race. In particular, it would prevent the development of more effective counterforce systems that might threaten the survival of retaliatory forces. The ban on all missile testing would further enhance stability by gradually reducing both sides' confidence in the counterforce capabilities of their own ballistic missile forces.

By covering production as well as deployment, the comprehensive freeze would prevent the stockpiling of weapons that could be rapidly deployed if the agreement were abrogated. Even if undertaken only as a hedge, such production would be destabilizing because the other side would see it as an indication of intent to break out of the agreement. Moreover, by stopping the production of nuclear materials for weapons purposes, the freeze would prevent further expansion of the stockpile of critical materials required for nuclear weapons. The ban on the fabrication of nuclear weapons would prevent stockpiled fissionable materials or old nuclear warheads from being used to make newer models. Freeze supporters also assert that the comprehensive freeze would not undercut European security because the freeze would preserve the present overall U.S.-Soviet strategic parity, which is the real determinant of European security.

Repair and Replacement. Supporters of the comprehensive freeze argue that the proposal can manage the practical problems of maintaining existing nuclear forces until these systems are removed by an agreed process of reductions. Aging delivery vehicles could be repaired with new parts or, if necessary, replaced with new delivery vehicles of the same type. These provisions would permit existing forces to continue operation without becoming more dangerous through modernization.

In this connection, it is argued that in practice aircraft and missiles can be maintained almost indefinitely by simply replacing parts. The B-52Gs and B-52Hs, the last of the B-52 series delivered to the U.S. Air Force in 1960-62, are expected to remain serviceable through the 1990s and even into the next century. The useful life of the planes is limited only by the availability of spare parts. Although some parts that are now being cannibalized from retired, older-model B-52s may run out, new production lines for these parts could be opened. In the same manner, missiles can be maintained for long periods by replacing worn-out parts, such as inertial guidance components and computers. In Forsberg's "Call to Halt the Nuclear Arms Race," ballistic missile submarines are specifically exempted from the freeze and their replacement is allowed provided they are retrofitted with existing missiles.

Specific provisions could avert a situation in which purely technical considerations arising from a decrease in the reliability of certain systems would determine the choice and method of reductions. For instance, the tritium component of thermonuclear weapons must be replaced periodically due to tritium's 12-year half-life. To this end the freeze could permit the safeguarded operation of sufficient reactor power to maintain the existing tritium inventory. But freeze supporters emphasize that a clear distinction must be made between maintenance and modernization in any such provisions for replacements.

In the final analysis, according to freeze supporters, the durability of existing forces is more of a factor in the postfreeze program of reductions that it is an obstacle to a freeze on new production. It would not be a real problem to maintain forces for 5 to 10 years, and 20 to 30 years is also manageable. In any case, it is clearly technically possible to maintain forces without innovation through replacement of parts, say freeze leaders. But they acknowledge that such maintenance may raise political difficulties depending on the systems involved and the timing of reductions.

Dual-Capable Systems. Freeze supporters recognize that dual-capable systems, such as tactical aircraft and cruise missiles, present a special problem since they can deliver both conventional and nuclear munitions. They argue, however, that there are a number of acceptable ways to deal with this problem within the comprehensive framework. The freeze defined in "Call to Halt the Nuclear Arms Race" would allow the continued production of these dual-capable systems with only conventional capability, although this raises problems with verification. An alternative arrangement would allow the production of dual-capable systems only as replacements for existing dual-capable aircraft on a one-for-one basis. Yet another suggested arrangement would exclude these vehicles from the freeze and control the nuclear warheads that can be carried by them.

Defensive Systems. Freeze supporters argue that it is not necessary to complicate the comprehensive nuclear freeze by including nonnuclear air defense and antisubmarine warfare systems, since these systems do not pose a serious threat to existing retaliatory strategic forces. They point out that ballistic missile defenses are adequately constrained by the SALT I ABM Treaty. In the view of freeze supporters, foreseeable advances in the technology of antiballistic missile systems, antisubmarine warfare, and air defense will do little to decrease each side's capacity for devastating retaliation.

Specifically, freeze supporters argue that existing ICBMs do not have to be improved to maintain deterrence, since ballistic missiles will be able to penetrate defenses in a retaliatory strike as long as the SALT I ABM Treaty remains in force. The Soviet Union has not yet initiated any programs that would threaten U.S. submarines, nor are there any new programs on the horizon. With regard to the competing technologies of strategic bomber penetration and antiaircraft defenses, improvements in nonnuclear penetration techniques, such as jamming, should permit existing bombers to hold their own against advances in nonnuclear air defenses in the event of a nuclear freeze. (Nonnuclear air defense and antisubmarine warfare systems are not in the comprehensive freeze as originally proposed, but they are included in the resolution proposed by the House of Representatives in May 1983.)

Negotiability. Freeze supporters assert that, contrary to the view of many of their critics, a comprehensive freeze could be negotiated relatively rapidly if both sides genuinely support the objective. They point out that the Soviet Union has formally endorsed the concept of a comprehensive freeze. They also note that many of the elements of a freeze have already been successfully negotiated in connection with the SALT II and Comprehensive Test Ban negotiations. In addition, because of the freeze's comprehensive nature, many of its definitions and provisions would be easier to formulate and agree upon than in the case of a more limited agreement, where permitted activities have to be defined with great care.

Critics of the Comprehensive Freeze

General Issues. The comprehensive nuclear freeze proposal has been criticized both by supporters of the START approach, which seeks to improve the strategic balance by deep reductions without qualitative restraints, and by supporters of the incremental approach, which permits modernization within specified limits, as characterized by the SALT process.

Supporters of the START approach argue that the comprehensive freeze at existing levels would be extremely dangerous to U.S. security. It would undermine strategic stability by locking the United States into a position of strategic inferiority with vulnerable retaliatory forces. The Soviet Union is now ahead of the United States in every static measure of strategic power except for total strategic warheads, according to these critics. By banning modernization the freeze would prevent the United States from correcting these dangerous deficiencies in its nuclear forces.

Moreover, implementation of the freeze would probably cause further deterioration of the U.S. strategic position in the future, according to critics. Under the freeze the Soviet Union, which already has the theoretical counterforce capability to destroy a large part of the U.S. land-based ICBM force, could threaten the entire U.S. strategic triad in the foreseeable future, contend the critics. The comprehensive freeze would limit the United States to its present strategic systems and capabilities while not constraining either Soviet nonnuclear air defense systems or nonnuclear antisubmarine warfare systems, thus making the U.S. strategic triad increasingly vulnerable over time. The results would be a progressive erosion of the U.S. deterrent relative to superior Soviet forces. In short, according to freeze critics, a nuclear freeze would leave the United States in a weakened position and make war more, not less, likely.

Other freeze critics who disagree with the Reagan Administration's assessment of the present U.S.-Soviet strategic relationship nevertheless share the concern about the increasing vulnerability of both sides' strategic forces. The freeze approach does not address the fundamental problem of the stability of the U.S.-Soviet strategic relationship, they argue, and would in fact prevent future efforts to improve the stability of this relationship.

Some freeze critics agree that the approach would in principle be in the overall interest of U.S. security but question whether it could be negotiated in a form that would, in the end, be acceptable in the United States. For this reason, they are concerned that it might not lead to a ratifiable agreement and would divert arms control efforts from more promising and practical goals.

Modernization. From the perspective of most critics of the freeze, the U.S. strategic force requires modernization to remain a stable deterrent. Consequently, a freeze, even if verifiable, would simply perpetuate accumulating problems and vulnerabilities by preventing essential corrective actions.

These critics argue that the freeze would terminate every current U.S. program designed to correct problems that have developed in the U.S. strategic posture as a result of the Soviet Union's large-scale arms buildup. At the same time, the freeze would not affect the Soviet programs that have the greatest potential for upsetting the strategic balance. For example, the freeze would bar the United States from developing a new survivable land-based system, but it would do nothing to eliminate the Soviet threat from large land-based missiles with accurate MIRVs that make these new U.S. developments necessary. The freeze would bar modernization of the U.S. strategic air force, but it would not block Soviet air defense programs. The freeze would prohibit the modernization of U.S. submarine-launched ballistic missiles, such as the extended-range Trident II missile, but it would do nothing to prohibit the development of Soviet antisubmarine warfare capabilities.

Europe. According to some critics, the comprehensive freeze would present a special problem in Europe, since by preventing the planned U.S. deployment of intermediate-range missiles, the freeze would place the NATO alliance in a militarily inferior position. The freeze would lock in the overwhelming Soviet advantage in intermediate-range nuclear missiles in Europe. At present there are some 600 Soviet intermediate-range nuclear missiles capable of striking U.S. allies and no comparable U.S. systems.

Reductions. Some critics of the freeze inside and outside the administration argue that, despite its apparently radical approach to nuclear arms control, the freeze does not go nearly far enough. According to these critics, it neither requires immediate reductions nor creates a framework that would encourage reductions in the future. Although the freeze movement calls for the prompt negotiation of reductions after a comprehensive freeze has been agreed upon, the reduction process is not built into the initial agreement. Such reductions could prove very difficult or impossible to negotiate subsequently, particularly if the Soviet Union is satisfied with the force levels frozen by the agreement.

In contrast, supporters of the START approach point out that the central objective of START is substantial reductions in strategic nuclear forces. Similarly, supporters of the incremental approach of SALT point out that the SALT II agreement actually went beyond the freeze by requiring significant reductions in Soviet strategic forces. Moreover, the equal aggregate ceilings in SALT, together with the various equal subceilings, provide a framework for a continuing process of reductions.

Defensive Systems. Many critics of the freeze argue that the proposal to permit nonnuclear air defense and antisubmarine warfare developments to go forward without any constraints while freezing all improvements in strategic offensive forces could prove extremely dangerous to U.S. security and increase the risk of war. These critics emphasize the importance of maintaining the retaliatory capabilities of the air and sea legs of the strategic triad under a freeze since the existing vulnerability of U.S. land-based ICBMs could not be reduced by new survivable land-based systems.

While their assessments of the urgency of the problem differ substantially, these critics point out that the ability of the present generation of strategic bombers to penetrate to their targets will certainly decline in the future as Soviet air defenses improve. Substantial further improvements in air defense can be imagined, particularly against the static air threat that would exist under the freeze. These critics therefore emphasize that the retaliatory capability of the air arm of the triad will deteriorate gradually—some would say rapidly—unless the United States upgrades the penetration capabilities of its strategic bombers. These capabilities can be substantially improved by equipping existing bombers with air-launched cruise missiles or by introducing improved new bombers such as the B-1 or Stealth into the force. These developments, however, would be prohibited by the freeze.

In the case of antisubmarine warfare, these critics point out that even if the threat is not great today, one cannot rule out major future improvements that would threaten the sea-based leg of the triad. A freeze on strategic offensive missiles would also stop deployment of the Trident II missile, a weapon whose increased range will greatly complicate the problem confronting Soviet antisubmarine warfare by allowing U.S. ballistic missile submarines to operate in much larger areas of the ocean.

These criticisms would be answered if air defenses and antisubmarine warfare were included in the freeze, as they were in the resolution proposed by the House of Representatives in May 1983. However, critics note that this would add substantial verification problems to the agreement and would greatly complicate its negotiability.

Negotiability. Many critics of the freeze, including some who endorse its objectives, question whether it could in fact be easily and quickly negotiated with the Soviet Union. They argue that the negotiations would inevitably be a long, drawn-out undertaking that would not produce concrete results for several years, during which time the arms race would continue. They point out that the experience of the SALT, START, and Comprehensive Test Ban negotiations demonstrates conclusively that developing the detailed language of a comprehensive freeze agreement would prove to be extremely complex and time-consuming. Among other things, the agreement would have to deal with definitions and provisions governing a number of important problems in gray areas that are inherently very difficult and on which proposals have not yet been clearly formulated even by freeze proponents. These problems include the handling of dual-capable systems, definitions of permissible repair and replacement for all types of systems, and the precise limitations on testing. These critics also emphasize that although it may be theoretically possible to define procedures that would permit adequate verification of the agreement, these procedures may prove unacceptable to the Soviet Union, and possibly to the United States as well, because of the degree of intrusiveness involved. In any event, the negotiation of mutually acceptable verification procedures would be a long and difficult process.

Those critics who support the incremental approach of the SALT process argue that a prolonged and possibly unsuccessful freeze negotiation is a poor alternative to the early ratification of SALT II or an updated version of that agreement. Moreover, they express concern that the freeze movement, by raising unrealistic expectations about the prospects of an early freeze agreement, will divert arms control efforts into a controversial and unproductive path while undercutting a public consensus in support of more limited arms control agreements that might be negotiated relatively quickly. Those critics who support the START approach express concern that the freeze movement, even if it does not lead to negotiations, will reduce the prospect of achieving the much more significant arms control objectives that they believe are necessary to improve U.S. security.

Verification of a Comprehensive Nuclear Freeze

Supporters of the Comprehensive Freeze

A great deal of the controversy surrounding the comprehensive nuclear freeze has focused on the verifiability of such a proposal. The issue involves both the question of the inherent verifiability of the approach and the broader question discussed in Chapter 2 of how much verification is enough.

Freeze supporters argue that the proposal meets the same criterion of "adequate" verification that has been used to judge other arms control agreements, such as SALT I and SALT II. The combination of existing U.S. intelligence assets, cooperative measures, and reasonable on-site inspection, coupled with the existing consultation process, could give the United States ample warning of any significant clandestine program to violate a freeze on the testing, production, and deployment of additional nuclear weapons and delivery systems. A freeze on the testing of nuclear weapons could be adequately verified by National Technical Means supplemented by cooperative measures and on-site inspection procedures already agreed upon by the United States, the Soviet Union, and the United Kingdom during the Comprehensive Test Ban negotiations. A freeze on the production of fissionable materials for weapons can be verified with high confidence by combining National Technical Means with extensively used safeguards established by the

International Atomic Energy Agency to monitor peaceful nuclear power facilities in other countries. Finally, note freeze supporters, many of the important elements of a comprehensive freeze have already been successfully incorporated in the SALT treaties, which the U.S. intelligence community and the Carter Administration judged to be adequately verifiable.

The comprehensive nature of a total freeze on all testing, production and deployment activities would facilitate verification, according to supporters. Any indication of deployment or production would signal a possible violation, so details of definition or complex quotas on production would not arise. The synergistic effect of various mutually reinforcing aspects of the comprehensive freeze would make it easier to verify than the total of its individual components. Freeze supporters argue that this synergistic effect would not apply to most partial freezes.

If all production and deployment of nuclear weapons and delivery systems were suspended, there would be many opportunities to detect continued production of both large and small nuclear systems, freeze supporters argue. Any nuclear weapon system has a long production and deployment process, including not only the production of warheads and a particular delivery system but also the production of ancillary support equipment, the training of forces to use the system, the provision of security and command and control that may be unique to particular systems, and the establishment of a chain of command that may also be unique to particular systems in the field.

As an example of the verifiability of important aspects of the proposal, freeze supporters cite the production of fissionable materials for weapons. Soviet production facilities for fissionable materials are well known and regularly monitored by U.S. intelligence. These installations, which include both dedicated production facilities and nuclear power reactors, are by their nature large and difficult to conceal. The closing of these production facilities could be monitored with high confidence using National Technical Means alone and with certainty by even superficial periodic on-site inspections. Operating nuclear power reactors could be monitored by the effective safeguard procedures of the International Atomic Energy Agency (IAEA), which are now applied worldwide in connection with the Non-Proliferation Treaty. Extensive experience has shown that these IAEA procedures, which involve periodic inspections as well as emplacement of secure seals and sensors, can successfully monitor inventories and give timely warning of possible diversions of fissionable materials from peaceful power programs. Freeze supporters note that the Soviet Union, which has historically rejected on-site inspection, has shown some signs of greater flexibility on this point, as evidenced by its recent voluntary move to accept some IAEA inspection of its peaceful nuclear power program. This move parallels similar voluntary arrangements made by the United States and the United Kingdom with the IAEA to indicate their acceptance of the safeguards, which legally apply only to nonnuclear weapons states under the Non-Proliferation Treaty.

Freeze supporters argue that it is well known where the central strategic Soviet systems (i.e., ICBMs, SLBMs, and strategic aircraft), as well as many other systems, are produced, and that National Technical Means, particularly if supplemented with modest cooperative measures and periodic on-site inspections, would have little difficulty determining whether production had stopped. The only issue is whether clandestine production could take place elsewhere and, if so, whether it would be in sufficient quantity to have any military significance. Freeze supporters assert that the United States' ability to monitor the production path of weapons would ensure that possible clandestine production would not occur in militarily significant quantities. The halt in production of all nuclear weapons and delivery systems under a comprehensive freeze would assist this monitoring process. As long as tactical and battlefield systems and associated nuclear warheads continue to be manufactured, the entire production chain for nuclear weapons will remain operational, making verification much more difficult.

With regard to dual-capable systems such as tactical aircraft, cruise missiles, and short-range ballistic missiles, freeze supporters argue that, although these systems individually present some very difficult verification problems, there are a number of practical ways of adequately dealing with them. For instance, production of some systems could continue but only with a conventional capability. In this connection, special provisions were successfully developed in SALT II requiring functionally related observable differences (FRODs) to differentiate strategic bombers capable of delivering nuclear weapons from similar versions of the same aircraft designed to perform different missions, such as reconnaissance. Similar provisions could differentiate nuclear and nonnuclear tactical aircraft, freeze supporters suggest. An alternative arrangement might be to allow continued monitored production of dual-capable systems but only as replacements for existing equipment on a one-for-one basis. A final possibility might be to exclude these vehicles from the freeze and rely on the freeze on nuclear materials and nuclear warheads to limit their military significance.

Concerning the problem of clandestine production activities that might give the Soviet Union a breakout potential, freeze supporters state that in the end there would be little to gain and much to lose in any clandestine attempt to violate an agreement banning production. Nuclear weapons made and stockpiled in secret without fully tested delivery systems would not contribute to nuclear deterrence. Moreover, without testing it would not be possible to develop new types of weapons or delivery systems that would be sufficiently reliable in a counterforce mission. Above all, the number of weapons that could be produced clandestinely would be very small compared with the size of the current arsenals, about 20,000 to 30,000 warheads on each side. Thus, it is highly unlikely, according to freeze supporters, that either party would see any real military advantage in trying to build a small number of additional nuclear weapons or delivery systems clandestinely.

Critics of the Comprehensive Freeze

The verifiability of a comprehensive nuclear freeze has been challenged not only by the Reagan Administration but also by many arms control analysts who supported the verifiability of the SALT I and SALT II agreements. In this instance, critics often have different standards for the acceptable level of verification. Administration criticisms are based on a perspective demanding higher verification standards than have been called for in the past. Other critics base their assessment on the same standards of adequate verification that were applied to SALT I and SALT II.

The administration has flatly stated that a freeze on all testing, production, and deployment of nuclear weapons and delivery systems could not be verified. In this view, it would not be possible to verify deployment of all types of delivery systems with acceptable confidence. With respect to production of nuclear weapons and delivery systems, the task of verification would become unmanageable. The possibility of clandestine activity would also seriously endanger national security, according to the administration. For example, even with very intrusive on-site inspection, confidence in verifying the ban on the fabrication of nuclear weapons would be very low. Confidence would also be low in the verification of many smaller nuclear delivery systems and the large range of dual-capable delivery systems. The administration also does not accept the verifiability of a ban on nuclear testing. In this regard it is presently challenging the adequacy of the verification provisions of the Threshold Test Ban Agreement, and it has rejected negotiations on a comprehensive test ban in part because of presumed difficulties in verification. Thus, the administration contends that the problem of verification alone is sufficient reason to oppose the nuclear freeze as proposed.

The practical result of a comprehensive nuclear freeze, according to the administration, would be that the United States, as an open society, would live up to a freeze, while there would at best be considerable doubt as to whether the Soviet Union would abide by the nonverifiable aspects of the agreement. In this connection, the recent charges of Soviet violations and possible violations of the SALT II Treaty and other agreements designed to be "adequately" verifiable raise serious questions as to probable Soviet actions under a freeze agreement containing many provisions that would be much more difficult to verify. In light of this experience, according to the administration, the United States simply cannot base its national security on trust in the Soviet Union.

Some arms control analysts who supported the verifiability of the SALT agreements and a comprehensive test ban share some of the technical concerns about verifying a ban on the production of nuclear delivery systems, particularly those involving the fabrication of nuclear weapons and the production of small and dual-capable systems. Even if theoretically possible, they argue, the technical measures needed to ensure adequate verification would prove to be so intrusive that neither side would agree to them. In any case, negotiating these detailed and intrusive verification measures would be so complex that the negotiations would be very protracted. These analysts point out that the negotiations on verification provisions in SALT I and II required very extensive discussions. Furthermore, they note, as of the final recess in the START negotiations the Reagan Administration had not been able to work out even within the U.S. government the verification procedures for its START proposal, which would be less demanding than those for a comprehensive freeze. Thus, it might require an inordinate amount of time to work out specific measures to assure adequate verification of a comprehensive freeze.

Many of these analysts agree with the administration that none of the proposed approaches to the difficult problem of dual-capable systems offers much promise of assuring adequate verification of a freeze on those systems. In this case, they argue, the inherent problems are so difficult that it may not be possible to resolve them without unrealistically intrusive and extensive inspection. In SALT II even the relatively straightforward problem of defining heavy bombers proved difficult, since the Soviet Union uses Bear and Bison aircraft for reconnaissance and various naval missions as well as for strategic bombing. The sides were finally able to agree on a complex system for determining which aircraft would count against the SALT ceilings, but it is by no means clear that the same approach could be applied effectively to tactical aircraft or cruise missiles. According to these analysts, the endless debate over whether the capabilities of the Soviet Backfire bomber make it a heavy bomber barely hints at the problems that would be involved in a freeze in defining which U.S. and Soviet tactical aircraft should be treated as having a potential nuclear role.

In summary, the verifiability of the freeze has been challenged in many respects and on several levels. Some critics argue that such an agreement simply cannot be verified to meet U.S. security requirements. Others believe that even if the technical verification measures could be worked out, the requirements would prove so intrusive that neither side would be willing to accept them. Moreover, any negotiations to reach a mutually acceptable compromise would probably be so protracted that an agreement would at best take a long time to achieve.

The Soviet Union and the Nuclear Freeze

The Soviet Union presented a comprehensive freeze proposal to the UN General Assembly in October 1983. The Soviet proposal called on all nuclear weapons states to stop, under effective verification, the buildup of all components of nuclear arsenals, including all kinds of delivery vehicles and weapons; to renounce the deployment of new kinds and types of such arms; to establish a moratorium on all tests of nuclear weapons and new kinds and types of nuclear delivery vehicles; and to stop the production of fissionable materials for the purpose of creating arms. The Soviet Union has also stated that this approach could initially be undertaken by the Soviet Union and the United States on a bilateral basis.

The Soviet Union stated that the proposal would allow for nuclear weapons already deployed to be replaced within the limits of the normal requirements of operation. Only tests of nuclear delivery systems already deployed would be allowed in connection with replacement and the normal requirements of operation. Concerning verification, the Soviet Union has stated that the freeze could be effectively monitored by National Technical Means, the Standing Consultative Commission, and, if necessary, additional cooperative measures. The proposal is based on the present nuclear parity between the superpowers, according to the Soviet Union, and is not an end in itself but rather a first step toward reductions. The Soviet Union emphasizes the point that the freeze must come before reductions. A freeze is important in the Soviet view because it can erect a barrier to more destabilizing deployments of first-strike weapons.

Shortly after the Soviet Union formally presented its freeze proposal, several Soviet arms control experts, in unofficial conferences on arms control, stated that a freeze on the maximum spectrum of systems should be considered initially; if this proved infeasible, narrower approaches could be discussed. They acknowledged that many complicated questions exist, such as verification of dual-capable systems and nonnuclear defense. But they argued that, with political will, leaders could resolve these issues.

The introduction of the Soviet freeze proposal at the United Nations was the first official Soviet endorsement of the comprehensive freeze approach, although the Soviet press, in its extensive coverage of the U.S. freeze movement, had earlier praised the proponents of an immediate nuclear weapons freeze and criticized the Reagan Administration for rejecting the approach. Prior to submitting its proposal at the United Nations, the Soviet Union had announced proposals for more narrow freezes, including freezes on the development and deployment of medium-range arms in Europe and of strategic arms in general for the duration of the INF and START negotiations. Soviet President Brezhnev announced the first of these proposals in February 1981, when he called for a moratorium on the "establishment" of new facilities in Europe for NATO and Soviet medium-range nuclear missiles. This moratorium would extend from the beginning of negotiations on the limitation or reduction of such facilities until a permanent treaty was concluded. On March 16, 1982, soon after the INF negotiations began, Brezhnev announced that he had imposed a unilateral moratorium on the deployment of "medium-range nuclear armaments" in the European part of the Soviet Union, specifically noting a "freeze" on deployment of SS-20 intermediate-range ballistic missiles.

These events were followed by President Brezhnev's call in May 1982 for the freezing of the strategic armaments of the Soviet Union and the United States at the beginning of the START negotiations. In START the Soviet negotiators coupled calls for such a strategic freeze with the traditional SALT approach to arms control. In his message to the UN Special Session on Disarmament in June 1982, Brezhnev stated that the Western freeze proposals "on the whole . . . go in the right direction." He also said that the idea of a mutual freeze on nuclear arsenals as a first step toward reductions "is close to the Soviet point of view." In October 1982, before the UN General Assembly, Soviet Foreign Minister Gromyko followed up on these remarks, characterizing the Brezhnev strategic moratorium as his country's "concrete response" to calls for a freeze on the existing level of nuclear arms. Shortly after his succession, Soviet President Yuri Andropov, in his November 22, 1982, speech to the Central Committee of the Communist Party and in his report "Sixty Years of the USSR," reiterated the call for a freeze while negotiations were in progress.

Finally, at the October 1983 UN General Assembly session, the Soviet Union formally embraced the comprehensive freeze approach in a formal resolution. The Soviet resolution was endorsed by 84 countries, with the United States and 18 other countries opposed and 17 countries abstaining. Soviet officials have stated that the call for a freeze is consistent with their positions in the INF and START negotiations and does not preclude other approaches to arms control. In the United States there has been relatively little reaction to the Soviet proposal in the United Nations. The administration and some critics of the freeze have dismissed it as a propaganda move designed to appeal to worldwide antinuclear sentiment. Others have viewed it as a possible first step toward formal discussion of a comprehensive nuclear freeze.

4 The Intermediate-Range Nuclear Force (INF) Negotiations

INTRODUCTION

The U.S.-Soviet Intermediate-Range Nuclear Force (INF) negotiations, which began in Geneva, Switzerland, on November 30, 1981, originated in NATO's decision in December 1979 to deploy 572 U.S. intermediate-range missiles in Europe. The NATO action, which responded to the Soviet deployment of a new generation of intermediate-range weapons (SS-20 missiles and Backfire bombers), was described as a two-track decision, since in parallel with the deployment of ground-launched cruise missiles and Pershing II ballistic missiles the United States was to seek to negotiate equal limits on these missiles with the Soviet Union.

From the outset of the negotiations the two sides had fundamental differences in their assessments of the threat and the systems that should be limited. The United States sought equal worldwide levels (originally zero) on the intermediate-range missile forces of the United States and the Soviet Union. The Soviet Union sought to ban further missile deployments and to limit the Warsaw Pact and NATO, including British and French forces, to equal levels of intermediate-range missiles (originally aircraft as well) in the European theater. In the two years of negotiations, the two sides revised many details of their respective positions but never resolved the underlying differences.

In response to the initial U.S. deployments of ground-launched cruise missiles and Pershing II missiles in Europe in November 1983, the Soviet Union, as threatened, walked out of the INF negotiations. The

United States has emphasized its willingness to continue the negotiations but refuses to meet the Soviet precondition of first removing its recently deployed missiles from Europe.

BACKGROUND

The Origins

Since the late 1940s, nuclear weapons have been stationed in Europe and have played an integral role in NATO's strategy to deter an attack by the Soviet Union. NATO's reliance on a strategy of nuclear deterrence evolved as an alternative to the high cost and heavy manpower required to maintain an adequate conventional force to deter the Soviet Union. Taking advantage of U.S. nuclear superiority, NATO's strategy relied on the threat of massive nuclear retaliation by the United States as the main deterrent to a Soviet attack in Europe. The first nuclear weapons deployed in Europe were on U.S. bombers stationed in Britain. In the early 1950s the United States began to deploy tactical nuclear weapons in Europe intended for use against the superior Soviet conventional forces.

In the early 1950s the Soviet Union began acquiring its own nuclear weapons. By the end of the decade it began deployment of liquid-fueled medium-range SS-4 and SS-5 missiles, which had ranges of 2,000 km and 4,000 km respectively, in the European part of the Soviet Union. These missiles were initially very vulnerable as they were unhardened and clustered. By the mid-1960s about 600 SS-4s and 100 SS-5s had been deployed.

In response to the Soviet medium-range missile deployment and the perceived Soviet advantages in intercontinental ballistic missiles, the United States in 1959 deployed Thor and Jupiter missiles, which had a range of 2,500 km, in Europe. This deployment was in fact intended to reassure the alliance that the United States was prepared to defend Europe. However, the U.S. Thor and Jupiter missiles were removed from Europe in 1962 as part of the settlement of the Cuban missile crisis.

With the withdrawal of these U.S. missiles from Europe, a debate arose within NATO as to the future strategy of the alliance. The United States wanted to place more emphasis on conventional defenses in Europe, while European governments wanted to continue to rely on extended nuclear deterrence to avoid the heavy financial and logistic burden of conventional defense. President Charles de Gaulle, one of the major critics of the U.S. position, believed that a shift toward conventional strength would only undermine an already questionable nuclear deterrent strategy that relied on the United States being willing to sacrifice New York for Paris. The debate culminated in 1967, after France withdrew from the integrated NATO defense, in the adoption of the so-called flexible response policy.

Flexible response was a compromise. The Western European governments accepted the need for a conventional response to Soviet aggression, and the United States agreed to retain in Europe a nuclear force suitable for use in a controlled escalatory fashion, should the Warsaw Pact launch an attack that could not be contained conventionally. Flexible response, which is still NATO's basic strategy of deterrence, requires NATO to be able to respond to a Soviet attack with a range of options, from conventional defense alone to the use of either tactical or long-range theater nuclear forces to the ultimate use of intercontinental nuclear forces. To implement this strategy the United States announced that 400 warheads on Poseidon ballistic missile submarines were being committed to the European theater. By this time, Great Britain and France had also begun developing independent nuclear weapons forces.

The Soviet break with China in the early 1960s and the Chinese demonstration of a nuclear capability in 1964 put further demands on Soviet theater systems. In response, the Soviet Union moved to modernize its vulnerable SS-4s and SS-5s and began targeting some of its new light ICBMs, the SS-11, on Europe and the Far East. During the SALT I negotiations the Soviet Union sought to include in the restraints on the U.S. side American forward-based systems (e.g., medium-range bombers and missiles located in Europe or on aircraft carriers in the European theater) and the British and French nuclear forces. The Soviet Union argued that any system that could hit the homeland of either the United States or the Soviet Union should be considered a strategic system. The United States refused even to discuss those systems in the negotiations on the grounds that they were not intercontinental systems.

In the SALT I Interim Agreement the Soviet Union was allowed a larger number of ballistic missile submarines and submarine-launched ballistic missiles than was the United States. Although the United States denied that the disparity implied any compensation for British and French forces, the Soviet Union in a unilateral statement to the accord indicated that it had received partial compensation for these forces. The Soviet Union further asserted in this statement that, if the NATO allies increased the number of their ballistic missile submarines during the period of the SALT I Interim Agreement to exceed the numbers that were operational or under construction as of May 26, 1972, the Soviet Union would have the right to a corresponding increase in the number of its submarines.

The Soviet modernization program continued throughout the 1970s. In 1974 the Soviet Union began deploying the Backfire bomber, a substantially improved medium-range bomber that was apparently intended to strengthen its theater nuclear capabilities. These developments, coming at a time when the Soviet Union had achieved nuclear strategic parity with the United States, raised further concerns within NATO about the imbalance in Europe and the credibility of U.S. deterrence strategy. The issue came to the forefront of the NATO policy debate in 1977, when the Soviets began deploying a new missile, the SS-20, to replace the aging SS-4s and SS-5s. The SS-20 was an intermediate-range (4,000 km) missile with three accurate MIRVed warheads on a reloadable mobile launcher. The SS-20 was to be stationed in the western Soviet Union to cover all of Western Europe, in the Far East to cover Asian targets, and also east of the Ural Mountains where it could reach either region.

West German Chancellor Helmut Schmidt was the first Western leader to publicize the threat posed to Europe by the Soviet deployment of the SS-20. In his now famous October 1977 London speech, the chancellor argued that, by codifying strategic parity, the SALT I agreement and the SALT II negotiations on long-range nuclear systems had neutralized the U.S. strategic nuclear guarantee for the defense of Western Europe. While earlier U.S. superiority in nuclear forces had compensated for the Soviet SS-4s, SS-5s, and SS-11s, in the new circumstances of U.S.-Soviet strategic parity nothing compensated for the SS-20s and Soviet superiority in conventional forces. Chancellor Schmidt called on NATO to respond to this growing disparity in the theater nuclear balance.

NATO's Dual-Track Deployment Decision

On December 12, 1979, after two years of extensive study and consultation, NATO unanimously decided to modernize its long-range theater nuclear forces by deploying 464 ground-launched cruise missiles (GLCMs) and 108 Pershing II ballistic missiles beginning in 1983. The Pershing II, with its very high accuracy, terminal guidance, and much greater range (1,800 km) than the Pershing IA (700 km), would be able to strike targets in the western portion of the Soviet Union in approximately ten minutes. The ground-launched cruise missile, which was also to be very accurate but flew at subsonic speed, had a range of 2,500 km, which allowed it to attack targets deeper within the Soviet Union than the Pershing II could reach. The NATO decision also provided for the withdrawal from Europe of 1,000 nuclear warheads of the shorter-range tactical nuclear weapons and the retirement of one existing nuclear weapon in Europe for every new longer-range weapon deployed.

NATO's deployment decision was called "dual track" because the alliance also pledged to pursue a parallel effort to obtain an arms control agreement with the Soviet Union to limit theater nuclear forces. The NATO communique stated that limitations on U.S. and Soviet long-range theater systems should be negotiated bilaterally in a step-by-step approach using the anticipated SALT III framework. The immediate objective of these negotiations was to establish verifiable, equal limits on U.S. and Soviet land-based long-range theater nuclear missile systems.

As NATO was nearing its modernization decision, Soviet President Leonid Brezhnev announced in October 1979 a series of arms control initiatives that included an offer to limit deployment of SS-20 missiles if NATO would defer its decision to deploy new missile systems. The alliance rejected Brezhnev's offer, saying a freeze was not enough. The initial Soviet reaction to NATO's dual-track decision was that it had destroyed any possibility for negotiations on theater nuclear systems. The possibility for any arms control talks was further dampened by the Soviet invasion of Afghanistan, after which President Carter requested that the Senate suspend consideration of the SALT II Treaty.

In July 1980, President Brezhnev reversed his earlier position and announced that the Soviet Union was prepared to enter negotiations at any time. The United States agreed, and preliminary talks between the United States and the Soviet Union began in mid-October 1980. The opening U.S. position, which reflected the NATO decision, was that attention should initially be given to establishing equal ceilings on long-range land-based theater nuclear missile systems. Those ceilings would cover the planned U.S. GLCMs and Pershing IIs and the Soviet SS-20s and older SS-4s and SS-5s. The Soviet proposal, which incorporated Brezhnev's call for a freeze on new deployments, took a broader approach, stating that all American systems capable of striking Soviet territory from European bases, such as the F-111 bombers stationed in Britain or aircraft on carriers in the European region, should be included, along with French and British strategic forces. The talks recessed after the U.S. elections in November, having simply established the positions of the two sides.

While the new Reagan Administration assessed its arms control and foreign policy goals, the grass-roots antinuclear movement in Europe, which opposed the deployment of new U.S. missiles, continued to gain popular support. Widespread European fears that the United States was moving toward a nuclear war fighting capability—which had been kindled by the deployment decision, the 1979 controversy over the neutron bomb, and the Carter Administration's 1980 announcement of a new flexible targeting strategy—were further fueled by the Reagan Administration's hard-line rhetoric and its failure to resume arms control negotiations promptly. At the end of February 1981, President Brezhnev in his report to the 26th Soviet Party Congress called on the United States to join in negotiations and proposed a U.S.-Soviet moratorium on deployment of new medium-range nuclear missile launchers in Europe and European parts of the Soviet Union. Brezhnev stated that this moratorium "could come into force immediately as soon as negotiations on this question commence and would be effective until a permanent treaty on limitation or, even better, on reduction of such nuclear facilities in Europe is concluded."

The INF Negotiations

Although the NATO alliance rejected Brezhnev's call for a moratorium while negotiations were in progress, the stage was now set for the INF negotiations. It is possible to reconstruct these negotiations in detail, since both sides released their positions and leaked information in an effort to influence the public debate on the NATO deployment decision. The details of the negotiations help illuminate the underlying differences between the positions of the two sides in this area.

Secretary of State Alexander Haig and Soviet Foreign Minister Andrei Gromyko announced on September 24, 1981, that the United States and Soviet Union would begin negotiations on November 30, 1981, in Geneva. The vague wording of the joint communique, which referred to talks "on those nuclear arms which were earlier discussed" between the two sides, made it apparent there was disagreement from the beginning about which systems the negotiations should address. Another important point of discord that became apparent before the negotiations started was the strikingly different assessments of the balance in Europe. President Brezhnev declared that there was at that time a rough balance of approximately 1,000 NATO and Soviet systems in Europe. President Reagan insisted that the Soviet Union had an "overwhelming advantage" in intermediate-range systems. The main differences in the two assessments involved the number and kinds of aircraft included by both sides and inclusion by the Soviet Union of British and French forces.

The fall of 1981 was a period of increased tensions in the NATO alliance. Demonstrations in Western Europe against the deployments became more frequent and numerous. European anxieties were fueled by such developments as President Reagan's statement that a nuclear war limited to Europe was possible, Secretary Haig's statement that NATO's strategy included nuclear demonstration shots, and the U.S. decision to produce the neutron bomb. Both in the United States and in Europe, dissenting views on the wisdom of the NATO modernization decision were becoming more common. Analysts argued that the Pershing IIs and GLCMs would inherently be extremely vulnerable because of their inability to maneuver in the densely populated, limited land area of Western Europe and because of their proximity to the Soviet Union. Some questioned whether the NATO deployments would be sufficiently survivable to provide a credible deterrent. Consequently, there were questions about the validity of the underlying premise that the deployment provided a link to U.S. strategic forces.

On November 8, 1981, in his first major address on arms control, President Reagan sought to develop popular support at home and abroad for the alliance's approach to the upcoming negotiations. He presented the so-called zero option proposal, under which the United States would be prepared to cancel its deployment of Pershing IIs and GLCMs if the Soviet Union would dismantle all of its SS-20, SS-4, and SS-5 missiles. The NATO allies and both Democrats and Republicans on Capitol Hill warmly welcomed the President's proposal. The Soviet Union immediately criticized the proposal because it "deliberately" overlooked U.S. submarine-based systems, forward-based aircraft systems, and French and British nuclear systems. A week later, Soviet President Brezhnev stated that the zero option proposal was inequitable to the Soviet Union. He reiterated Moscow's proposal for a moratorium on new deployments of intermediate-range missiles in Europe and said that the Soviet Union was willing to make "radical" cuts in its forces.

As the first round of the INF talks opened in Geneva, there were mass protests against U.S. and Soviet arms policies in Denmark, Switzerland, Italy, West Germany, and Romania. The NATO ministers responded by endorsing the arms control process, stating that they planned to go ahead with the deployment of the new weapons in Europe should the negotiations fail. President Reagan asserted that it was the deployment decision that had brought the Soviet Union to the negotiating table, and that the antinuclear demonstrations in Europe "were bought and paid for by the Soviet Union."

President Brezhnev then proposed a two-thirds reduction in NATO and Soviet medium-range nuclear systems. This would amount to a reduction of about 600 systems from the 1,000 systems that the Soviet Union estimated each side to have. Brezhnev also called for the "real zero option," which he defined as the removal of all tactical and medium-range nuclear arms, including British and French arms. He stated, however, that if the West was not ready for "radical decisions" the Soviet Union would settle for deep cuts. President Reagan rejected Brezhnev's plan and announced that the United States had presented a draft treaty incorporating the proposed zero option.

The Soviet Union responded to the U.S. announcement with a statement from the Soviet news agency Tass that labeled the U.S. draft as "absurd" and detailed the Soviet draft position in the negotiations. The Soviet proposal sought to limit medium-range missiles, which the Soviet Union defined only as systems in Europe west of the Ural Mountains, forward-based nuclear-capable aircraft, and British and French systems. The main feature of this proposal was a call for phased reductions from the current balance, under which both sides would reduce to totals of 300 by 1990. The U.S. State Department in turn said that the Soviet proposal was unacceptable, because it would not prevent more SS-20s from being deployed in the European part of the Soviet Union, would permit unlimited deployment of SS-20s in the Asian part of the Soviet Union, and would not permit the United States to deploy intermediate-range missile systems in Europe.

By the end of the first round of negotiations in Geneva, it was clear that the sides differed on four fundamental issues: the United States sought to eliminate all intermediate-range nuclear missiles, whereas the Soviet Union sought to stop planned U.S. missile deployments but was not prepared to give up all of its existing intermediate-range missiles in return; the United States sought to limit only U.S. and Soviet systems, whereas the Soviet Union sought to include the British and French forces; the United States called for global limits on intermediate-range missiles, whereas the Soviet Union proposed to limit only systems in or "intended for use" in Europe; and the United States sought to limit only intermediate-range missile systems, whereas the Soviet Union also wanted to include nuclear-capable aircraft in a single aggregate ceiling. The sides disagreed on a number of other significant issues. These issues included the following: Soviet refusal to consider U.S. proposals for limits on shorter-range missile systems; U.S. insistence that reduced systems be destroyed as opposed to Soviet offers to reduce systems by a combination of destruction and withdrawal from range of Europe; and U.S. desire for a treaty of unlimited duration as opposed to the Soviet position that the treaty would have to be renewed in 1990.

As the first round of the Geneva negotiations recessed, President Brezhnev announced in March 1982 a unilateral freeze on new medium-range missiles in the European part of the Soviet Union. He also said that some missiles already in place would be removed in the course of 1982. While restating hopes for agreement, Brezhnev issued a warning, which was to be repeated throughout the course of the negotiations, that the deployment of U.S. missiles in Europe capable of striking targets within the Soviet Union would compel it "to take retaliatory steps that would put the other side, including the United States itself, in its own territory in an analogous position."

The White House rejected Brezhnev's unilateral moratorium, calling it "neither unilateral nor a moratorium," and charged that the proposal sought to maintain Soviet superiority, divide the West, and secure "unchallenged hegemony" for the Soviet Union over Europe. The NATO allies joined in dismissing Brezhnev's initiative and noted that existing deployments of SS-20s east of the Urals could also be targeted on Europe. President Brezhnev publicly responded to the NATO allies' concern in a nationally televised speech in May 1981, in which he pledged that missiles withdrawn from the European part of the Soviet Union would not be redeployed east of the Ural Mountains within range of the Western European nations.

The "Walk in the Woods" Formula

In the second phase of the formal negotiations, which ran from May 2 through July 20, 1982, little progress was made on the major issues, although the Soviets did make a number of minor changes and additions to their proposal. Meanwhile, secret informal discussions were taking place between Ambassadors Paul Nitze and Yuli Kvitsinsky in July 1982 on the major issues in the negotiations. The informal agreement that may have emerged from their discussions later became known as the "walk in the woods" formula. Reportedly, the formula contained the following provisions: the United States and the Soviet Union would each be limited to 225 intermediate-range missile launchers and aircraft in Europe (including the eastern slope of the Urals); each side would be limited to a subceiling of 75 intermediate-range missile launchers in Europe (including the SS-20s on the eastern slope of the Urals at Verknyaya Salda that could reach Europe), with the United States to deploy only cruise missiles (not Pershing IIs) within its sublimit; the Soviet Union would be limited to 90 intermediate-range missile launchers in the eastern Soviet Union outside the range of Europe; each Soviet ballistic missile launcher would carry no more than three warheads; each U.S. GLCM launcher would carry no more than four missiles, each with one warhead; all intermediate-range missile systems in excess of the limits would be destroyed; limited aircraft would be the U.S. F-111 and FB-111 and the Soviet Backfire, Badger, and Blinder; U.S. and Soviet shorter-range INF missiles would be limited to existing levels; and appropriate verification measures would be negotiated within three months.

Much controversy was eventually to surround the origin of the walk-in-the-woods formula, the role of each negotiator in its preparation, and the question of which side rejected it first. According to U.S. Ambassador Nitze, he and Soviet Ambassador Kvitsinsky agreed to develop jointly, without official commitment, a complete package of reciprocal concessions that would resolve all of the principal outstanding issues if accepted by both governments. Ambassador Kvitsinsky, on the other hand, maintains that Nitze unilaterally advanced a package deal along the lines of the walk-in-the-woods proposal, which Kvitsinsky agreed to send to Moscow but told Nitze would be unacceptable to the Soviet Union. Whatever the facts on the origin of the walk-in-the-woods formula, neither the United States nor the Soviet government embraced the concept. Nitze reports that in September 1982 Kvitsinsky returned to Geneva with instructions to reject the walk-in-the-woods formula. The U.S. press reported that after much debate within the administration, President Reagan had previously rejected further exploration of the walk-in-the-woods compromise, as a result of strong opposition from the Department of Defense.

After the recess of the second round, Ambassador Nitze announced that there had been "no progress on the central issue" of which weapons to include in an INF agreement. Relations between the two negotiating sides deteriorated as the United States charged the Soviet Union with violating its self-imposed moratorium on SS-20 deployments and the Soviet Union denied that it had deployed new SS-20 missiles west of the Ural Mountains.

The Soviet Offer to Match British and French Missiles

Shortly after the death of Leonid Brezhnev in November 1982, the new head of the Soviet government, Yuri Andropov, announced a new Soviet initiative. This proposal called for the Soviet Union to reduce the number of its medium-range missiles in Europe to about 160, equal to the number of French and British missiles, provided the United States did not deploy the 572 GLCMs and Pershing IIs. The new Soviet leader said that this meant that the Soviet Union "would reduce hundreds of missiles, including dozens of the latest missiles known in the West as the SS-20s." When making this offer, Andropov added that there must also be an accord on reducing to equal levels each side's number of medium-range nuclear-capable aircraft stationed in Europe. Soviet Foreign Minister Gromyko elaborated that the Soviet Union was prepared to negotiate on the basis of "mutuality" a reduction of its shorter-range SS-21, SS-22, and SS-23 nuclear weapon systems targeted on Western Europe. He also confirmed that "some rockets could be completely destroyed and some could be redeployed behind a line in Siberia where they could no longer hit targets in Western Europe."

The Reagan Administration rejected Andropov's proposal, saying that it would still leave a Soviet monopoly on intermediate-range missiles in Europe and deny the United States "the means to deter the threat." The NATO allies also quickly rejected Andropov's offer, and the French Foreign Minister said that he was "shocked" that the Soviet Union would attempt to include the French nuclear arsenal in U.S.-Soviet arms control talks since the French nuclear force was "independent." By the end of November 1982, when the third session of the formal negotiations ended, the two sides remained far from agreement.

The series of Soviet initiatives, combined with the continuing political demonstrations in Europe and the pending elections in several European countries, put pressure on the United States to appear more forthcoming in the INF negotiations. Pressure from NATO for the United States to advance a position less stringent than the zero option peaked in mid-January 1983 when former U.S. Arms Control and Disarmament Agency Director Eugene Rostow revealed that Ambassadors Nitze and Kvitsinsky had developed the informal walk-in-the-woods compromise the previous summer. Initially, the White House informed the press that the informal agreement was inadequate and could not have served as the basis for an accord. Several days later, the White House informed the press that the United States had not closed the door quite as firmly as Moscow, explaining that Secretary of State George Shultz had sought to signal Gromyko on September 28 that while the understanding was inadequate the informal channel should remain open. Gromyko rejected reports that the U.S. and Soviet negotiators had earlier reached any tentative agreement at Geneva.

To reassure NATO about the sincerity of the U.S. interest in an agreement, Vice President George Bush publicly read a letter in Europe from President Reagan to President Andropov inviting Andropov to meet Reagan to sign a treaty banning all intermediate-range land-based missiles "whenever and wherever he wanted." President Reagan himself later acknowledged that the letter was not a new proposal but rather a response to the "vast" Soviet propaganda effort "to discount the legitimate" U.S. proposal. President Andropov quickly rebuffed the offer, saying that the conditions for the meeting were totally unacceptable.

The U.S. Interim Agreement Proposal

On March 31, 1983, President Reagan proposed an "interim agreement" in which the United States would substantially reduce its planned deployment of Pershing IIs and ground-launched cruise missiles, provided the Soviet Union reduced the number of warheads on its intermediate-range missiles worldwide to an equal level. This announcement followed the reelection of West German Chancellor Helmut Kohl after a campaign in which one of the major issues was the deployment of U.S. missiles on German soil. President Reagan said that his ultimate goal was still to eliminate all intermediate-range nuclear weapons in Europe. Officials also noted that only the zero option would prevent U.S. deployments from beginning in December. The NATO allies acclaimed the President's proposal, which was actually the initial position called for in the December 1979 dual-track deployment decision, and called on the Soviet Union for a quick and constructive response. Foreign Minister Gromyko held a lengthy press conference in which he gave a detailed explanation as to why the proposal was unacceptable. Tass called Gromyko's statement an "unambiguous rejection," whereas the State Department referred to it as an "unconstructive initial Soviet reaction."

President Andropov once again captured the headlines on this unprecedentedly high-profile negotiation with another apparent Soviet initiative on May 4, 1983. Andropov stated that the Soviet Union was now prepared to reach agreement "on the equality of nuclear potentials in Europe both as regards delivery vehicles and warheads with due account for the corresponding armaments of Britain and France. . . . The same approach would be applied also to the aviation systems of this class deployed in Europe." The Soviet President noted in his speech that this approach would leave fewer medium-range missiles and warheads in the European part of the Soviet Union than before 1976, when the Soviet Union did not have SS-20 missiles. The U.S. government welcomed Soviet willingness to negotiate ceilings on nuclear warheads as well as missiles in Europe but once again emphasized that it could not accept renewed Soviet demands for equality with NATO's combined nuclear forces.

As the deployment date drew nearer without progress in the INF negotiations, President Reagan and the NATO allies reaffirmed several times during the summer of 1983 that the deployment was the only way to prod the Soviet Union into serious negotiations. Meanwhile, in the fifth round of the formal sessions, the Soviet negotiators threatened to respond with countermeasures should NATO proceed with its planned deployments. Despite the Soviet rejection of the interim measure and the inflammatory tone of the formal sessions, President Andropov again attempted to display some flexibility in August 1983. He stated that if a mutually acceptable agreement were achieved, the Soviet Union, in reducing the number of medium-range missiles in the European part of the Soviet Union to equal that of the British and French missiles, would "liquidate" all of the missiles to be reduced. Andropov emphasized that this would considerably reduce the total number of SS-20s, not just shift them to the Asian part of the Soviet Union. The U.S. government responded by saying that, despite his attempts to appear flexible, Andropov was not addressing the two fundamental problems in the negotiations: the right of the United States to deploy new missiles to achieve parity, and the need for global limitations.

The Soviet downing of a Korean airliner at the end of August 1983 deeply disrupted overall U.S.-Soviet relations. The Reagan Administration decided, however, to proceed not only with the INF negotiations but with an announcement of a new U.S. proposal. At the United Nations on September 26, 1983, President Reagan announced that, within the context of an interim agreement providing the United States and the Soviet Union with the right to equal numbers of intermediate-range missile warheads globally, the United States was prepared to consider a commitment not to offset the entire worldwide Soviet INF missile deployment with U.S. INF missile deployments in Europe. The United States would retain the right to deploy missiles elsewhere to meet the equal worldwide ceiling. The United States would also be prepared to negotiate the mix of deployed Pershing IIs and GLCMs and to explore equal verifiable limits on specific types of U.S. and Soviet land-based aircraft. Soviet officials dismissed the new U.S. initiative, complaining that it failed to address the fundamental issues of the British and French systems and the planned deployment of new U.S. missiles in Europe.

By mid-October 1983, senior Soviet spokesmen warned that the Soviet Union would suspend negotiations if the U.S. missiles were deployed in Europe. On October 26, President Andropov announced a new "flexible" Soviet proposal, disclosing that Moscow was prepared to cut down to "about 140" the number of Soviet missiles in the European part of the country. In addition, on entry into force of an agreement covering Europe and assuming no change in the "strategic situation" in Asia, the Soviet Union would unilaterally halt deployment of additional

SS-20 missile systems in the eastern Soviet Union. If the United States deferred the INF deployment in Europe indefinitely, the Soviet Union would unilaterally scrap its remaining SS-4 missiles over a two-year period.

The State Department initially held that there was little new in Andropov's latest proposal, but a few days later President Reagan commented that the United States would study the new Soviet proposal and address it in Geneva. On November 14, 1983, as U.S. missiles arrived in Britain, the President announced that the United States would agree to a global warhead limit of 420 on U.S. and Soviet intermediate-range missiles. Tass responded that any arms control plan that included the deployment of U.S. missiles was unacceptable.

The "Walk in the Park" Proposal

In mid-November Ambassadors Nitze and Kvitsinsky engaged in a final controversial exchange. According to Nitze, Kvitsinsky informed him on November 13 that if Washington proposed equal reductions in Europe of 572 warheads on intermediate-range missiles from the present Soviet force level and the planned U.S. deployment force level, Moscow would accept the proposal. This formula would have resulted in no U.S. deployments and a reduction of 572 warheads on the Soviet side. This would have left the Soviet Union with about 140 SS-20s with 420 warheads in range of Europe. This was roughly the same number of missiles and warheads that the Soviet Union attributed to French and British nuclear forces. The proposed formulation would have avoided specifically referring to those forces. Nitze informed the Soviet ambassador that he could not imagine the United States transforming the Soviet proposal into a U.S. proposal but would inform Washington immediately. The United States in turn informed its NATO allies. On November 17, according to Nitze, the Soviet Union began informing NATO governments that Nitze, not Kvitsinsky, had proposed the mutual reductions by 572 warheads and that Washington was likely to reject the proposal. According to Kvitsinsky, on the other hand, Nitze initiated the discussions of equal U.S. and Soviet reductions of 572 warheads. Kvitsinsky indicated that the Soviet Union would seriously consider the initiative if formally presented. Kvitsinsky claims that West Germany leaked the modified proposal and ascribed it to the Soviet Union so that it could reject the proposal.

In any case, the West German Bundestag voted on November 23, 1983, to reaffirm the U.S. missile deployment, assuring the first round of deployments in Europe. President Andropov followed the vote with the announcement that Moscow was withdrawing from the negotiations. He also announced that the Soviet Union had decided to deploy additional nuclear weapons "in ocean areas and seas" near the United States and to accelerate preparatory work on the deployment of Soviet "operational-tactical" nuclear missiles in East Germany and Czechoslovakia. The Soviet delegation officially walked out of the negotiations in Geneva on November 23, 1983. President Reagan and the NATO alliance expressed their regret that the Soviet Union had discontinued the negotiations and reaffirmed their desire to continue them. The Soviet Union set the removal of the new U.S. missiles as the precondition for resuming the talks.

SUMMARY OF THE U.S. AND SOVIET INF POSITIONS AS OF NOVEMBER 1983

The draft U.S. INF proposals as of November 23, 1983, reportedly included the following elements:

The draft zero-zero treaty, which would be of unlimited duration, would ban all U.S. and Soviet ground-launched nuclear ballistic and cruise missiles with ranges between that of the Pershing II (1,800 km) and 5,500 km. All such existing missiles, their launchers, and other agreed support equipment and structures would be destroyed. New types of missiles with these ranges would be prohibited. The practical effect of this measure would be to eliminate all SS-20, SS-4, and SS-5 missiles and all Pershing IIs and GLCMs.

The draft treaty also called for collateral measures that would limit shorter-range missiles with ranges between those of the Soviet SS-23 and SS-12/22 (500 to 1,000 km) to the number deployed as of January 1, 1982. Excess inventories of such missiles would be destroyed, although modernization and replacement, within certain qualitative limits, would be permitted on a one-for-one basis. The combined effect of these provisions would be to ban any ground-launched nuclear missiles with ranges greater than that of the SS-12/22. The United States agreed to discuss limits on the Pershing I in the negotiations.

The U.S. interim proposal called for equal global levels of warheads on U.S. and Soviet intermediate-range missiles. Specifically, the United States proposed in November 1983 a global warhead ceiling of 420 warheads on these missiles. Also, within the context of an agreement providing the right to equal global levels, the United States stated that it was prepared to consider a commitment not to offset the entire worldwide Soviet missile deployments with U.S. missile deployments in Eu-

rope, although it would retain the right to deploy U.S. INF missiles up
to the ceiling elsewhere; to negotiate the mix of Pershing IIs and
GLCMs to be deployed; and to explore equal and verifiable limits on
specific types of U.S. and Soviet land-based aircraft.

The draft Soviet INF proposal as of November 1983 reportedly called
for a moratorium on future deployments of all "medium-range" systems in Europe and a treaty of unlimited duration with the following
elements and options:

• A ban on U.S. medium-range missiles in Europe and reductions of
Soviet SS-20 missile launchers within range of Europe to approximately 140; or alternatively, elimination of *all* (U.S., Soviet, British,
and French) "medium-range" and tactical nuclear weapons in Europe,
otherwise referred to as the "real zero option."

• Dismantling or destruction as the primary means of reduction for
excess systems. However, a certain percentage of excess systems could
be withdrawn from the "zone of reduction or withdrawal," which would
be west of the line 80 degrees east (near Novosibirsk in Siberia) to
latitude 57 degrees north and then to the mouth of the Lena River. One
excess missile associated with each excess launcher would also be
"liquidated."

• A unilateral halt of the deployment of additional SS-20 missile
systems in the eastern Soviet Union, assuming no change in the "strategic situation" in Asia.

• Discussion of specific numbers and types of aircraft to be limited in
the context of equal levels of NATO and Soviet aircraft with a "medium
radius" of action. (Soviet representatives have indicated that they were
prepared to agree to a level of 300 to 400 aircraft. This would include
aircraft on U.S. carriers in the European area and the French Mirage IV
as the only non-U.S. aircraft under the NATO aggregate.)

• Unilateral destruction of remaining Soviet SS-4 missiles over a
two-year period if the United States would agree to defer indefinitely
the deployment of intermediate-range missiles in Europe.

• Certain quantitative unilateral constraints on systems with
ranges between 500 and 1,000 km in a separate protocol.

Originally, the Soviet Union had proposed that if the United States
did not deploy the U.S. missiles, it would agree to equal limits of 600
"medium-range" missiles and aircraft "in or intended for use" in Europe by the end of 1983 and 300 by the end of 1990. British and French
forces would be counted under the U.S. ceiling. A subceiling would limit
Soviet missiles in Europe to 162, the current number of British and
French missiles. If the level of British and French missiles increased in

quantity or quality, the subceiling would be increased to allow the Soviet Union to maintain equality in missiles with Britain and France.

THE MAIN ISSUES SURROUNDING THE INF NEGOTIATIONS

The European Nuclear Balance: Different U.S. and Soviet Perspectives

The underlying issue in the INF negotiations is the different assessments of the balance of nuclear forces in Europe by the United States and the Soviet Union. These differing views determined each side's position on the scope of the negotiations. The United States and NATO believe that there is a significant imbalance favoring the Soviet Union, which was the main reason for NATO's 1979 decision to have the United States deploy 464 GLCMs and 108 Pershing IIs in Europe. The Soviet Union believes that there is presently a regional balance, which was the main Soviet rationale for opposing any U.S. deployments in Europe. Analysts have argued for years about the actual number of nuclear systems in Europe. The following sections describe the official U.S. and Soviet assessments of the balance of intermediate-range nuclear systems since the INF negotiations began in 1981.

The U.S. View

The U.S. perspective, which its NATO allies share, is that since the mid-1970s there has been a growing disparity in the nuclear balance in Europe. The most important factor that has contributed to the growing threat to Western Europe and to the instability of the European military balance has been the continuing deployment of Soviet intermediate-range systems, in particular the new mobile solid-fueled SS-20 missile. Since 1977 the Soviet Union has deployed some 378 SS-20s, each with three accurate 150-kt MIRVed warheads. More than 240 of these SS-20s can be targeted on Western Europe. The United States and its NATO allies consider the SS-20 to be a new generation of longer-range, qualitatively improved intermediate-range missiles, not simply an improvement on the older, fixed Soviet SS-4 and SS-5 systems previously deployed in Europe. Also, unlike the SS-4s and SS-5s, the SS-20s can be reloaded, which further complicates the balance. The United States has estimated that as of January 1, 1984, Soviet warheads on all intermediate-range missiles including those in the Far East totaled approximately 1,300, not including the SS-20's reload capacity.

In the U.S. view, until the deployment of the U.S. Pershing II and

GLCM systems began in the late fall of 1983, the United States and its NATO allies had no systems that posed a comparable threat and thus provided a deterrent against a Soviet threat of attack or actual attack on Western Europe. This situation led to NATO's 1979 dual-track decision to modernize its intermediate-range nuclear forces while pursuing arms control to reduce the threat posed by these Soviet systems. The United States and its allies were concerned that without the U.S. deployment in Europe or without an agreement that would decrease the Soviet threat, the NATO alliance would have to rely upon U.S. strategic forces to respond to a Soviet attack with intermediate-range missiles. The alliance would also have to rely solely on these U.S. strategic forces to deter a Soviet threat of attack to attain political goals. Under such circumstances the United States and its allies feared that deterrence would be undermined, since the Soviet Union might miscalculate in a crisis that an attack limited to Europe would not produce a strategic response from the United States.

Further disrupting the balance in Europe in the U.S. view was Soviet development and deployment of a new generation of shorter-range ballistic missiles (the SS-22s and SS-23s) with ranges from 500 to 1,000 km. Once the U.S. Pershing IA missiles are replaced by Pershing IIs, the United States will have only a small number of operational short-range (110 km) Lance missiles. In the U.S. view, the shorter-range Soviet missile systems pose an increasing threat to the survivability of NATO's air bases and seaports, command, control, and communications facilities, and key nuclear and conventional forces. In the mid-1970s the Soviet Union also markedly enhanced its theater nuclear aircraft capabilities when it started to deploy the Backfire bomber. In the U.S. view, all of these recent Soviet deployments have increased the gap between NATO's nuclear capabilities and those of the Soviet Union.

President Reagan countered the Soviet claim of approximate parity in the overall European balance, including missiles and aircraft, with the claim that the Soviet Union had an overwhelming six to one advantage. According to estimates provided by the U.S. State Department in 1981, total U.S. intermediate-range nuclear systems numbered about 560, compared with 3,825 Soviet systems (see Table 1). The U.S. estimate included 2,700 Soviet fighter-bombers that U.S. analysts believe can deliver nuclear weapons or can be readily converted to that status. The Soviet Union does not include these bombers in its estimates. The United States also pointed out that the Soviet Union claimed that parity existed in Europe from 1979 through 1982, during which time the number of Soviet missile warheads on land-based intermediate-range missiles increased from 800 to 1,300. The United States also criticized

TABLE 1   U.S. and Soviet Views of the INF Balance in 1981

U.S. COUNT			
U.S.		**Soviet**	
Missiles	0	SS-20 missiles	250
F-111 fighter-bombers	164	SS-4s and SS-5s	350
F-4s	265	SS-12s and SS-22s	100
A-6s and A-7s	68	SS-N-5s	30
FB-111s (in U.S. for		TU-26 Backfire bombers	45
use in Europe)	63	TU-16 Badgers and TU-22 Blinders	350
		SU-17, SU-24, and MIG-27	
		fighter-bombers	2,700
TOTAL	560		3,825

SOVIET COUNT			
Western		**Soviet**	
U.S.		Land-based missiles	
Fighter-bombers		(SS-20s, SS-4s, SS-5s)	496
(F-111s, F-4s, A-6s,		Submarine missiles	
A-7s, FB-111s)	555	(SS-N-5s)	18
Pershing IA missiles	108	Medium-range bombers	
		(Backfires, Badgers, Blinders)	461
British			
Polaris missiles	64		
Vulcan bombers	56		
French			
Land-based intermediate-range			
ballistic missiles	18		
Submarine missiles	80		
Mirage 4 bombers	33		
West German			
Pershing IA missiles	72		
TOTAL	986		975

SOURCE: The *New York Times*, November 30, 1981, p. A12.

the Soviet calculations for including the independent British and French nuclear systems.

The Soviet View

The Soviet Union has continued to insist that rough parity in intermediate-range weapons existed in Europe throughout the late 1970s and early 1980s until it was disrupted by the U.S. deployment of Pershing IIs and GLCMs. In 1979, according to Soviet calculations, the rough parity included approximately 1,000 systems apiece for the Soviet Union and the NATO countries (see Table 1). The Soviet estimates included on their side the SS-20, SS-4, and SS-5 missiles and intermediate-range bombers; on the NATO side the Soviet estimates included the U.S. forward-based nuclear force (FB-111 bombers, F-111 and F-4 fighter-bombers, A-6 and A-7 carrier-based aircraft, and the Pershing IA missiles), together with the nuclear forces of Britain and France (ground-based French S-2 and S-3 missiles, British Polaris and French M-20 submarine-based ballistic missiles, and Vulcan, Buccaneer, and Mirage bombers). At the end of 1983, then Soviet Chief of the General Staff Marshall Ogarkov presented slightly different estimates that included 938 Soviet launchers (465 bombers and 473 missiles) and 857 NATO launch vehicles (695 aircraft and 162 British and French missiles). Soviet spokesmen pointed out that NATO delivery systems have a range of 1,000 to 4,000 km and thus can reach targets within the Soviet Union, making them analogous in combat potential to Soviet SS-20 missiles.

The Soviet Union has argued that the SS-20 is simply a replacement on a one-for-one or one-for-two basis for the older SS-4 and SS-5 missiles, whose service lives have expired. To counter the arguments about the destabilizing nature of the upgraded SS-20s, Soviet spokesmen have noted that even though they carry three warheads, their combined yield (three times 150 kt) is less than that of one old SS-4 and SS-5 warhead (around 1 Mt). Consequently, Soviet spokesmen have repeatedly stated that the process of replacing obsolete missiles has decreased both the total number of Soviet "carriers" and the total yield of the warheads these systems could deliver. At the end of 1983 the Soviet Union stated that it had reduced its missile launchers from 600 to 473 and that the SS-5 missile had been totally withdrawn. President Andropov charged that U.S. assessments of an imbalance pretend that 1,000 medium-range U.S. and NATO nuclear systems in the European zone do not exist.

The Soviet government has argued that the deployment of new U.S. intermediate-range systems will disrupt the existing rough parity by giving NATO a major advantage in both the number of missile launchers and associated missile warheads. With these new deployments, according to the Soviet Union, the United States is starting another exceptionally dangerous round in the arms race.

The Soviet Union has also emphasized the charge that the new U.S. missiles dramatically change the nuclear balance since they can reach Soviet territory in a very short time whereas similar Soviet missiles cannot reach the United States at all. In the Soviet view, these new U.S. systems, particularly the Pershing IIs, undercut the foundation of strategic stability because they threaten the Soviet command and control systems. More generally, the Soviet Union has argued that these new U.S. deployments in Europe are part of a larger U.S. program designed to give the United States the capability to launch a preemptive attack against the Soviet Union. In the Soviet view, the Pershing II, in the context of the MX missile, the Trident II D-5 missile, anti-satellite weapons, the new command and control systems, and the new ballistic missile defense initiative, will produce a destabilizing shift toward U.S. preemptive superiority.

The Main Differences Between the U.S. and Soviet INF Proposals

The main differences between the U.S. and Soviet proposals center on three main issues: (1) the systems to be included in the negotiations, (2) the geographic scope of the negotiations, and (3) the treatment of third-country nuclear forces.

The U.S. Approach

The five criteria guiding the U.S. approach to the negotiations were that an agreement (1) must entail equal limits and rights for the United States and the Soviet Union, (2) should address only U.S. and Soviet systems, (3) should apply to INF missiles regardless of location and should not shift the security problem in Europe to the Far East, (4) should not weaken the U.S. contribution to NATO's conventional deterrence and defense, and (5) must be verifiable.

The U.S. View: Systems to be Covered. In the view of the United States and its NATO allies, the chief source of the destabilizing imbalance in Europe has been the new SS-20 missile. Therefore, the opening U.S. position, the zero option, sought to eliminate the entire class of intermediate-range land-based missiles, together with their launchers and certain support structures and equipment. It also banned testing and production of new intermediate-range missile systems. The practical implication of the zero option was that the United States, which had no system comparable with the SS-20, SS-4, or SS-5 missiles, was willing to forego the deployment of its Pershing IIs and GLCMs if the Soviet Union would eliminate its existing intermediate-range systems.

The U.S. interim proposal covered the same land-based missiles. Throughout the negotiations the United States insisted that it must have equal rights to deploy intermediate-range systems to offset any Soviet deployments if these systems were not to be eliminated. The informal walk-in-the-woods formula suggested that the United States might be prepared to forego deployment of the Pershing II missiles while proceeding with the deployments of the GLCMs in exchange for certain Soviet concessions, but the U.S. government officially rejected that approach and continued to argue that both systems were necessary to offset the Soviet advantage in this area. As long as the Soviet Union maintained an SS-20 force, the United States argued, the right to deploy a mix of Pershing IIs and GLCMs was essential to provide flexibility in delivery systems and to hedge against the possible loss of either system.

Even though the United States would not forego the right to deploy either Pershing IIs or GLCMs, it did modify its position, announcing that it was willing to negotiate the mix of the planned deployment level of the Pershing IIs and GLCMs. Toward the end of the negotiations the United States also acknowledged the Soviet concern about aircraft by indicating a readiness to explore limits on specific types of U.S. and Soviet aircraft. But the two sides continued to disagree over what aircraft should be counted and how aircraft should be included in the negotiations. The U.S. assessment gave the Soviet Union a five-to-one advantage in this area, whereas Soviet assessments gave NATO an advantage.

The U.S. approach on intermediate-range missiles also included collateral restraints on shorter-range nuclear missiles. The original U.S. position included limits on Soviet missiles with ranges between those of the Soviet SS-23 and SS-12/22 (500 to 1,000 km) at the levels deployed as of January 1, 1982. The net effect of the U.S. zero option proposal would have been to ban any ground-launched nuclear missiles with ranges greater than that of the Soviet SS-12/22. The United States did not include limits on the shorter-range U.S. Pershing IA missile (700 km), on the grounds that shorter-range Soviet systems could fulfill the missions of Soviet intermediate-range missiles to a much greater extent than the Pershing IA could fulfill the mission of U.S. intermediate-range systems. However, in June 1983, after the Soviet Union agreed to consider limitations on these shorter-range systems, the United States agreed to consider restraints on the Pershing IA.

The U.S. View: Geographic Scope. The U.S. proposal called for worldwide limits on intermediate-range missiles because the range, mobility, and transportability of the SS-20s made these missiles a potential threat to the NATO allies even if deployed in the Far East. The U.S. zero option solved this problem by completely eliminating these systems. Under the interim proposal the main reason for global limits was to ensure that the security threat to Europe was not shifted to the Far East. In September 1983 the United States modified its position by stating that under an agreement providing for equal global ceilings, the United States was prepared not to match the entire worldwide deployment of Soviet intermediate-range missiles with U.S. missiles in Europe and offered to explore the level of these European deployments.

With regard to the Soviet proposal, the United States argued that, with the range of 5,000 km it calculated for the SS-20, the line behind which the Soviets proposed to deploy SS-20s would still have them within range of Western Europe. In addition, the Soviet offer to freeze intermediate-range systems in Asia was subject to a unilateral Soviet assessment of the strategic situation there and thus did not constitute a real limitation.

The U.S. View: Third-Country Forces. The U.S. position dealt only with U.S. and Soviet systems. It excluded any limitation on or compensation for third-country forces. In response to the Soviet argument that British and French nuclear forces should be included, the United States maintained that these were independent forces of two sovereign nations that were not parties to the negotiations. The United States also argued that it does not determine or control the composition or employment of these forces, which are national minimum deterrents. They are for the most part submarine-based and differ in role and characteristics from the land-based intermediate-range missiles under discussion at Geneva. The U.S. negotiators argued that, unlike the U.S. systems, the British and French systems were not intended to deter attacks on other nonnuclear NATO countries. The U.S. negotiators also argued that the Soviet effort to include those forces violated the fundamental principle of equal rights and limits between the United States and the Soviet Union. Finally, top-level U.S. government officials noted that, while the Soviet Union had previously sought in other arms control negotiations to obtain compensation for British and French forces, it had in the past found it possible to enter into agreements with the United States even though its demands for compensation were rejected.

General U.S. Criticisms of the Soviet Position. The United States consistently maintained that the basic problem with the Soviet negotiating position was that the Soviet Union would not acknowledge the U.S. right to parity in the area of intermediate-range nuclear force missiles. Until that right was acknowledged, there could be no agreement. The United States argued that the last official Soviet offer of a ceiling of 420 warheads, with an implicit ceiling of 140 SS-20 launchers capable of targeting Europe and a freeze on Soviet systems in Asia, still matched Soviet forces with British and French forces while not allowing the United States the right to deploy any intermediate-range missiles. In the U.S. view, the Soviet Union would maintain a destabilizing unilateral advantage that could be used for political intimidation in a crisis in Europe.

The United States also criticized the Soviet Union for its general approach to the negotiations. The Soviet Union was viewed more as fighting a propaganda war to stop U.S. deployments in Europe and disrupt the NATO alliance than as negotiating in good faith. The United States also held the image of Soviet flexibility in the negotiations to be a misperception cleverly developed by the Soviet Union, since in reality all of the Soviet proposals were merely elaborations of the original proposals presented at the first session of the negotiations.

The Soviet Approach

The main thrust of the Soviet approach has been that there is a balance of medium-range missiles in Europe between NATO and the Soviet Union and that the subject of the negotiations should be limitations on the nuclear arms of the Soviet Union and all NATO forces threatening the Soviet Union. Only this approach will assure the Soviet Union equality and equal security in Europe. The Soviet Union argued that the SS-20 had not disrupted the balance and was in fact simply a modernized replacement of older systems whose service lives had come to an end. The deployment of new U.S. missiles, however, constituted a buildup of new nuclear weapons capabilities. This buildup therefore undermined the very purpose of the talks and made the negotiations meaningless in the Soviet view. The Soviet Union emphasized that no Soviet systems in Europe can target the United States, whereas the new U.S. systems can hit the Soviet Union in about ten minutes.

In the Soviet view, theater forces are directly related to the overall strategic balance that was implicitly agreed to in SALT I and SALT II. In these agreements, overall capabilities were pronounced equal. Soviet advantages in intercontinental ballistic missiles and theater forces offset U.S. advantages in strategic bombers, forward-based systems threatening the Soviet Union, allied forces, and the overall quality of U.S. forces, as exemplified by its submarine-launched ballistic missiles. Any revisions of the SALT approach would have to be reciprocal and equal. Present systems would be traded for present systems; future systems would be traded for future systems.

The Soviet View: Systems to be Covered. From the outset of the negotiations the Soviet Union maintained that if the sides did not agree to remove all nuclear weapons from Europe, then an agreement should provide for equal reduced levels of both missiles and aircraft between the Soviet Union and NATO, not just the United States. Since a balance of theater forces already existed in Europe, the Soviet Union called for a ban on the deployment of new U.S. intermediate-range missiles. The Soviet Union argued that both the Pershing IIs and the GLCMs would disrupt the existing balance, but it focused particular attention on the dangerous and destabilizing aspects of the Pershing IIs. Although in the walk-in-the-woods formula the Soviet negotiator appeared to be agreeing to allow the United States to deploy GLCMs in Europe within a larger set of trade-offs, the Soviet Union officially rejected this approach and reaffirmed its insistence on no U.S. deployments.

The Soviet Union argued that its proposals were seeking approximate parity at lower levels of medium-range systems. In the Soviet view, it is a false distinction to separate nuclear weapons on aircraft from those on missiles. The Soviet Union argued that it cannot ignore the thousands of nuclear weapons on U.S. aircraft in the European zone and on U.S. aircraft carriers, just as it cannot ignore the other nuclear weapons on NATO delivery systems.

Although toward the end of the negotiations the United States agreed to consider aircraft, the Soviet Union argued that U.S. accounting of aircraft turned the NATO's real 50 percent advantage over Soviet medium-range aircraft into a fivefold advantage for the Soviet Union. The Soviet Union insisted that the United States was counting, along with Soviet medium-range bombers, a large number of tactical fighter-bombers that have never carried and cannot, as now configured, carry nuclear weapons. At the same time, it asserted that the United States excluded whole categories of U.S. forward-based aircraft capable of striking the Soviet Union. Consequently, the initial Soviet approach in the negotiations was to call for a ban on new U.S. deployments and a reduction to equal levels of U.S. and Soviet medium-range missiles and aircraft in or intended for use in Europe, with British and French systems being counted under the U.S. totals. This position eventually evolved to include a subceiling for missiles and to allow warheads to be the unit of account to compare Soviet forces with existing British and French forces. The resulting warhead ceiling of 420 would have required reducing Soviet SS-20 launchers within range of Europe to 140.

The Soviet View: Geography. The Soviet Union argued that the negotiations should not include missiles or aircraft out of range of European targets, since its missiles in the Far East pose no threat to Europe. These missiles are intended to protect Soviet security in the Far East. By the end of the negotiations, the Soviet Union proposed to freeze the number of its SS-20 missiles in Asia contingent on the "strategic situation." The Soviet Union stated that this geographic distinction would be assured by its proposed "zone of reduction and withdrawal" behind which, it asserted, the SS-20s could not hit European targets. In defining the zone the Soviet Union claimed that the SS-20 has a range of 4,000 km, while the United States claimed that it has a range of up to 5,000 km.

The Soviet View: Third-Country Forces.   By the end of the negotiations, the Soviet claims for compensation for British and French forces became the central issue of disagreement. The Soviet Union argued that it could not be expected to ignore more than 400 warheads on British and French sea- and land-based missiles that are aimed at the Soviet Union and its allies. Moreover, these warheads are likely to increase from the present level to 1,200 by 1990 and to 2,000 by the end of the century if planned programs are carried out. Under such circumstances, the Soviet Union argued that it was impossible to exclude these missiles from the count of NATO weapons in Europe threatening the Soviet Union. In addition, the Soviet Union claimed that its proposals addressed European concerns about the deployment of the SS-20s by reducing the warheads on Soviet intermediate-range systems to below the level that existed in 1976.

The Soviet Union has not accepted the rationale that the British and French systems are independent, given British participation in NATO and vigorous French support for the decision to deploy U.S. missiles. Soviet representatives also argued that it defies logic for the United States to present the modernization decision as a NATO mandate while claiming that NATO armaments should not be counted in the negotiations.

General Soviet Criticisms of the U.S. Proposals.   The Soviet Union criticized the U.S. proposals generally on the grounds that they called for inequitable and disproportionate reductions in Soviet forces. In the Soviet view, the zero option was, in effect, an attempt to impose unilateral disarmament on the Soviet Union, since it would have to scrap all of its medium-range missiles while the United States and its NATO allies would retain all of their nuclear weapons in this category.

The U.S. interim proposals, in the Soviet view, also ran counter to the principle of equal security. The interim proposal would allow the deployment of U.S. Pershing IIs and GLCMs in Europe, which would pose a new strategic threat to the Soviet Union and disrupt the existing balance in Europe. Thus, by proposing a trade between future U.S. systems and present Soviet systems with no constraints on British and French forces, the U.S. proposals contradicted the principles of equality and equal security and the overall balance incorporated in SALT I and SALT II.

The Soviet Union also criticized the general U.S. approach to the negotiations, charging that the United States simply used the negotiations as a smokescreen for the U.S. missile deployments in Europe. In this context, Soviet spokesmen criticized the United States for revealing and distorting the private negotiations for propaganda purposes.

Verification

Little has been said publicly about either side's approach to the verification requirements of its respective proposals. It is not clear whether any specific proposals have actually been made. Since initiating the INF negotiations with the Soviet Union, the U.S. government has stated standards for verification that go well beyond previous U.S. requirements for adequate verification (see Chapter 2). In any case, the limitation on INF systems raises several extremely difficult verification issues. Both the U.S. zero option and the U.S. interim proposal would apply only to cruise missiles armed with nuclear missiles; they would permit such missiles with conventional warheads. Unless extremely intrusive inspection is permitted, it is not clear how this distinction will be verified. Although a zero global level has been said to be easier to verify, this advantage disappears if nonnuclear missiles of the same type are permitted.

The U.S. ban on excess or reload missiles, which complements the ban on launchers and missiles in the zero option proposal and the reduction of missiles and launchers in the interim proposal, would appear to require intrusive on-site inspection of Soviet and U.S. production and storage facilities.

Limitations on dual-capable aircraft with conventional as well as nuclear capabilities present another set of difficult verification problems. Effective procedures in this area would require extensive cooperative measures. Besides verifying the limitations of aircraft within an agreed zone, there will be the problem of monitoring similar aircraft stationed outside the zone of limitation.

The Deployment and the Future of NATO

The termination of the INF talks caused by the Soviet walkout at the start of the deployment of U.S. systems in Europe raises the issue of whether on balance NATO's 1979 deployment decision increased or decreased the cohesiveness of the NATO alliance.

The U.S. and other NATO governments have emphasized that NATO's decision to proceed with the deployments, despite an intense four-year campaign by the Soviet Union to split the alliance, has greatly strengthened the alliance and reinforced NATO's fundamental commitment to collective security. The deployment represents a major victory, according to these governments, because the Soviet Union made the deployment a test of its ability to influence security decisions within the alliance. Moreover, some observers argue that the unprecedented, detailed consultations within NATO during the conduct of the negotiations strengthened the cohesiveness of the alliance.

On the other hand, there is a growing belief in some quarters in Europe and the United States that the political problems created by the U.S. missile deployment may have actually reduced the cohesiveness of the NATO alliance. From this perspective, the situation may become progressively worse as deployments continue in stages over the next few years. The collapse of the negotiations contributed to what appears to be a widespread view in Europe that the new U.S. missiles will increase the threat of Soviet nuclear attack rather than deter it. Also, the process of deployment provides the Soviet Union with a continuing political target, which focuses public attention on the issue of whether the NATO nations should seek more independence from U.S. policy.

The United States and the current NATO governments have been broadly criticized by opposition parties for not having tried harder to achieve agreement in the negotiations. For example, the Social Democrats in Germany, the Labour Party in Britain, and the Labor Party in Norway all supported the Soviet proposal to match British and French systems. Some European opposition leaders who originally supported the 1979 dual-track decision, believing that it would produce negotiated reductions in the numbers of nuclear weapons in Europe, are now calling for a delay in the deployments and a return to the negotiating table. Denmark has opposed the deployment, both Greece and Belgium have shown little support for the decision, and Norway and the Netherlands are having difficulties maintaining support for the deployment. These political developments suggest that the deployment decision may in the long run cause serious political problems within the alliance.

Integrating the START and INF Negotiations

In searching for a constructive approach to resuming the INF negotiations, the possibility of folding these negotiations into the START negotiations has been considered. The United States has left open the option of merging the negotiations at a later date, but the Soviet Union has shown little interest in this approach.

Critics of this approach argue that the two negotiations are dealing with separate systems that have different purposes and should therefore be treated separately. Two separate imbalances of forces need to be redressed, one in Europe and one between the strategic arsenals of the superpowers. Therefore, merging the talks may simply result in compromising the establishment of a balance in one area for successful negotiations in another. This is a special concern of the Europeans, who fear that their interests may be compromised in the process of negotiating the strategic portion of the overall agreement. It is further argued that joining the two negotiations may complicate the arms control process to the point where no agreement is possible.

Those who support integrating the two negotiations argue that the initial premise of NATO's 1979 decision was to conduct European intermediate-range negotiations within the context of the next round of strategic arms limitations talks. In the real world, they state, the systems being discussed in the two negotiations are inherently intermingled. Therefore, the issues of each negotiation are in fact part of the same conceptual framework. Supporters of this approach also argue that an agreement is more likely to be negotiated if it is part of a larger package deal, since merging the two negotiations will provide more areas for possible trade-offs.

5 Strategic Defensive Arms Control: The SALT I Anti-Ballistic Missile (ABM) Treaty

INTRODUCTION

The Strategic Arms Limitation Talks (SALT) began in November 1969 under the Nixon Administration with the goal of limiting ballistic missile defense systems and strategic offensive nuclear systems. On May 26, 1972, after two and a half years of negotiations, Presidents Richard Nixon and Leonid Brezhnev signed the SALT I Anti-Ballistic Missile (ABM) Treaty and the Interim Agreement on Strategic Offensive Arms. The ABM Treaty, which is of unlimited duration, obligated both sides not to undertake a nationwide ballistic missile defense and severely limited each side's deployment and development of ballistic missile defense systems. Recently, the renewed U.S. interest in nationwide ballistic missile defense has raised serious questions about the future of the ABM Treaty. (See Chapter 2 for a detailed discussion of the SALT I and SALT II limits on strategic offensive arms.)

BACKGROUND

The Origins

In the early 1950s, in response to the development by the Soviet Union of nuclear weapons and long-range strategic aircraft, the United States embarked on a major program of nationwide air defense. The system was to consist of interceptor aircraft and surface-to-air missiles (SAMs) together with a complex network of radars, computers, and communication for the ground control of intercepts. However, with the advent of intercontinental ballistic missiles (ICBMs), which could readily destroy the air defense system and against which no defense was deployed, continued buildup of air defense appeared futile. It was finally stopped in the early 1960s, and most of the system was later dismantled. Today only skeletal remains exist, leaving the United States with essentially no air defense capability.

In contrast, the air defense effort in the Soviet Union has been continuous and intensive. The primary component of the Soviet system is a collection of SAM batteries consisting of over 10,000 interceptor missiles of different designs deployed throughout the Soviet Union. New and upgraded SAM systems continue to augment or replace those already deployed. Despite these Soviet efforts, the U.S. military remains confident that its strategic aircraft, flying at low altitudes and employing defense suppression tactics and electronic countermeasures, can effectively penetrate Soviet defenses. Moreover, by upgrading the bomber force it is believed that this penetration capability can be maintained into the foreseeable future. Given this assessment of the limited effectiveness of Soviet air defense and the complete vulnerability of the Soviet Union to ballistic missile attack, continued Soviet emphasis on air defense remains something of a puzzle. This is the component of the U.S. and Soviet strategic force posture where there is the largest asymmetry.

The United States began studying defensive measures against ICBMs in the mid-1950s. The Army developed the Nike-Zeus system, which was the forerunner of several ABM systems later considered for deployment, the Nike-X, Sentinel, and Safeguard. Although the Nike-Zeus system "worked" in a narrow technical sense, in that it could destroy one or a few nuclear warheads arriving at a low rate, it was judged incapable of handling a massive attack, especially given the devices to aid penetration that the Soviet Union would have been able to incorporate by the time the system was operational. For that reason, President Dwight Eisenhower refused to approve production and deployment of the system.

The Johnson Years

Despite Eisenhower's decision against early deployment of an ABM system, research and development continued on ballistic missile defenses. The Nike-Zeus system evolved into the Nike-X system, which included a new high-performance short-range interceptor, phased-array radars to replace mechanically steered radars, and more advanced data processing. Nike-X promised substantially higher performance than the Nike-Zeus, but it too remained only a development program. The anticipated penetration technology and destructive powers of nuclear warheads were too much for the system to handle. The Soviet Union also continued its vigorous program of research and development, and in 1964 it paraded a large ABM interceptor missile in Moscow.

By 1964, Secretary of Defense Robert McNamara began to argue that stability in the strategic relationship between the United States and the Soviet Union should rest on a capability of "assured destruction." That is, the United States should be able to destroy a large fraction of Soviet cities and industry after absorbing an all-out Soviet first strike. It was recognized that the Soviet Union would inevitably match this capability, given the inherent destructive power of nuclear weapons. According to McNamara, each side would be deterred from launching a nuclear attack if it was certain that the opposing side would have, even after an attack, a retaliatory force that could deliver unacceptable damage to the attacker's society.

In 1966 the Army and the Joint Chiefs of Staff, supported by a growing congressional interest in ABM defense, attempted to obtain production authority for the Nike-X system. Those opposed to ABM deployment argued that the Soviet Union would respond to the potential degradation of its ability to threaten U.S. urban targets by increasing its offensive forces or resorting to various countermeasures. Either of these responses would be much less costly than the U.S. defensive systems. Secretary McNamara argued that the best way for the United States to penetrate future Soviet ABM defenses would be to upgrade U.S. offensive forces by deploying the new Poseidon and Minuteman III missiles with multiple independently targetable reentry vehicles (MIRVs), which were then under development in the United States. The previous plan had been to equip missiles with penetration aids, such as chaff and decoys. While there might be some element of doubt as to whether such penetration aids would be effective, there was no doubt that a heavy MIRV attack would overwhelm the type of defense then technically feasible.

Military proponents of the U.S. deployment argued that the deployment of an ABM system around Moscow strongly suggested that Soviet strategy was based on developing a strategic nuclear force with a capability beyond that required for assured destruction. They also argued that the Soviet Union would not be able to afford the arms race that McNamara had stated would ensue from an ABM deployment by the United States. Even if the Soviet Union attempted such an arms race, they claimed, the United States could stay ahead if it took the initiative on the ABM deployment.

When President Lyndon Johnson appeared to be leaning towards developing an ABM system, largely in response to congressional pressure to match the Soviet deployments, Secretary McNamara offered a compromise. It involved including the production and procurement money for an ABM system in the budget but withholding the funds pending efforts to explore ABM limitations with the Soviet Union. If stability relied on assured destruction, which defensive deployments might be perceived to threaten, then arms limitations on these systems became an essential part of the stability equation. President Johnson agreed to the compromise, and the strategic arms limitation process was set in motion.

In his January 1967 budget message, President Johnson called for intensive development of the Nike-X system but stated that he would "take no action now" if the Soviet Union was willing to begin negotiations on mutual limitations on ABM systems. Soviet Premier Alexei Kosygin stated in response to a letter from President Johnson in March 1967 that he agreed to bilateral discussions, but that the discussions should be on "means of limiting the arms race in offensive and defensive nuclear missiles." Shortly afterward, the two governments announced that they agreed in principle to begin discussions on both offensive and defensive systems, at an unspecified future date. When President Johnson and Premier Kosygin met at Glassboro, New Jersey, in June 1967, the United States again tried to establish a timetable for negotiations. Despite Johnson's and McNamara's arguments about the need to constrain defensive systems and about the link between defensive systems and an accelerating offensive arms race, Kosygin still appeared reluctant to limit defensive systems.

With President Johnson unable to move the Soviet Union to the negotiating table and unable to withstand congressional pressure for ABM deployment, Secretary McNamara announced on September 8, 1967, the decision to deploy the Sentinel ABM system. The Sentinel ABM was described as a limited or "thin" system oriented primarily toward defending the U.S. population against a potential Chinese missile attack in the 1970s and against an accidental launch. The announcement came at the end of a long speech that focused largely on the infeasibility of a "thick" ABM system designed to protect U.S. cities against an all-out Soviet attack. McNamara emphasized that a nationwide ABM system would be very expensive and would only accelerate the arms race. The speech marked the beginning of an unprecedented public debate over U.S. strategic doctrine that did not subside until the signing of the SALT I ABM Treaty in 1972.

Nine months after the United States announced its decision to deploy the Sentinel system, the Soviet Union indicated that it was willing to proceed with negotiations on strategic arms limitations. On July 1, 1968, at the signing of the Non-Proliferation Treaty, President Johnson announced that the United States and the Soviet Union had agreed to enter "in the nearest future" into discussions on the limitation and reduction of both offensive and defensive strategic weapons. A joint U.S.-Soviet announcement that the talks would start on September 30, 1968, was scheduled for August 21, 1968. But when the Soviet Union invaded Czechoslovakia on the day of the announcement, August 21, the United States postponed the talks indefinitely. In November 1969 the Soviet Union tried to reestablish contact and accepted the objectives proposed earlier by the United States. Despite the Soviet interest and President Johnson's personal efforts, until the final days of his term, to revive the talks, he was unable as a lame duck president to commit his successor to a resumption of the talks.

The Nixon Years: ABM and SALT

The approval of the Sentinel system, combined with the collapse of the scheduled arms talks, aroused unexpectedly intense public opposition to the deployment of ABM interceptor missiles armed with nuclear warheads outside major cities. In response to the growing criticism of the Sentinel system and another Soviet offer to begin arms control negotiations, the new Nixon Administration announced in February 1969 a temporary one-month halt to the deployment of Sentinel pending a review of U.S. strategic policy. This action was in sharp contrast to candidate Nixon's criticisms of arms control and calls for nuclear superiority during the 1968 presidential campaign.

The Nixon Administration declared a doctrine of strategic "sufficiency" as the basis for its review of U.S. strategic forces. This doctrine set forth the following criteria: maintenance of an effective strategic retaliatory capability to deter surprise attack by any nation against the United States; preservation of stability by reducing the vulnerability of U.S. strategic forces and thereby minimizing the Soviet Union's incentive to strike first in a crisis; prevention of a strategic relationship that would permit the Soviet Union to inflict significantly more damage on the U.S. population and industry than U.S. forces could inflict on the Soviet Union in retaliation; and defense of the United States against small-scale nuclear attacks or accidental launches.

Referring to these criteria, President Nixon announced in mid-March 1969 the decision to deploy the Safeguard ABM system. This system employed the same technical components as the Sentinel system, but its new primary mission was to protect some Minuteman ICBMs, some

Strategic Air Command bases, and the National Command Center in Washington against a possible preemptive counterforce attack by the Soviet Union. Defense of the population against a small Chinese attack was retained as a secondary mission. Thus, the Nixon Administration concluded, as had the Johnson Administration, that nationwide ABM systems using available technology could not protect the U.S. population from a heavy attack.

The Safeguard system, like the previous Sentinel system, consisted of three main subsystems: radars, missiles, and computers. The perimeter acquisition radar (PAR) detected and predicted the trajectory of an incoming warhead while it was still several thousand miles from its target. The missile site radar, which was responsible for battle management, took over the tracking of the incoming missile from the PAR and guided the defending missiles to the point of intercept. In a period of about 10 to 20 minutes, computers would have to interpret the radar signals, identify potential targets, distinguish between warheads and decoys, eliminate false targets, track incoming objects, predict trajectories, allocate and guide interceptor missiles, and aim and detonate interceptor missiles when they got within range of a target. All of this would have to be accomplished in a very hostile environment involving radar blackout from offensive and defensive nuclear explosions as well as enemy countermeasures. The system employed two nuclear interceptor missiles: the Spartan, with a multimegaton warhead, and the Sprint, with a relatively small (few kiloton) enhanced-radiation warhead. The Spartan, with a range of several hundred kilometers, was designed to intercept warheads outside the atmosphere, where its high yield would permit misses of substantial distance. The Sprint, with a range of 40 km and a very high acceleration, was designed to intercept missiles at low altitudes after chaff and light decoys had been stripped away by the atmosphere.

The Nixon Administration's decision to go ahead with Safeguard intensified public and congressional debate. The administration argued that the Safeguard system was necessary because of the Soviet development of the ABM system around Moscow and the growth of the Soviet strategic offensive arsenal. Moreover, the Safeguard deployment was not provocative and would not induce a Soviet response large enough to upset the strategic balance, since Safeguard was primarily a limited defense of strategic forces, not cities. Finally, the administration argued that the funding of the Safeguard program would induce the Soviet Union to negotiate ABM limitations.

Many members of the scientific community joined the ABM debate. In general, they argued against the decision to deploy the Safeguard system, as they had against the proposed earlier Sentinel system. They argued that the system was unlikely to perform according to specifications against a real attack. The radars, missiles, and computers in the system were at the limits of existing technology, and extraordinary coordination would be required among those subsystems with a very short reaction time in the poorly understood environment induced by multiple nuclear explosions. Scientists also argued that Soviet ballistic missiles could easily penetrate the Safeguard system by a variety of tactics or overwhelm it by sheer numbers. The weakest links in the system were the few soft radars on which the system completely depended. These radars would be highly vulnerable to nuclear attack. In addition, measures to confuse the system, such as decoys, chaff, and jamming, and measures to blind the radars, such as nuclear blackout, would be particularly effective at high altitudes. At lower altitudes, incoming missiles could use several effective tactics to escape destruction, especially when targeted against cities and other targets. With all these options, it was argued, the attacker could confidently overcome the system at a much lower cost than that of the defensive system itself.

With regard to its mission, there was almost unanimous agreement that Safeguard, both in its proposed deployment or in an expanded future configuration, could not defend the U.S. population against a heavy Soviet attack. The only mission for which a Safeguard-type ABM system might have significant capability would be to defend the hardened U.S. ICBM force, which at that time was not threatened by Soviet ICBMs. Even if this threat developed, detractors argued, it was by no means clear that an ABM system would be the right response since there were alternative, less costly ways of maintaining the survivability of the land-based leg of the triad if this were in fact necessary. In any case, the Safeguard system as proposed could not be counted on to contribute much because its large soft radars were themselves extremely vulnerable to attack.

Scientists who opposed the Safeguard deployments also argued that there was no escape from the strategy of mutual deterrence, which relied on the certainty of being able to deliver a crippling retaliatory blow. Any of the proposed ABM systems would probably have little effectiveness in actual combat, they argued. However, the greater uncertainty over what constituted a secure deterrent would force an intensified arms race at higher and more destructive levels. Upgrading the ABM system would cost much more than implementing effective offensive countermeasures, including the deployment of more reentry vehicles. The U.S. decision to develop MIRVs, which was in large part a response to the Soviet Moscow ABM deployment, illustrated dramatically how improved defenses could stimulate increases in offensive arms.

The administration barely won approval for the first phase of the Safeguard system with a 51 to 50 tie-breaking vote in August 1969. In October 1969, as both superpowers faced the prospects of costly defensive deployments that might lead to an even more costly race in offensive strategic arms, the United States and the Soviet Union agreed to begin the Strategic Arms Limitation Talks on November 17, 1969.

The U.S. position in the SALT I negotiations was that the agreements had to limit both offensive and defensive forces and that any limits had to be verifiable. In the first year and a half of negotiations, the parties were able to reach an agreement on ABM limitations but were unable to reach a comprehensive agreement on offensive systems. The Soviet Union then sought to restrict the negotiations to anti-ballistic missile systems and to defer limits on offensive systems. The United States held that to limit ABM systems but allow the unrestricted growth of strategic offensive systems would be incompatible with the basic objectives of SALT and that it was essential to take first steps in the control of offensive strategic arms. The long deadlock was finally broken when it was agreed to conclude a permanent treaty limiting ABM systems and to agree separately to certain interim limitations on offensive systems while continuing negotiations to achieve a more comprehensive, long-term treaty on offensive strategic arms. On May 26, 1972, Presidents Nixon and Brezhnev signed the SALT I ABM Treaty of unlimited duration and the five-year Interim Agreement on Strategic Offensive Arms.

In the ABM Treaty the United States and the Soviet Union agreed not to deploy ABM systems for national or regional defense. Within this broad constraint, the sides agreed that each would have only two ABM deployment areas (later amended to one). Precise quantitative and qualitative limits were imposed on the ABM systems that could be deployed, along with restraints on radars and interceptor missiles. The treaty specifically prohibited the development, testing, or deployment of sea-, air-, space-, or mobile land-based ABM systems and their components, since such systems might provide the base for a nationwide defense. The treaty contained extensive provisions, paralleling those in the SALT I Interim Agreement, to facilitate its verification by National Technical Means (NTM), which included the satellite reconnaissance systems of both sides. These provisions included bans on interference with these systems and on deliberate concealment measures to impede verification by NTM. The treaty also provided for a U.S.-Soviet Standing Consultative Commission to deal with compliance questions and to promote the implementation of the treaty.

The signing of the SALT I ABM Treaty formalized the mutual recognition that deterrence based on the assured destruction of an attacker's society was the basis of security in the nuclear age. The treaty also signified the two sides' agreement that effective measures to limit ballistic missile defense systems would help curb the ongoing strategic offensive arms race and decrease the risk of nuclear war.

The U.S. Senate demonstrated the broad consensus that had evolved on these difficult and emotional issues by voting 88 to 2 to advise ratification. On July 3, 1974, Presidents Nixon and Brezhnev signed a protocol to the SALT I ABM Treaty that limited each side to a single site instead of the two sites permitted in the original treaty. By 1975 the United States had deactivated the one Safeguard installation it had built at Grand Forks, North Dakota, because it was judged to have little military utility by itself.

Throughout the 1970s, both sides continued significant research on ballistic missile defense technology within the constraints of the treaty. At the same time, both the United States and Soviet Union began to improve the counterforce capabilities of their missile forces, thereby making active defense of ICBM silos a more relevant consideration. This turned the U.S. research and development program for ballistic missile defense increasingly toward the problem of hard-point defense of silos. This was inherently a much simpler technical problem than the defense of populations. Since only a very small area had to be defended, intercepts could be made at low altitudes, and relatively high leakage rates were acceptable given the low value of individual targets. Despite these research activities, there was little pressure to revise the treaty. The communique released at the end of the first scheduled review conference stated: "The parties agree that the treaty is operating effectively, . . . serves the security interests of both parties, decreases the risk of the outbreak of nuclear war, facilitates progress in the further limitation and reduction of strategic offensive arms, and requires no amendment at this time."

The Strategic Defense Initiative

In the early 1980s the issue of ballistic missile deployment and the future of the ABM Treaty once again became a major security issue. Efforts to develop a survivable mode for the deployment of the MX missile focused renewed attention on the potential of ABM systems for last-ditch, hard-point defense. Hard-point defense was considered as a possible supplement to the racetrack deployment proposed by the Carter Administration, and it appeared that the MX "dense pack" basing mode proposed in the early stages of the Reagan Administration might involve a hard-point defense component. The early Reagan Administration defense budgets substantially increased research and development for hard-point defense systems, which had continued after the ABM Treaty and the demise of the Safeguard system. The joint communique of the 1982 ABM review conference, while ensuring continuation of the treaty, signaled less enthusiasm for the treaty than at the conference five years before. In the summer of 1982, Secretary of Defense Caspar Weinberger added to the speculation over the future role of ballistic missile defense when he announced the administration's underlying strategic concepts. These indicated the possibility of an increased focus on ballistic missile defense.

On March 23, 1983, President Reagan reopened the ABM debate on the national level in a major address referred to as the "Star Wars" speech. He called for a major technological effort to develop a defense against strategic nuclear missiles that would eventually make nuclear weapons "impotent and obsolete." He stated that, consistent with the obligations of the ABM Treaty, he was instigating a "comprehensive and intensive effort to define a long-term research and development program to begin to achieve our ultimate goal of eliminating the threat posed by strategic nuclear missiles." He called upon the scientific community to use their talents to find ways "to intercept and destroy strategic ballistic missiles before they reached our own soil or that of our allies." Shortly after the President's speech, Secretary of Defense Weinberger stated in a news conference that he was confident that American science could achieve a total defense of the United States.

The Soviet reaction to the President's speech was immediate and extremely critical. Soviet President Yuri Andropov stated, "At first glance this may seem attractive to uninformed people. . . . In fact, the development and improvement of the U.S. strategic offensive forces will continue at full speed and in a very specific direction—that of acquiring the potential to deliver a nuclear first strike." The Soviet President went on to unveil a new arms control proposal to prevent an arms race in space.

In testimony on the Strategic Defense Initiative program for fiscal year 1985 that was to implement the President's proposal, administration officials were less clear about whether the program was to be directed specifically at a nationwide defense effort. They suggested that although the purpose of the Strategic Defense Initiative was to assess technologies for a highly effective, multitiered defensive system, intermediate versions of a less comprehensive ballistic missile defense might be useful in assuring the survival of the U.S. deterrent force.

Administration officials also said that a definitive statement of the goals of the program could only be made after an initial research effort.

To determine the technical feasibility of the President's long-term goal, the Defense Department formed two study groups of scientists and national security experts. Their reports formed the basis for the proposed Strategic Defense Initiative in fiscal year 1985. One of the study groups, the Defensive Technologies Study, concluded that emerging technologies held substantial promise. It recommended a long-term research and development program to develop options that could guide future decisions concerning ballistic missile defense.

The Strategic Defense Initiative, which emerged from these studies, is designed to explore the feasibility of a multitiered system that could engage ballistic missiles and warheads along their entire trajectories, including the boost, post-boost, mid-course, reentry, and terminal phases. Funding for fiscal year 1985 is $1.74 billion (88 percent from the Department of Defense, 12 percent from the Department of Energy), which is to be focused on five technologies that have been determined to offer the greatest promise for ballistic missile defense. The 1985 request also includes an additional funding increment of about $250 million to augment these technologies and exploit other new technological opportunities. For the five-year period 1985–89, an estimated $24 billion will be required.

One purpose of the program is to demonstrate at an early time key technologies needed for a highly effective, multitiered, low-leakage ballistic missile defense that could prevent all but a small fraction of the attacking force from reaching targets. The knowledge gained from these demonstrations would support a decision in the early 1990s on whether to proceed with deployment. The administration has emphasized that since the program is a research and technology effort, it can be carried out for the next several years within existing treaty constraints.

The proposed Strategic Defense Initiative has been divided into five technical areas: (1) surveillance, tracking, and acquisition; (2) directed energy weapons; (3) kinetic energy weapons; (4) systems analysis and battle management; and (5) support programs. The largest amount of research dollars will be used to attack perhaps the most challenging problem for the system: surveillance, tracking, and acquisition.

In the directed energy weapons program, four basic technologies have been identified as potentially capable of countering an enemy attack: space-based chemical lasers, ground-based chemical lasers, space-based particle beams, and directed energy from nuclear explosions. The goal of the directed energy program is to bring the most promising of these concepts, which might be used against missiles in the boost and post-boost phases, to technical maturity in the 1990s. Among the technologies to be examined are excimer short-wavelength lasers, free electron lasers, neutral particle beams, and lasers pumped by nuclear explosions.

A major effort will also be directed at development of kinetic energy projectiles. These projectiles collide at extremely high velocities with warheads or with post-boost vehicles that have not dispensed all of their reentry vehicles. Kinetic energy weapons may also engage reentry vehicles that reach the terminal phase of their flight.

The Strategic Defense Initiative, as proposed by the President and developed in the fiscal year 1985 budget, has stirred considerable opposition, both domestically and abroad. Once again important members of the scientific community have questioned the technical prospects of the program and emphasized the tremendous cost of such systems. Much of the criticism echoes the earlier criticisms of previous ABM approaches, namely, that the proposed defensive systems are fundamentally unable to defend successfully against a large number of extremely destructive nuclear weapons, and that the proposed defensive systems could be countered at much less cost by the offense. In addition, critics argue that these systems, whatever their actual capabilities, would have a very destabilizing effect on the strategic relationship and would accelerate the offensive arms race. Although the technology has changed, the fundamental criticisms are very similar to those in the early debates over the Nike-Zeus, Nike-X, Sentinel, and Safeguard systems.

A significant new element of the 1980s debate is the impact of the Strategic Defense Initiative on the SALT I ABM Treaty, which is the cornerstone of the arms control process. Although the Reagan Administration has asserted that its immediate program is consistent with the ABM Treaty, critics point out that the program will inevitably collide with the treaty in only a few years if the administration pursues its stated goals.

By the summer of 1984 it was clear that the Strategic Defense Initiative would be an issue in the presidential campaign. The Democratic platform attacked the Star Wars proposal in very strong language. It asserted that "our best scientists agree that an effective population defense is probably impossible" and that "this trillion-dollar program would provoke a dangerous offensive and defensive arms race." Candidate Walter Mondale was on record as being strongly opposed to the program.

PROVISIONS OF THE SALT I ABM TREATY

The SALT I ABM Treaty (Appendix C), which is of unlimited duration, obligates the United States and the Soviet Union not to deploy ABM systems for "defense of the territory of its country," not to provide a base for such a defense, and not to deploy ABM systems for defense of any individual regions except within the very restrictive limits defined in the treaty. The treaty originally provided that each side could have two ABM deployment areas, one to protect its capital and another to protect an ICBM launch area. This was subsequently amended in 1974 to allow each side to deploy ABMs at only one of those sites, whichever it chose.

Precise quantitative and qualitative limits were imposed on the ABM system that may be deployed. At each site there could be no more than 100 interceptor missiles and 100 launchers. The number and characteristics of associated radars were specified in detail. Further deployment of radars to give early warning of a strategic ballistic missile attack were not prohibited, but such radars had to be located along the periphery of a nation's territory and oriented outward so that they could not also serve as battle management radars in an ABM defense of the interior.

Qualitative improvements of ABM technology were severely limited. For example, the parties agreed not to develop, test, or deploy ABM launchers capable of launching more than one interceptor missile at a time; existing launchers could also not be modified to give them this capability. Systems for rapid reload of launchers were similarly prohibited. These provisions, which were clarified by agreed statements, also banned interceptor missiles with more than one independently guided warhead.

The treaty specifically prohibited development, testing, or deployment of sea-, air-, space-, or mobile land-based ABM systems and their components, since these systems could provide the base for a nation-wide defense. Moreover, an agreed statement to the treaty made clear that if future technology produced ABM systems "based on other physical principles and including components capable of substituting for ABM interceptor missiles, ABM launchers, or ABM radars," specific limitations on such systems would be developed to fulfill the treaty's basic obligation not to deploy nationwide ABM systems.

The treaty also prohibited the upgrading of air defense systems to give them a capability against ICBMs and SLBMs.

The treaty included provisions to facilitate its verification by National Technical Means (NTM), which include the two sides' satellite reconnaissance systems and other systems to collect technical intelligence. The treaty specifically banned interference with NTM used for verification and deliberate concealment measures that impeded verification of the treaty by NTM.

The treaty also provided for a Standing Consultative Commission to consider compliance questions and other problems under the treaty as well as to develop additional procedures for its implementation. While the treaty is of unlimited duration, provisions were made for its review at five-year intervals.

THE MAIN ISSUES SURROUNDING THE SALT I ABM TREATY

Reopening the ABM Debate

The central issue involving the 1972 SALT I ABM Treaty today is whether or not the United States should move toward a defense-oriented strategic policy involving a nationwide ballistic missile defense system. The ABM Treaty was specifically designed to prevent such systems. While President Reagan's proposed Strategic Defense Initiative is a research program that does not involve any immediate compliance issues under the ABM Treaty, the goal of the program is completely contrary to the treaty's fundamental obligation. Moreover, the initiative itself would appear to raise questions under the treaty in the not too distant future.

Supporters of the Strategic Defense Initiative

General Issues. Some supporters of the Strategic Defense Initiative see it as the first step in a program to achieve President Reagan's goal of a highly effective ballistic missile defense. Such a defense would permit the United States to move away from the present offense-oriented strategy based on the threat of retaliation to a defense-oriented strategy. Other supporters see it as a more modest technical development effort directed at strengthening the existing strategy of deterrence with a ballistic missile defense system.

Both groups see the initiative as a timely response to several major developments that challenge the basic premises on which the SALT I ABM Treaty was based. These developments include growing dissatisfaction with a strategic policy of deterrence based on the threat of retaliation and mutually assured destruction, various scientific and technical developments that suggest the possibility of an effective ballistic missile defense, and evidence of a major Soviet effort to develop a ballistic missile defense system as part of a broader defensive military effort. Both groups argue that, if feasible, an effective defensive system would strengthen deterrence, increase stability, and reduce reliance on the present threat of assured destruction.

Reassessment of the U.S. Offense-Dominated Strategy. In his March 23 speech, President Reagan stated that it was a sad commentary on the human condition that even with major arms reductions it would still be necessary to rely on the "specter of retaliation" for deterrence. He stated that current technology has attained a level of sophistication where it is reasonable to embark on a program to counter the Soviet missile threat with defensive measures. If successful, these measures would make these weapons "impotent and obsolete" so that people could live secure in the knowledge that their security did not rest upon the threat of U.S. retaliation to deter a Soviet attack, a policy that is not only morally repugnant but increasingly lacking in credibility. It is argued that a strategy that increases reliance on defensive systems would offer a new basis for managing the long-term relationship with the Soviet Union.

Supporters of the initiative also emphasize that it is necessary to go beyond past strategic assumptions about deterrence. In today's strictly offense-dominated U.S.-Soviet confrontation, the continuing growth of the Soviet ballistic missile threat could force the United States to make ever more difficult improvements in its offensive forces to assure a survivable retaliation force. On the other hand, a new balance between offensive and defensive forces resulting from a ballistic missile defense could enhance deterrence against deliberate attack and provide greater safety against accidental use of nuclear weapons or unintended nuclear escalation. The extent to which these possibilities can be realized depends on how present uncertainties about technical feasibility, costs, and Soviet responses are resolved.

Supporters of the initiative argue that a U.S. defense against ballistic missiles will enhance deterrence even in its early phases. It will reduce Soviet confidence in the success of a preemptive counterforce attack, since weapons could not be counted on to destroy high-priority strategic targets. Defenses could also protect critical U.S. command and control centers, which would increase the credibility of the U.S. deterrent and reduce the military utility of a preemptive attack. In addition, the U.S. effort could moderate the development of future offensive systems, according to supporters. A defensive system against ballistic missiles would not have to be completely leak-proof to attain these objectives and enhance deterrence. Supporters of the initiative also argue that U.S. research and development on ballistic missile defense might cause the

Soviet Union to shift its emphasis from destabilizing ballistic missiles, with their short flight time, to air-breathing forces, which take much longer to reach targets, thus allowing more time for decision. The Soviet Union would also have to devote an increased portion of its research and development efforts to developing countermeasures and new types of delivery vehicles.

Supporters of the initiative argue that deterrence would be enhanced not only by an advanced, low-leakage multitiered defense system but by intermediate versions of a ballistic missile defense system. Such intermediate systems could not provide the protection available from a complete multitiered system, but they could play a useful role in defeating limited attacks and deterring large attacks. A defensive system assuring the survival of a significant number of U.S. nuclear weapons would greatly add to the deterrence against Soviet attack.

Emerging Technologies. The main reason for reconsidering ballistic missile defense at this time, according to most supporters of the Strategic Defense Initiative, is the emergence of technologies that may make an effective ballistic missile defense feasible. However, unless it undertakes a long-term research and development program, the United States will never be certain about the possibilities. At the time of the earlier ABM debate, the United States had no effective way of intercepting missiles during the boost phase, the means were not available to discriminate confidently between warheads and sophisticated decoys, computers could not manage thousands of simultaneous engagements, and the terminal defense with nuclear warheads at low altitudes involved unacceptable collateral damage to the defended area. Today, according to initiative supporters, directed energy systems—such as high-powered chemical lasers, X-ray lasers pumped by nuclear explosions, particle beams, and "hypervelocity" kinetic energy weapons— appear to offer promising kill mechanisms against missiles in the boost phase. Precision sensors can make it possible to discriminate warheads from decoys, chaff, and debris. New computers and electronic advances make it possible to manage thousands of engagements simultaneously. And precision sensors permit "hit to kill" intercepts without requiring nuclear warheads. While acknowledging that it is not known today how effective and reliable such systems can be made, supporters of the initiative assert that these new technologies, which have seen very rapid advances in recent years, provide a compelling rationale for reexamining the technical issues associated with achieving an effective ballistic missile defense.

The initiative's supporters argue that it will be extremely difficult to design countermeasures against a multitiered defensive system because components of the system will engage attacking missiles in all phases of their flight. The initiative calls for research into surveillance systems that would observe a potential target using infrared, visible, and radar radiation. A decoy, for example, might easily be constructed to simulate a warhead to a single sensor, but a decoy that could simulate real warheads to a variety of sensors would be almost as heavy and sophisticated as an actual warhead. The combination of different weapons and sensors in three or more layers would at the minimum drive an opponent to extremely expensive, and significantly less capable, missiles. Supporters of the initiative argue that this is a positive result in its own right. Such questions and options played a central role in determining the recommended research and technology program.

The Soviet Ballistic Missile Defense Program. One of the main reasons for the Strategic Defense Initiative cited by its supporters is the advanced state of the Soviet ballistic missile defense program. For a number of years the Soviet Union has been pursuing some of the new technologies that are now believed to hold promise for an effective defense. It is argued that unilateral Soviet deployment of an advanced ABM system, when added to the Soviet Union's already extensive air defense and civil defense programs, could sufficiently reduce U.S. retaliatory capabilities to a point where the credibility of U.S. deterrence would be brought into question. Thus, the U.S. research effort will provide a necessary and vital hedge against the possibility of a one-sided Soviet deployment.

Reagan Administration officials have testified that decisions on the deployment of defensive systems do not rest solely with the United States. Soviet history, doctrine, and programs—including the active program to modernize the existing ABM defense around Moscow, which is the only operational ballistic missile defense in the world—all indicate that the Soviet Union may deploy an ABM system when it deems that such a deployment would be advantageous, they argue. Since long-term Soviet behavior cannot be predicted reliably, the United States must be prepared to deploy its own ABM defense. A U.S. research and development program on ballistic missile defense that provides a variety of deployment options will help resolve the many uncertainties the United States would now confront in making such a decision.

Some administration officials have stated that the Soviet Union is as much as ten years ahead in certain aspects of high-technology ballistic missile defense systems. Among the worrisome Soviet programs is a very large research program on directed energy weapons, including chemical lasers that could be either ground- or space-based. Administration officials have estimated that Soviet space-based ABM systems could be tested in the 1990s, but that they will probably not be operational until after the turn of the century. Officials have also noted that the Soviet capacity to launch payloads into space, although less sophisticated, is significantly greater than the U.S. capacity. In view of these ongoing Soviet programs, supporters of the Strategic Defense Initiative argue that an increased U.S effort to develop defense technologies should not intensify the Soviet effort since a very vigorous Soviet development program is already under way.

Arms Control. According to some supporters of the Strategic Defense Initiative, the premise of the SALT I ABM Treaty that limitations on ABM defenses would curb the growth in offensive arms has been disproven. The Soviet offensive buildup continued after the ABM Treaty. Some supporters also argue that the U.S. option to deploy an ABM system would increase U.S. leverage on the Soviet Union to agree to mutual reductions in offensive nuclear forces. In turn, such reductions could reinforce the potential of defensive systems to stabilize deterrence. The reductions that the United States has proposed in the START negotiations would be very effective in this regard.

Critics of the Strategic Defense Initiative

Critics of the President's Strategic Defense Initiative argue that it holds out a false and dangerous hope that effective nationwide defenses can be developed against nuclear weapons. In reality, nothing has happened scientifically, technologically, militarily, or politically to change the underlying strategic reality codified in the SALT I ABM Treaty, they say. Attempts to develop nationwide defenses against ballistic missiles would still weaken deterrence and destabilize the military balance, increasing the likelihood of nuclear war. Consequently, despite its promise of a more humane and moral approach to strategic policy, President Reagan's proposal will actually lead to a more dangerous relationship with the Soviet Union and an increase in the arms race. Furthermore, within a relatively short time, critics point out, the Strategic Defense Initiative will come into conflict with the provisions of the SALT I ABM Treaty and probably other arms control agreements as well. This could lead to the collapse of the entire arms control framework, the product of two decades of negotiations, long before work ever begins on an ABM system.

At the same time, many critics agree that the United States should pursue some research programs on defensive systems as a hedge against Soviet technological breakthroughs. They emphasize, however, that this research should be carefully limited to stay within the constraints of the SALT I ABM Treaty. It should not be undertaken in the context of seeking a nationwide ballistic missile defense, which is clearly contrary to the underlying objective of the ABM Treaty.

The Realities of Assured Destruction.   The present offense-dominated strategy based on the prospect of assured retaliatory destruction is inherent in the millionfold increase in the power of nuclear weapons over conventional high explosives, the critics argue. This strategy is not the result of a policy choice or of a lack of determination to pursue defensive technology. Rather, the unprecedented power of nuclear weapons and the vast stockpiles of such weapons in the world put impossible technical demands on a defensive system to defend an inherently vulnerable urban society. Nuclear weapons have made the nuclear superpowers into mutual hostages. No matter how unpleasant this relationship may be, it is a fact of life in the nuclear age. According to the initiative's critics, emerging advances in technology have not changed this fundamental situation.

The Technical Infeasibility of a Nationwide Defense.   Critics of the Strategic Defense Initiative emphasize that a nationwide defense must be extremely effective because any weapons that leaked through would destroy the target being defended. For example, if the United States were able to deploy a ballistic missile defense system that through remarkable improvements in defense technology was 95 percent effective against the present Soviet threat, a Soviet attack could still deliver several hundred ICBM warheads on U.S. cities. Thus, any conceivable system would in practice almost certainly fail unless both sides first reduced their offensive forces to relatively low numbers.

To overcome such a defense, the offense could concentrate its firepower against targets of its choice while a national defense would have to protect all major urban areas. The offense could also increase the size of its attack by adding more missiles to its force or by increasing the number of MIRVed warheads on existing missiles. The apparent size of an attack can be increased by incorporating large numbers of decoys in the missile payload. The offense can conceal the location of attacking warheads from radar sensors with chaff and from infrared sensors with balloons or other devices. It can also try to blind defensive sensors with precursor nuclear explosions or employ a wide variety of unpredictable jamming techniques.

The offense can also attack individual vulnerable components such as key radars or other sensors. In the case of satellite-based systems, this tactic can even be carried out in advance of hostilities, possibly without the knowledge of the defense. One such countermeasure could be to deploy space mines that would follow critical elements of the space defense system and detonate on command. Finally, the offense always has the option of simply circumventing the defense by introducing or emphasizing an entirely different mode of attack. These different attack modes could employ either ballistic missiles or air-breathing vehicles such as cruise missiles, which a defensive system oriented toward ballistic missiles would not cover. A separate air-breathing threat would have to be countered with an entirely different defensive system, which would be extremely complex and expensive. Moreover, such a system could still be overwhelmed by cruise missiles of advanced design that were equipped to facilitate penetration. Thus, whether the missile defense system is a space-based laser involving hundreds of satellites at a cost of many hundreds of billions of dollars or pop-up nuclear-pumped X-ray systems with mind-boggling technical requirements, the fact remains that a sophisticated offense has a vast array of techniques to overwhelm or circumvent these systems at much lower cost.

Some critics acknowledge that it would be technically possible to design a complex system to defend hardened redundant targets such as missile silos. In the case of missile silos a significant leakage is acceptable, since only a fraction of the silos need survive to constitute an effective deterrent. These critics question, however, whether such an expensive system would be cost effective compared with other approaches to ensuring the survivability of deterrent forces. Moreover, they contend that, whatever its intended purpose, such a system would be perceived as an effort to achieve a nationwide defense and would have a serious destabilizing effect. It would certainly not be permitted under the SALT I ABM Treaty.

In all of these cases, critics point out, the offense could negate at much lower cost whatever reduction in damage the defense might achieve simply by increasing its penetration capability.

The Destabilizing Effects of Nationwide and Intermediate Defenses. The most important danger of a major national effort to achieve an effective ballistic missile defense, according to critics, is that it will be an unprecedented stimulant to the arms race. If the United States mounts a nationwide defense, the Soviet Union will quickly follow suit. The Soviet research and development program in this area is comparable with the U.S. program, and it is inconceivable that the Soviet Union would surrender this area of strategic military development to the United States. In turn, the competition in offensive systems would accelerate as both sides deployed forces capable of penetrating anticipated future defensive systems. The decision to deploy MIRVed warheads on the Minuteman III and Poseidon missiles, which was sparked by the threat of a Soviet ballistic missile defense in the late 1960s, is the most impressive historical example of this action-reaction cycle.

These critics also argue that a nationwide ABM defense, regardless of its actual effectiveness, would substantially decrease crisis stability. Such a system would inevitably be seen by the other side as part of an effort to achieve a first-strike capability. If the ABM system was, or was perceived to be, effective, it would permit the possessor to launch a preemptive counterforce attack confident that the other side's retaliation would not inflict unacceptable damage. Even if such a system did not appear capable of defending effectively against a full-scale, coordinated attack, it might be believed, or perceived, to be capable of effectively handling a reduced and poorly coordinated retaliation after a massive preemptive counterforce attack. In a major crisis that might escalate to nuclear war, critics argue that both sides would be under greatly increased pressure to preempt if either side had a nationwide ABM system. The situation would be most unstable if both sides had such systems. On both sides, the capabilities of each side's system would tend to be judged less than perfect by the knowledgeable possessors and exaggerated by the other side. Critics point out that this strategic situation would create the maximum pressure for preemption in a crisis.

The Soviet Ballistic Missile Defense Program. Critics of the Strategic Defense Initiative state that its advocates have overstated the magnitude and significance of the Soviet ballistic missile defense effort. In any event, some critics continue, the magnitude of the Soviet effort is largely irrelevant to the U.S. decision on whether to undertake the initiative. Critics point out that the Soviet Union has traditionally invested considerably larger effort in air defenses than has the United States. Yet, despite the vast amounts invested in Soviet air defense, the U.S. Air Force is confident that it can penetrate these defenses using a variety of penetration tactics together with air-launched cruise missiles. Therefore, even if the Soviet program were as extensive as some claim, this does not mean that the United States should imitate such a cost-ineffective effort.

According to critics of the initiative, most independent assessments conclude that the Soviet effort, while technically more advanced in some selected respects, is technologically considerably inferior in many respects to that of the United States. Soviet scientists are carrying out technologically advanced research on certain high-powered laser techniques, but they appear to be behind in those techniques that the United States considers most promising. They have also shown interest in particle beam work, but there is no evidence that such work has proceeded beyond the laboratory stage.

It is true, critics acknowledge, that the old Soviet ABM deployment around Moscow is being replaced by a system using more modern interceptors and that the associated radars are being modernized. This is being done, however, without violating the ABM Treaty. In the ABM system permitted for Moscow, the number of interceptors would be totally inadequate to protect the city against a U.S. retaliatory strike.

There is no reason to believe, according to these critics, that the Soviet Union has a near- or intermediate-term capability to deploy area defenses rapidly. In this regard, they note that even if the recently discovered Krasnoyarsk radar proves to be a violation of one of the provisions of the ABM Treaty, it would, when complete, add little, if anything, to a nationwide ballistic missile defense capability against a U.S. retaliatory strike. Like the other Soviet large phased-array radars, it is extremely vulnerable to attack using a variety of precursor or defense suppression tactics. Thus, critics maintain that the present Soviet ballistic missile defense effort is not even remotely a threat against the U.S. deterrent. Even if it were a legitimate source of concern, they emphasize, the appropriate response would be further improvements in offensive penetration at relatively low cost rather than imitation of the Soviet effort.

Arms Control. Critics state that the Strategic Defense Initiative will put the United States on a short-term collision course with the violation or termination of the SALT I ABM Treaty, the cornerstone of arms control efforts to date. Despite reassurances by the Reagan Administration, they assert that the Strategic Defense Initiative will come into direct conflict with the the ABM Treaty in only a few years. At that time the United States would presumably withdraw from the treaty rather than operate in clear-cut violation of its provisions. In view of the limited prospects for technical success or strategic stability from the pursuit of defensive systems, these critics argue that termination of the ABM Treaty is an unacceptably high price to pay for this exploratory program.

Critics of the Strategic Defense Initiative assert that it will actually accelerate the arms race and further undercut the arms control regime built over the past 20 years. In addition to the ABM Treaty, other arms control casualties of the program could be the Outer Space Treaty, which prohibits the stationing of weapons of mass destruction in space, and the Limited Test Ban Treaty, which prohibits the testing of nuclear weapons in space. These treaties would certainly be incompatible with any serious efforts to pursue the technology of X-ray lasers pumped by nuclear explosions. Any hopes for negotiating a comprehensive test ban agreement or an anti-satellite accord would also be lost. In short, these critics argue, the Strategic Defense Initiative will lead to the destruction of the entire arms control framework developed over the past two decades.

Verification and Compliance

Since the signing of the SALT I Treaty, the United States and the Soviet Union have raised a number of questions about the other side's compliance with the treaty (see Chapter 2 for a full discussion of these compliance issues). During the SALT II ratification hearings the Carter Administration reported that all compliance questions relating to the SALT I ABM Treaty had been satisfactorily resolved when referred to the Standing Consultative Commission. On January 23, 1984, President Reagan submitted a classified report to Congress entitled "Soviet Non-Compliance with Arms Control." In his transmittal message the President charged that the Soviet Union has almost certainly violated the ABM Treaty. The report stated that a new phased-array radar under construction near Krasnoyarsk in central Siberia "almost certainly constitutes a violation of legal obligations under the ABM Treaty of 1972 in that in its associated siting, orientation, and capability, it is prohibited by the Treaty." The Soviet Union has officially stated that the Krasnoyarsk radar is for space tracking, but the United States does not accept this explanation because of the radar's technical characteristics and location.

For its part, the Soviet Union publicly released a diplomatic note on January 29, 1984, charging the United States with specific violations of the SALT I ABM Treaty. Among the charges were that the United States was developing both a mobile and space-based ABM system, was working on multiple warheads for ABM interceptors, was building and upgrading new large phased-array radars on its coasts (Pave Paws) that, despite their asserted early warning function, cover large areas of the United States and could serve as battle management radars for a future U.S. ABM system, and was incorporating ABM capabilities in the intelligence radar on Shemya Island. The United States has rejected all of these charges as being without technical or legal merit.

6 Anti-Satellite (ASAT) Arms Control

Since Sputnik was launched in 1957, satellites have played an important role in the military programs of both the United States and the Soviet Union. Today, satellites serve a wide variety of extremely important security functions, including early warning of strategic attack, intelligence on the current and projected military threat, precision navigation and targeting, communications for command and control, and verification of arms control agreements. The critical importance of satellites to U.S. national security has focused special attention on the evolving threat posed by anti-satellite (ASAT) systems designed to attack satellites. The problem is complicated by the interaction of anti-satellite and anti-ballistic missile (ABM) developments. Ballistic missile defense systems have an inherent ASAT capability; ASAT technology can contribute to ballistic missile defense development; and space-based ABM systems would be vulnerable to ASAT systems.

The United States and the Soviet Union both initiated ASAT programs in the early 1960s. The United States maintained a direct-ascent nuclear-armed ASAT system until the mid 1970s. The Soviet Union has worked intermittently on a coorbital nonnuclear ASAT system that is now considered to be operational. The United States is on the threshold of testing a new dedicated nonnuclear ASAT system with considerably greater capabilities than the existing Soviet ASAT system.

There is a long history of arms control agreements relating to space. In 1963 the Limited Test Ban Treaty prohibited nuclear tests in space. In 1967 the Outer Space Treaty prohibited stationing weapons of mass destruction in space. In 1972 the SALT I ABM Treaty prohibited inter-

159

ference with satellites that helped verify the agreement. In 1978 the Carter Administration initiated negotiations on an ASAT agreement with the Soviet Union. An agreement was not reached, and there have been no further negotiations since 1979. The Soviet Union has continued to advocate such an agreement and in 1983 presented to the United Nations a detailed draft treaty banning weapons in space, ASAT systems, and the use of force against satellites. The Reagan Administration has formally taken the position that a ban on ASAT systems would be contrary to U.S. security interests. Nevertheless, in the summer of 1984 the U.S. government accepted a Soviet invitation to discuss the ban on weapons in space in Vienna in September 1984 but indicated it would discuss other arms control proposals as well. As of September 1, 1984, the two sides had been unable to agree on an agenda for the meeting.

BACKGROUND

The Origin of the ASAT Program

The earliest U.S. studies of specific systems designed to attack satellites were commissioned by the U.S. Air Force in the late 1950s. They focused on two basic approaches: a "killer satellite" interceptor, which would be placed in orbit and then maneuvered to its target, and a direct-ascent interceptor, which would intercept the target when it passed overhead. In 1960, three years after the Soviet Union's successful launch of Sputnik, the U.S. Air Force began research and development on the first U.S. anti-satellite program. The program (designated SAINT), which never reached the test phase and was canceled in 1962, involved the concept of a coorbital interceptor that could inspect and destroy a target. The U.S. ASAT program was then temporarily incorporated into the Army's anti-ballistic missile program, since it was recognized that the Nike-Zeus ABM test facilities on Kwajalein Island in the Pacific could also serve as a nuclear-armed direct-ascent anti-satellite system against satellites that came within a range of a few hundred kilometers. From 1964 to 1967 a few Nike-Zeus interceptors were deployed there as an anti-satellite system.

In response to what appeared to be an emerging threat of a Soviet system for orbital nuclear bombardment, the U.S. Air Force also resumed its ASAT mission. In 1964 several intermediate-range Thor rockets, which were modified for an anti-satellite mission, were deployed on Johnston Island in the Pacific. Although the Thor system had considerably greater range than the Nike-Zeus, its kill mechanism, a high-yield nuclear warhead, meant that its use in peacetime or conventional war would risk collateral damage to friendly satellites at great distances, as well as nuclear escalation. When the anticipated threat of Soviet orbital nuclear weapons never materialized, U.S. interest in an anti-satellite weapons capability faded and the Thor missiles, which had been extensively tested, were finally retired in 1976. The Thor system reportedly could still be restored to operational status on relatively short notice.

The early Soviet efforts on ASAT weapons probably began in 1964 with the establishment of a division of the strategic defense forces whose mission was that of "destroying the enemy's cosmic means of fighting." By 1967, preliminary tests of a Soviet ASAT system had begun. The multiton Soviet ASAT system is launched by a modified version of the Soviet Union's early large ICBM (the SS-9). The ASAT itself is a coorbital interceptor that uses an active radar to home in on its target within two orbits after launch and destroys its target with a nonnuclear warhead. The initial test program of the interceptor from 1968 to 1972 was judged by the United States to have been successful.

During the 1960s and early 1970s, the United States and the Soviet Union concluded a number of arms agreements limiting the militarization of outer space. The Limited Test Ban Treaty of 1963 bans the testing of nuclear weapons in space. The Outer Space Treaty of 1967 prohibits the deployment of nuclear weapons and other "weapons of mass destruction" in space. The 1972 SALT I ABM Treaty bans the development, testing, or deployment of ballistic missile defense systems in space. Both SALT I and the 1979 SALT II Treaty also ban interference with any satellite providing National Technical Means to verify those agreements. Although the cumulative effect of these agreements restricts many weapon systems in space and protects intelligence satellites used for verification purposes, none of the agreements explicitly limits the development and deployment of dedicated anti-satellite systems unless they involve the deployment of nuclear weapons in space.

The Ford-Carter Years

In 1976 the Soviet Union resumed the testing of its ASAT system using the same technology upgraded to permit intercept on the first orbit instead of the second orbit. In response to this development, the Ford Administration in its final days directed the initiation of a new ASAT program for the stated purpose of deterring use of the Soviet system. The directive also called for a study of arms control options.

The Carter Administration undertook a two-track approach to the ASAT problem. It sought to negotiate an agreement limiting such systems while concurrently developing the new ASAT system. Moving quickly to initiate the arms control process in this area, the Carter Administration agreed with the Soviet Union in March 1977 to establish a U.S.-Soviet working group on ASAT as one item on the agenda of arms control issues the two countries would explore. Three rounds of ASAT negotiations were held between the United States and the Soviet Union during 1978–79. The first session in Helsinki in May 1978 was preliminary in nature and devoted to discussion of the scope of a possible agreement. Two subsequent rounds of ASAT talks were held in Bern, Switzerland, from January 23 to February 16, 1979, and in Vienna, Austria, from April 23 to June 17, 1979.

The U.S. approach to the negotiations was to seek a treaty banning attacks on all satellites and to establish an agreed moratorium lasting for a year or so on the testing of ASAT systems. The moratorium would provide time to negotiate a more detailed treaty on ASAT testing and deployment that would deal with the verification problems involved.

The draft texts developed in these negotiations have never been made public, and there are varying opinions on how close to agreement the two sides came in the negotiations. Although important progress was made, it seems clear that several important issues remained unresolved. Among these issues was whether the treaty would apply only to U.S. and Soviet space objects or to those of other countries as well. The United States considered this an important issue since unless nonsignatories were covered the agreement appeared to legalize attacks on third parties. The sides also disagreed on whether ASATs could be used for self-defense against "hostile" acts. The United States objected to a formulation that could again legalize the use of ASATs. The problem was complicated by ambiguities as to what activities were covered by the concept of "hostile" acts. The Soviet Union also reportedly proposed language that might limit the Space Shuttle because of its inherent ability to rendezvous and capture or interfere with satellites. The U.S. position was that the shuttle was neither an ASAT nor an ASAT launch platform. In addition, there was a fundamental unresolved issue within the U.S. government as to whether a moratorium on the testing of ASAT systems should include all potential ASAT systems, including those using directed energy kill mechanisms, or simply those systems using direct-ascent and coorbital interceptors.

In June 1978, a year after the United States declared the Soviet ASAT operational, President Carter summarized his Presidential Directive on National Space Policy, stating that "while the United States seeks verifiable, comprehensive limits on anti-satellite capabilities and use, in the absence of such agreement, the United States will vigorously pursue the development of its own capabilities." The ASAT system chosen for development, which is the system currently in the testing phase, was an air-launched miniature homing vehicle delivered by a small two-stage rocket carried on an F-15 fighter aircraft. The rocket is guided by an inertial guidance system to intercept a satellite whose orbital parameters have been determined by ground-based sensors. In the final engagement, the miniature homing vehicle uses infrared sensors to home in on the satellite and destroys it on impact.

By 1978 the Soviet Union had encountered difficulties in its attempts to upgrade its low-altitude ASAT system to permit it to attack its target on the first orbit. The Soviet Union had also begun testing a more advanced interceptor with an optical homing device, which would be less vulnerable to countermeasures such as evasive maneuvering and jamming. All six of the tests of this improved system failed. Just prior to the opening of the ASAT negotiations in May 1978, the Soviet Union announced that it would undertake a unilateral testing moratorium on ASATs.

By June 1979 the ratification of the SALT II Treaty had taken overriding priority in the arms control planning of the U.S. government, and it was decided not to press ahead with the uncertain ASAT negotiations, which involved unresolved policy issues, until the SALT II ratification was completed. Nevertheless, the United States and the Soviet Union did agree in a joint communique at the signing of the SALT II Treaty in Vienna "to continue actively searching for mutually acceptable agreement in the continuing negotiations on anti-satellite systems." After the Soviet Union's invasion of Afghanistan in December 1979, the United States made no effort to resume the bilateral ASAT negotiations.

The Reagan Years

The policy agenda of the new Reagan Administration called for a complete review of arms control policy and objectives and consequently put any further movement on these issues, including the ASAT negotiations, on the back burner. In March 1981 the Soviet Union, which had never previously destroyed a target with an ASAT system, successfully performed a complete operational test of its ASAT system using a radar homing device. Although the technology used in the Soviet system was still significantly inferior to the proposed U.S. program, the test sparked increased interest both in the press and in the government in ASAT research and development.

Several months after the Soviet ASAT test, Foreign Minister Andrei

Gromyko on August 11, 1981, submitted to the UN General Assembly a draft treaty banning the deployment of any weapons in outer space. He explained that the proposed treaty would preclude the stationing of weapons in outer space that were not already covered by the definition of "weapons of mass destruction." The Soviet draft treaty had a limited impact on ASAT development. While it would have banned space-based ASATs, it did not appear to restrict ground-based or air-launched systems, such as the U.S. system currently under development. The draft treaty obligated the parties to the agreement "not to place in orbit around the earth objects carrying weapons of any kind, install such weapons on celestial bodies, or station such weapons in outer space in any other manner, including on reusable manned space vehicles of an existing type or of other types" that parties may develop in the future. The parties also undertook "not to destroy, damage, disturb the normal functioning or change the flight trajectory of space objects" of other parties.

The U.S. government did not react favorably to the Soviet draft treaty. The UN General Assembly, however, approved the draft treaty and referred it to the First Committee. The Committee on Disarmament was instructed to include on its agenda for negotiations the Soviet draft treaty as well as the question of negotiating agreements to prevent an arms race in outer space. The United States was the only country that opposed the establishment of a working group on the subject, stating that immediate progress could not be expected and that the area had to be approached with extreme care. Without U.S. cooperation, the activity in the Committee on Disarmament stagnated.

By the early 1980s the United States already had an active program for developing the technology for more sophisticated ASAT systems, such as ground-based and space-based lasers. Tests of an airborne gas dynamic laser for use against tactical missiles had been conducted by the Air Force, and the Defense Advanced Research Projects Agency was funding a space-based laser program, involving the Alpha 2-MW infrared chemical laser, the Talon Gold pointing and tracking system, and the Large Optics Demonstration Experiment (LODE). By this time the Soviet Union also had an active high-powered laser research and development program. The press reported that the Soviet Union had, along with its coorbital interceptor system, ground-based test lasers with probable ASAT capabilities. It was also speculated that the Soviet Union was conducting research and development in the area of space-based laser ASAT weapons.

In early 1983 top administration officials explained that the U.S. rationale for developing ASAT weapons was largely to deter the Soviet

Union from using its capability. With the administration's position on space weapons and with the upcoming U.S. F-15 ASAT test scheduled for the fall, domestic pressure began to build in Congress and in the arms control community for movement in the area of space arms control.

On March 23, 1983, in a major address to the American people, President Reagan escalated the entire debate on weapons in space by calling on the scientific community to support a major technological effort to develop a defense against strategic nuclear missiles that would eventually make these systems "impotent and obsolete." The president's speech added fuel to the debate over ASAT arms control. Due to the overlap of technologies, the new Strategic Defense Initiative suggested that the administration was unlikely to agree to an ASAT ban that would restrict ballistic missile defense developments. While the President asserted that the program should be consistent with the provisions of the SALT I ABM Treaty, the new initiative raised many questions about technical developments that could be related to either ballistic missiles or ASAT systems.

Soviet President Yuri Andropov quickly followed the Reagan Administration's announcements with further calls for arms control negotiations on these issues. In response to a petition from a group of American scientists to ban weapons in space, Andropov stated that the United States and the Soviet Union were approaching a crucial time when failure to negotiate a ban on weapons in outer space would make an extension of the arms race into outer space inevitable. The U.S. State Department indirectly responded to Andropov's call for a ban on weapons in space by noting that the Soviet Union was the leader in developing an ASAT interceptor and that the Soviet arms control initiatives were in fact efforts to maintain a monopoly in this area.

By mid-May, with no movement by the U.S. administration in this area, the Union of Concerned Scientists presented a draft treaty for arms control in space at a hearing of the Senate Foreign Relations Committee. The draft treaty included a ban on the testing of all anti-satellite weapons. Its supporters emphasized that an ASAT treaty would clearly be in the U.S. interest because the United States is more dependent on space systems than is the Soviet Union. They also voiced the opinion that if the United States proceeded with tests of its air-launched ASAT weapon, verifying restraints on these systems would be made much more difficult.

At the same hearings, the Reagan Administration stated that it was not considering negotiations on ASAT in the near term. The Director of the U.S. Arms Control and Disarmament Agency, Kenneth Adelman, told the Senate committee that the United States, which was on the verge of testing its anti-satellite weapon in space, had no plans to resume negotiations with the Soviet Union to limit such weapons. Adelman said, "We should not rush into negotiations on these subjects unless we are ready with verifiable proposals that will enhance national security." Adelman stated that "there are difficult technical problems, including verification, that constitute fundamental obstacles to progress in this area." Among his key concerns were Soviet ASAT capabilities, which he said had created an asymmetry that "is a serious obstacle to achieving an equitable space arms control agreement."

Throughout the summer of 1983, a number of resolutions were introduced in Congress on both sides of the space weapons issue. The most significant action came on July 26, 1983, when the Republican-controlled Senate unanimously approved Senator Paul Tsongas's (D-Mass.) amendment to the fiscal year 1984 Department of Defense authorization bill stating that no funds could be obligated or expended to test any explosive or inert anti-satellite warheads against objects in space unless the President determines and certifies to the Congress that (1) the United States is endeavoring in good faith to negotiate with the Soviet Union a mutual and verifiable ban on anti-satellite weapons and (2), pending agreement on such a ban, testing of explosive or inert anti-satellite warheads against objects in space by the United States is necessary to avert clear and irrevocable harm to the national security.

The fiscal year 1984 request for the ASAT system was $225 million. This included $19.4 million for components for the first production line version of the ASAT. On August 5, 1983, the House and Senate conferees included the Tsongas amendment in the defense authorization, making it law until September 1984. The conferees also specified that the $19.4 million for procurement for the first ASATs could not be obligated unless the President submitted a report to Congress on U.S. ASAT policy and arms control plans no later than March 31, 1984.

Two weeks later, Soviet President Andropov announced for the first time Soviet willingness specifically to ban all ASAT systems. At a meeting in the Kremlin with a group of Democratic senators in August 1983, Andropov called on the United States to negotiate a complete prohibition on the testing and deployment of any space-based weapons for hitting targets on earth, in the air, or in outer space. He stated that the Soviet Union was also prepared to agree to prohibit the testing and development of all new anti-satellite systems and to eliminate all existing anti-satellite systems. In addition to these proposals, Andropov stated that the Soviet Union "assumes the commitment not to be the first to put into outer space any type of anti-satellite weapon." Shortly after Andropov's announcement, Foreign Minister Gromyko, in a letter to the UN Secretary General, made public a new Soviet draft treaty on space weapons entitled the Treaty on the Prohibition of the Use of Force in Outer Space and from Space Against the Earth. The draft treaty incorporated Andropov's proposals but excluded the unilateral moratorium on testing, which was transmitted separately.

At the United Nations the new Soviet draft treaty was referred to the First Committee. The United States was the only member of that committee to vote against a compromise resolution that would have established an ad hoc working group on outer space with a view toward undertaking negotiations. The U.S. position was that it supported the establishment of a working group on outer space to address a broad range of space arms control issues before any conclusions could be drawn about pursuing negotiations in the Committee on Disarmament. However, the United States did not favor having a working group undertake negotiations.

On January 21, 1984, the U.S. Air Force conducted the first test of the U.S. F-15 ASAT over the test range at Vandenberg Air Force Base in California. The missile was fired at a point in space rather than at an actual target, so that it did not violate the Tsongas amendment. Nevertheless, the Soviet news agency Tass criticized the U.S. government's test on January 24, 1984, stating that the "tests of the anti-satellite system carried out by the United States are an open challenge to the U.N. resolutions directed against the arms race in outer space."

A month after the U.S. test, during a hearing on the fiscal year 1985 defense budget, Richard DeLauer, Under Secretary of Defense for Research and Engineering, told the House Armed Services Committee that "ambitious tests" were planned during the coming year to demonstrate the capability of the F-15 ASAT system. DeLauer also disclosed that work had begun on a comprehensive study to select a "follow-on system with additional capability to place a wider range of Soviet satellite vehicles at risk" and that research on the Strategic Defense Initiative will also include an anti-satellite component.

The test launching of the U.S. F-15 ASAT, closely following the collapse of the START and INF negotiations, was viewed by critics of the administration's approach to arms control as a major step toward a situation in which it would be impossible to negotiate an ASAT agreement. Joining congressional voices for movement on space arms control, Democratic presidential contender Walter Mondale in February 1984 proposed that the United States initiate a temporary moratorium on the testing of ASAT systems along with a six-month moratorium on underground nuclear testing to break the impass on arms control talks with the Soviet Union. Mondale also pledged that if elected he would vigorously move forward on negotiations to reach an ASAT treaty.

As the administration was finalizing its report on ASAT arms control for Congress in March 1984, the new Soviet leader, Konstantin Chernenko, made his first appeal for negotiations on the militarization of outer space. In a speech carried by Tass, Chernenko stated that the United States could prove its "peaceableness" by concluding an agreement renouncing the militarization of outer space. This appeal was reiterated by Soviet officials in Moscow during a visit by several U.S. senators in the same month.

In his official ASAT report to Congress on March 31, 1984, President Reagan formally rejected the comprehensive ban proposed by the Soviet Union. President Reagan stated in the transmittal letter that "no arrangements beyond those already governing military activities in outer space have been found to date that are judged to be in the overall interest of the United States and its allies." The report stated that the factors that impede the identification of effective ASAT arms control measures include significant difficulties of verification, diverse sources of threats to U.S. and allied satellites, and threats posed by Soviet targeting and reconnaissance satellites that undermine conventional and nuclear deterrence. The President cautioned Congress that even though the executive branch would continue to study space arms control in search of selected limits on specific activities in space, he did not believe it would be productive to engage in formal negotiations. The report concluded that verification problems were profound and that the Soviets had a "destabilizing advantage" with the anti-satellite weapons they already had.

Following the release of the President's report, Soviet leader Chernenko, while renewing calls for negotiations on space, stated in an interview in *Pravda* on April 9, 1984, that "bluntly and frankly, they do not want to reach an agreement. But going so far as to make a mockery of common sense, they express readiness to talk with us with the sole aim of agreeing that accord on this issue is impossible. It is thus that the people in Washington understand political dialogue and talks in general."

In response to these developments, Republicans and Democrats in the Senate and House rallied around resolutions calling for space arms control. A version of Senator Larry Pressler's (R-S.Dak.) resolution, which called for a temporary halt in the ongoing U.S. effort to develop an ASAT, and resolutions challenging the Tsongas amendment, which was to expire in September, provided the vehicles for debate. In the House a new umbrella organization, the Coalition for Peaceful Uses of

Outer Space led by Representative George Brown (D-Calif.), was formed to try to eliminate funding for ASAT testing from the fiscal year 1985 military requests. Democratic presidential candidate Mondale once again criticized the administration for its space policy and outlined a five-point proposal that went beyond his earlier call for a "temporary" moratorium on testing anti-satellite weapons and negotiations with the Soviets "to get a verifiable ban" on these weapons. This proposal included a reaffirmation of the U.S. commitment to the 1972 ABM Treaty; a temporary moratorium on the testing and deployment of all weapons in space; and, following that, negotiation of a "verifiable treaty blocking weaponry in the heavens." Other critics of the administration's report argued that it was simply a laundry list of problems facing ASAT arms control and failed to compare the advantages to U.S. national security of an ASAT ban with an unconstrained ASAT race. The NATO Defense Ministers, meeting in April for NATO's Nuclear Planning Group session, also reportedly expressed skepticism and anxiety about U.S. military plans for space, and several allies urged the U.S. administration to enter into negotiations with the Soviet Union to forestall an arms race in space.

On May 28, 1984, the House of Representatives approved by a vote of 238 to 181 a ban sponsored by Representative Brown on the further testing of U.S. ASAT weapons until the Soviet Union resumed testing. Following the House vote, Soviet leader Chernenko once again called on the United States to negotiate "without delay" a pact banning the use of anti-satellite weapons. The Soviet leader said the Soviet Union would maintain its "unilateral moratorium" on launchings of anti-satellite weapons as long as the United States abstained "from placing in space anti-satellite weapons of any type," which he said also covered "test launchings of anti-satellite weapons." Chernenko also renewed Andropov's offer to "liquidate" all existing anti-satellite systems as part of an agreement.

The U.S. State Department responded that Washington was ready to "talk" about anti-satellite weapons but not "negotiate" and that the government would not engage in formal negotiation on an issue where it believed there was no reasonable chance of verification. The following day, July 12, 1984, the Senate passed a compromise amendment to the fiscal year 1985 military authorization bill that was less restrictive than the previous year's version of the Tsongas amendment but more restrictive than the challengers to that amendment desired. Among other provisions, the amendment stated that no funds could be obligated or expended to test any explosive or inert anti-satellite warheads against objects in space unless the United States was endeavoring in good faith to negotiate with the Soviet Union a mutual and verifiable agreement with the strictest possible limitations on anti-satellite weapons consistent with the national security interests of the United States. The previous year's language had called for a ban on these systems. The compromise was approved on a 61 to 28 vote.

In a prime time news conference following the Senate action, President Reagan stated that the United States had not "slammed the door" on ASAT negotiations. Subsequent news stories quoted an unnamed White House official as stating that the United States intended to present ASAT treaty proposals to the Soviet Union within a month. According to press reports, White House officials also stated that four options were being considered for a limited agreement on ASAT weapons: limiting each nation to one type of satellite interceptor, banning ASATs that could destroy high-altitude satellites, confidence-building measures involving the exchange of information about each other's weapons in space, and an agreement under which both nations would agree not to interfere with each other's satellites.

Following the President's press conference, Soviet Ambassador Anatoli Dobrynin presented a note to Secretary of State George Shultz on June 29, 1984, proposing formal negotiations between the United States and the Soviet Union in Vienna in September on an agreement to prevent "the militarization of outer space," including a ban on space weapons and the mutual renunciation of anti-satellite systems. President Reagan responded almost immediately, stating that the United States was prepared to hold wide-ranging arms control talks with the Soviet Union, including discussions seeking agreement on "feasible negotiating approaches" to limits on anti-satellite systems, but would not allow the agenda to be restricted to the militarization of outer space. The President stated that the United States would also discuss "mutually agreeable arrangements under which negotiations on the reduction of strategic and intermediate-range nuclear weapons can be resumed."

The Soviet Union rejected the U.S. response to the proposal for talks as "totally unsatisfactory," but emphasized that the offer to open negotiations in September on preventing "the militarization of outer space" remained open. There followed a series of diplomatic and public exchanges on the nature and scope of the proposed meeting in which each side insisted it was prepared to meet but accused the other side of attempting to manipulate the agenda for political purposes and questioned the seriousness of the other side's interest in the meeting. While at first the White House took the position that the meeting was definitely on and the U.S. delegation would be in Vienna on September 17 to start the talks, by the end of July both sides discounted the likelihood that the meeting would actually start on that date or before the U.S. presidential election. It was reported that President Reagan offered in a letter to Soviet leader Chernenko to delay the start of the proposed talks until after the November election to keep the issue out of the presidential campaign.

After a month of diplomatic and political maneuvering, there were reportedly still several basic differences separating the approaches of the two sides to the proposed September meetings. The most fundamental difference was that the Soviet Union insisted on limiting the talks to the militarization of outer space and space-based weapons, while the United States refused to discuss space weapons other than ASATs and insisted on the right to raise other subjects that it considered related to the substance of the meeting. The subjects the United States wanted to discuss included the general problem of offensive weapons and the specific question of the resumption of the START and INF negotiations. With regard to the character of the meetings, the Soviet Union wanted to identify the talks as representing a commitment to the negotiating process on arms control of space-based weapons and objected to the U.S. formulation that cast the talks as simply seeking "agreement on feasible negotiating approaches." Finally, the Soviet Union reportedly refused to accept U.S. efforts to limit the stated agenda of the talks to anti-satellite weapons as opposed to the Soviet proposal to deal with all space-based weapons. The Soviet approach would have broadened the talks to deal explicitly with the U.S. Strategic Defense Initiative as well as dedicated ASAT systems.

SUMMARY OF U.S. AND SOVIET POSITIONS ON ASAT ARMS CONTROL

The U.S. Position

President Reagan reported to Congress on March 31, 1984, that no arrangements or agreements beyond those already governing military activities in outer space have been found to date that are judged to be in the overall interest of the United States and its allies. The factors that impede the identification of effective ASAT arms control measures include major verification problems, existing threats to U.S. and allied satellites, and threats posed by Soviet targeting and reconnaissance satellites. The President reported that, notwithstanding these difficulties, the United States would continue to study space arms control in search of selected limits on specific types of space systems or activities that could satisfactorily deal with the problems outlined.

In connection with the talks proposed for September 1984 in Vienna, the U.S. government has undertaken a major internal review of its approach to arms control of space-based and ASAT weapons. According to press reports, this review includes assessment of four options for a less comprehensive approach to the ASAT issue: limiting each nation to one type of satellite interceptor, banning ASATs that can destroy high-altitude satellites, confidence-building measures involving the exchange of information about each other's weapons in space, and an agreement under which both nations would agree not to interfere with each other's satellites. There is no indication as to which of these positions the United States will advance if the Vienna meeting takes place.

The Soviet Position

The Soviet offer of a unilateral testing moratorium, announced by Andropov in August 1983, states, as reported by Tass, that the Soviet Union "assumes the commitment not to be the first to put into outer space any type of anti-satellite weapon, that is, imposes a unilateral moratorium on such launchings for the entire period during which other countries, including the U.S.A., will refrain from stationing in outer space anti-satellite weapons of any type."

The Soviet 1983 draft Treaty on the Prohibition of the Use of Force in Outer Space and from Space Against the Earth is comprehensive in scope and of unlimited duration. The treaty prohibits both the use of force or the threat of its use against space objects and the use of force or the threat of its use by space objects against targets in space, in the atmosphere, and on earth. The treaty specifically prohibits testing and deployment of any space-based weapons intended to attack targets on earth, in the air, or in space. The treaty would obligate the signatories not to destroy, damage, or disrupt the normal functioning of other states' space objects or to change their flight trajectories. It specifically requires the elimination of all existing anti-satellite systems and prohibits the testing or development of new anti-satellite systems. It also specifically prohibits the testing or use for "military" purposes, including anti-satellite purposes, of any manned spacecraft. The Soviet draft treaty, which is a multilateral treaty, states that verification would be provided by National Technical Means and calls for consultation and cooperation with regard to those means.

THE MAIN ISSUES SURROUNDING ANTI-SATELLITE ARMS CONTROL

The U.S. View

Although President Reagan has recently announced that the United States is willing to discuss with the Soviet Union feasible negotiating approaches to constraints on anti-satellite systems, he has taken a strong public position against a comprehensive ban on anti-satellite systems, saying that it would not be in the overall security interest of the United States and its allies. According to the administration, the problems facing ASAT arms control, which more than offset the potential benefits, include the lack of effective verification, the Soviet potential for breakout, the problem of defining ASAT systems, and the risks of disclosing sensitive information. The administration also argues that a U.S. ASAT capability is necessary for U.S. security and to maintain deterrence.

Verification. The Reagan Administration argues that it is not possible to verify a comprehensive ASAT ban. The verification problem is particularly serious in the case of ASAT systems since a small number of satellites serve critical U.S. security needs. Consequently, even very limited cheating on ASAT limitations could pose a very large risk to the United States. The administration emphasizes that it would be extremely difficult, if not impossible, to verify that the current operational Soviet ASAT interceptor had been eliminated. This verification problem is complicated by the fact that the interceptor is launched by a space booster, the SS-9, that is also used for a number of other space launch missions and would presumably be retained by the Soviet Union. Since it is not known how many ASAT interceptors or SS-9 boosters are available, the Soviet Union could maintain a covert supply of interceptors that could be quickly readied for operational use with little risk of detection by the United States. SS-9 boosters could then be diverted from other missions to launch the interceptors.

The verification problem is also complicated by the inherent difficulties in defining an ASAT system, since it can be a by-product of systems developed for other missions. This creates problems in specifying what systems or tests should be prohibited. According to the administration, the fact that ASAT capabilities are inherent in some systems developed for other missions or could be developed in an undetected or surreptitious manner makes it impossible to verify compliance with a truly comprehensive testing limitation that would prohibit tests of all methods of attacking satellites. Government witnesses have testified that test bans on more limited classes of ASAT systems may be verifiable, but the breakout threat from limited bans is a very serious problem.

As further examples of the verification problems of ASAT arms control, the government asserts that ground tests of a ground-based laser ASAT weapon would be easy to conceal and that space tests of such systems could be difficult to detect. Moreover, although circumstances might be suspicious, it would be extremely difficult as a practical matter to determine whether an orbiting satellite contained a weapon. Finally, while there would normally be little question that a satellite had been destroyed or damaged, it could be difficult, or impossible, to verify the source of the attack.

Breakout. The Reagan Administration has argued that the tremendous importance of a few critical U.S. satellites creates a strong incentive for the Soviet Union to maintain a capability to break out of any agreement. This breakout potential could exist even if the Soviet Union actually destroyed all of its existing ASAT interceptors, since it would retain the capability to produce and redeploy relatively quickly a system in which it could have confidence. If prior to a ban the United States had not tested its own ASAT system, the Soviet Union alone would possess such proven technology. Under a strict ASAT ban, the Soviet Union could change the basic character of its ASAT program. For example, under the guise of space rendezvous and docking operations, which the Soviet Union routinely conducts, spacecraft could be developed to detonate next to another nation's spacecraft.

Definition. The Reagan Administration has emphasized that a central problem inherent with ASAT arms control is the difficulty in defining an ASAT or a space weapon for arms control purposes. There are technologies and systems designed for purposes other than ASAT missions, even some with little or no ASAT capabilities, that may be difficult to exclude from an ASAT definition. Likewise, there are technologies and systems with a possible ASAT application that might not be included in an ASAT definition. If the survivability of satellites is a main concern, then ASAT capability relates to all systems capable of damaging, destroying, or otherwise interrupting the functioning of satellites. Such systems would include:

• Maneuvering spacecraft, such as the coorbital interceptor operationally deployed by the Soviet Union.

• Direct-ascent interceptors, such as the miniature homing vehicle system being developed by the United States. This category would also include exo-atmospheric ABM interceptors and intercontinental ballistic missiles with nuclear payloads. The Soviet nuclear-armed ABM interceptors would have such ASAT capabilities.

• Directed energy weapons, such as lasers and particle beams, whether ground-based or space-based. The United States has stated that current Soviet ground-based test lasers have probable ASAT capabilities.

• Electronic countermeasures of sufficient power output to damage or interrupt satellite functions. The United States considers this a current Soviet ASAT capability.

• Weapons that could be carried on the Space Shuttle or space stations.

Clearly, there are many different types of systems that could be used to destroy satellites, and many space activities have capabilities inherently useful for ASAT purposes. For example, the rendezvous and docking operations routinely conducted by the Soviet Union could be used to conceal development of one or more types of ASAT techniques. Restricting the definition of what is an ASAT weapon could simplify an agreement and make it easier to verify, but it could make such an agreement ineffective in achieving its purpose of protecting satellites.

Disclosure of Information. The administration has also pointed out that while the establishment of cooperative measures might diminish the difficult verification problems associated with ASAT arms control, these measures could cause additional problems. Cooperative measures meant to enhance verification of an ASAT arms control agreement might require access to U.S. space systems that the Soviet Union alleged to have ASAT capabilities. This could create an unacceptable risk of compromising sensitive security information.

U.S. Military Requirements and Deterrence. The Reagan Administration has argued that ASAT limitations could undermine deterrence. Since the Soviet Union has an operational ASAT capability and the United States does not, the current situation is viewed as destabilizing. For example, if during a crisis or conflict the Soviet Union were to destroy a U.S. satellite, the United States would lack the capability to respond in kind and would either have to accept a major loss in capabilities or escalate the conflict. The administration argues that to counter Soviet satellites by attacking their ground facilities would be an uncertain alternative to an ASAT capability and would certainly risk escalation of a conflict. Thus, it is argued that a U.S. capability to destroy satellites clearly responds to the need to deter such Soviet attacks on U.S. satellites.

The U.S. government also argues that the United States must be able to protect its forces against the threats posed by Soviet satellites. Specifically, a comprehensive ASAT ban would afford a sanctuary to existing Soviet reconnaissance satellites designed to target U.S. naval and land forces. The absence of a U.S. ASAT capability to prevent Soviet targeting from satellites could be seen by the Soviet Union as a substantial factor in its ability to conduct a successful conventional attack on U.S. and allied forces. It might also offset the deterrent effect of superior U.S. and allied naval warfare capabilities. Conversely, Soviet uncertainty over the availability of satellites to target naval forces would decrease Soviet confidence in its ability to attack U.S. naval forces, thereby adding to deterrence and stability. In this manner, a U.S. ASAT capability would help deter a conventional conflict.

U.S. Evaluation of Soviet Initiatives.   The Reagan Administration has taken the position that the Soviet draft treaty and the proposed moratorium are unacceptable. It argues that the motives behind the Soviet initiatives are suspect. The timing of the Soviet offer suggests that it is designed to curtail the testing of the new U.S. ASAT program, thereby leaving the Soviet Union with a unilateral advantage in ASAT capability. Moreover, in addition to its operational ASAT system, the Soviet Union currently has other systems with potential ASAT capabilities that would not be constrained by the Soviet moratorium, which deals only with space-based systems. The proposed moratorium, for example, would not affect tests of ground-based lasers in an ASAT mode.

Furthermore, according to the administration, a test moratorium would not necessarily cause the Soviet operational system to atrophy. After a hiatus of several years in ASAT testing, the administration points out, the Soviet Union was able to resume testing of its ASAT system without any apparent degradation in performance. Research and development programs, such as the U.S. ASAT program, would pay a much higher price for a moratorium on testing, and even a short delay in the test program would delay the time that the U.S. ASAT could be operational.

With regard to the draft treaty itself, the Reagan Administration argues that it lacks effective verification provisions since it provides for nothing beyond National Technical Means of verification, which are deemed inadequate. Specifically, the draft treaty does not indicate how the elimination of the operational Soviet ASAT system would be verified. Moreover, the prohibition on the destruction, damaging, and disruption of other states' space objects could pose verification problems.

The administration argues that the draft also does not deal with residual ASAT capabilities. For example, dismantling of the Soviet coorbital ASAT system would still permit the Soviet Union to use its Galosh ABM interceptor missiles in an anti-satellite role. In addition, the draft treaty proposes that "piloted" spacecraft not be used for "military" purposes. Since the term "military" does not appear elsewhere in the draft and would appear to cover such activities as reconnaissance and communications as well as weapons, the administration suggests that this provision is intended to constrain the Space Shuttle, which is the primary U.S. launch system for national security as well as civil space missions.

The Soviet View

The 1983 Draft Treaty Constraints. The Soviet government has proposed that U.S.-Soviet talks on space should deal not only with ASAT systems but also with the broader question of the militarization of space as addressed in the Soviet 1983 draft treaty. The Soviet draft treaty specifically calls for a ban on the testing and development of new anti-satellite systems and for the elimination of all existing anti-satellite systems. The Soviet Union has also stated that its unilateral commitment not to be the first to put into space any type of "anti-satellite weapon" is still in force as a first step toward a total ban on anti-satellite weapons. The Soviet Union has called on the United States to declare a similar moratorium on its activities before the opening of official arms control negotiations on space.

The 1983 Soviet draft is broader in scope and more precise in definition and terms than the 1981 draft, which Soviet Foreign Minister Gromyko referred to as simply an extension of the 1967 Outer Space Treaty. The parties to the draft treaty undertake "not to destroy, damage or disrupt the normal functioning of other states' space objects, nor change their flight trajectories." Along with the more comprehensive constraints on ASAT systems, which were not included in the 1981 draft, the new Soviet proposals cover a range of activities, including a prohibition on the use, threat of use, testing, and deployment of any space-based weapons against targets in space, in the atmosphere, or on earth.

Verification. The verification provisions in the 1983 Soviet draft treaty call for the use of National Technical Means of verification and state that the parties will undertake to consult and cooperate with each other in resolving any questions that may arise with regard to the objectives of the treaty or its observance. Soviet leader Chernenko has asserted that a moratorium on anti-satellite weapons tests is verifiable. He stated on June 12, 1984, "The Soviet Union is convinced that monitoring a freeze on anti-satellite weapons tests is possible, and, moreover, is extremely reliable above all through the national technical means the sides have at their disposal. . . . Effective monitoring of the sides' compliance with a moratorium on orbital anti-satellite weapons could be ensured by the means both sides have at their disposal for tracking objects in space. As for suborbital anti-satellite systems, then apart from the aforementioned facilities it would also be possible to enlist the use of other radioelectronic facilities of the United States and of the Soviet Union that are stationed on the ground, in the world's oceans and in space." Chernenko continued that consultations, the exchange of information, or possibly other forms of cooperation could be found to deal with uncertain situations. Given real interest in finding effective solutions, he said, any questions relating to the militarization of outer space could be successfully resolved during negotiations.

Soviet analysts point out the impossibility of trying to solve the problem of verification without even discussing it within a negotiating framework. The U.S. position on verification has been criticized by the Soviet Union as a means of undercutting negotiations to permit the U.S. space weapons program to proceed.

Soviet analysts argue that the technical difficulties involved in the verification of an ASAT agreement are no more difficult than the technical difficulties involved in the verification of any other arms control agreement. They point out that ASAT weapons would be much easier to control before the systems are deployed rather than after they are deployed by both sides. Concerning the verification problems presented by residual ASAT capabilities, Soviet analysts acknowledge that there are various ways of destroying satellites, including such clumsy procedures as docking with enemy satellites. Nonetheless, they argue that there is no problem in verifying whether or not a satellite has been destroyed and that it is much easier to see what is happening in space than it is to see what is happening on earth from space.

Arguments for ASAT Arms Control. Soviet analysts have argued that the matter of banning anti-satellite weapons is urgent because the deployment of such weapons would destabilize the strategic situation.

By being able to destroy early warning satellites, such weapons would have the capability to blind the other side. Soviet analysts note that an attack on a single satellite, or even the failure of a satellite, during a time of tension, could have grave consequences. Some Soviet analysts emphasize that an attempt to destroy the opponent's satellites would be regarded as the first step to a nuclear war.

Soviet spokesmen have emphasized that urgent steps are necessary before the militarization of space becomes irreversible. The U.S. ASAT systems are viewed as the first step toward a more comprehensive U.S antimissile system. Because of the link between ASAT and ABM developments, the Soviets have urged comprehensive bans on the militarization of outer space. Anti-satellite systems are also connected by Soviet analysts to the buildup of U.S. strategic offensive systems, including the MX, the Trident II missile, the Stealth bomber, and Pershing II and cruise missiles. These analysts assert that the U.S. ASAT program is part of a larger move by the United States to develop a destabilizing first-strike capability.

U.S. Supporters of ASAT Arms Control

Although there are different opinions on what may be the best formula for ASAT arms control, domestic supporters of ASAT arms control agree that the strictest possible limitation on anti-satellite weapons is in the national security interest of the United States. The following arguments are usually highlighted in support of urgent ASAT arms control efforts.

The Importance of Satellites to U.S. Security. Domestic supporters of ASAT arms control emphasize that satellites are vital to U.S. national security and strategic stability and that their survival can best be protected by strict ASAT arms control limitations. To the extent that critical satellite systems were still considered at risk under an ASAT agreement, survivability measures such as hardening and redundancy should be incorporated into future satellite systems. Conversely, unrestricted ASAT development endangers the survival of all U.S. satellite systems. While satellite systems are also important to the Soviet Union, there is little question that the United States is more dependent on these systems. In a political crisis, both superpowers would depend on satellites to assess the actions of the other side and of the rest of the world as well. These systems serve a critical role in deterring either side from attempting a preemptive strike in a severe crisis, since the possibility of effective surprise would be greatly reduced. If hostilities occurred, the information and communications from satellites would be vital to efforts to keep the crisis from escalating to a nuclear catastrophe. If an attack occurred, satellites would provide early warning and permit the launching of retaliatory forces.

Since satellites play such an important role in deterrence, domestic supporters point out, ASAT systems are inherently destabilizing. ASAT systems will inevitably be viewed by the other side as supporting a first-strike strategy. By having the capability to shoot down crucial military satellites at the outset of a preemptive first strike, a potential adversary could be perceived as planning to degrade the effectiveness of an opponent's retaliatory forces in a crisis.

Some of these domestic supporters point out that ASAT arms control is also in the U.S. security interest since the United States is more dependent on military communication satellites than is the Soviet Union. U.S. military forces are spread around the globe and require secure long-range communications. More than 60 percent of long-haul U.S. military communications are now transmitted via satellites, and it is argued that there are no alternative facilities that provide a satisfactory replacement for this satellite system. In contrast, Soviet military forces are mainly on or near to the Eurasian land mass and can readily communicate by a variety of ground-based and airborne facilities in addition to satellites. Indeed, one can argue that the Soviet Union uses its satellites as a backup for its ground-based and airborne communications and intelligence gathering systems, while the converse is true of the United States.

These domestic supporters also argue that ASAT arms control is in the U.S. interest because the United States will be less able to adapt to the costs of an unrestrained ASAT competition. The United States operates with fewer satellites than does the Soviet Union since U.S. satellites are much more sophisticated and long-lived than the Soviet counterparts. Moreover, although many U.S. satellites are currently secure in high orbits, this situation will not long be the case if ASAT technology continues unconstrained. In contrast, Soviet satellites for the most part are relatively short-lived systems in low earth orbit. Soviet practice is to replace satellites frequently and maintain more in orbit. The Soviet Union is therefore better situated to deal with a race for satellite redundancy. Since ASAT technology strongly favors the attacker over the defending satellite, critics argue that advocates of ASAT development have lost sight of the cost to the United States of developing defensive capabilities for satellites. Adding survivability features such as armor plating, antijamming devices, evasive maneuvering capabilities, or shielding against lasers, or developing active defenses for a satellite, will markedly increase the weight, and hence the cost, of future satellites while providing marginal protection against improved ASAT weapons. Deterring a Soviet ASAT attack on high-value U.S. satellites also would not be achieved by the sure response of destroying a single Soviet satellite, supporters argue, as that would be a price the Soviet Union is willing to pay.

Time Urgency for ASAT Arms Control. Domestic supporters of ASAT arms control emphasize that there is now a unique opportunity for agreement since the tested but rudimentary technology of the Soviet ASAT system and the untested but more advanced technology of the U.S. F-15 ASAT system do not provide either side with a really threatening ASAT capability. When fully developed the U.S. F-15 ASAT system will pose a much more serious threat to satellites. The Soviet Union must be expected to move quickly to duplicate the more advanced technology of the U.S. system. Of particular concern is the fact that once tested and deployed, the small size and mobility of a jet-launched miniature homing vehicle system will present a very difficult verification problem for future ASAT agreements. This will take on particular significance given the greater capabilities of the F-15 ASAT system.

The result of an unrestrained ASAT weapons development, according to domestic supporters of ASAT arms control, would be an extremely expensive, destabilizing arms race in space from which neither side would gain any security advantage. Currently, all anti-satellite weapons deployed or undergoing field tests have a maximum altitude of several thousand kilometers or less. Hence, they could attack satellites only in low or highly elliptical orbits. Since the early warning, navigation, attack assessment, and communications satellites essential to the U.S. strategic forces are all in very high orbits, they are not at risk in the near term. The Soviet Union faces a somewhat greater potential threat, since some of its essential communications satellites and all of its early warning satellites are currently in highly elliptical Molniya orbits that the new U.S. F-15 ASAT system could attack from bases in the general area of the orbits' perigees.

These domestic supporters argue that it is logical to constrain the Soviet program now, while the system is slow, only marginally reliable, and capable of attacking only a few satellites at a time. The technology of the U.S. F-15 ASAT system, on the other hand, would potentially present a prompt threat against a large component of deployed satellites. Achievement of this capability, which could represent a significant military threat if fully exploited, constitutes the crossing of an important threshold leading to an arms race in space.

Arguments Against the Military Utility of ASATs. Domestic supporters argue that, contrary to the administration's assessments, development of ASAT weapons will not improve but rather undermine deterrence. If both countries achieve an ability to destroy the opponent's early warning, communications, and navigation satellites, they argue, each would have considerable incentive to initiate such an attack during a time of acute crisis. The existence of significant ASAT capabilities on both sides would therefore decrease crisis stability and increase the threat of war. If past experience is any guide, supporters argue, the U.S. ASAT program, far from discouraging the Soviet Union from further ASAT developments, will only stimulate these developments.

Domestic supporters also challenge the administration's rationale that ASATs are needed to deny a sanctuary to Soviet ocean reconnaissance satellites capable of guiding aircraft and submarines to important U.S. naval ships. It is argued that the Soviet Radar Ocean Reconnaissance Satellite (RORSAT) is not an extremely threatening system and that the capability to destroy Soviet reconnaissance satellites is neither necessary nor sufficient to protect the U.S. fleet. Countermeasures against such a system, such as jamming from ships, are available and fairly straightforward.

Verification. Supporters of ASAT arms control do not accept the administration's conclusion that a comprehensive ASAT agreement poses impossible verification problems. They argue that when the verifiability of a comprehensive ASAT ban is examined in detail, the risks to U.S. security of possible violations are small compared with the dangers inherent in unlimited development of ASAT capabilities.

Supporters emphasize that the United States has a very effective and redundant intelligence system for keeping track of Soviet activities in space. Moreover, these capabilities will increase significantly in the future. As evidence of past capabilities, they point out that the United States has successfully monitored Soviet ASAT activities for the last 15 years even though the Soviet Union has never acknowledged the existence of its ASAT program.

Supporters argue that the prohibition on attacks on satellites can be verified with high confidence, since the operation of U.S. satellites is closely monitored and sensors can diagnose the cause of failure. While direct evidence of the source of the attack might not always be immediately available, circumstantial evidence would be overwhelming, since no country other than the Soviet Union will have the capability or motivation to undertake such attacks against U.S. satellites in the foreseeable future.

Supporters argue that, while problems exist, a ban on dedicated ASAT systems can be adequately verified. Specifically, a ban on the further testing of the existing coorbital systems could be verified with high confidence. Such tests are easily identified and would be monitored from launch to intercept, as they have been in the past. Any effort to upgrade this system to give it a high-altitude capability would be particularly obvious, since it would require an extremely large booster to get the heavy payload into a geosynchronous orbit. Without additional tests it would not be possible to upgrade the capability of this system significantly.

Supporters acknowledge that the complete elimination of the existing Soviet ASAT system cannot be verified with high confidence since some payloads might not be destroyed. But they argue that if launchers for the specialized SS-9 boosters are eliminated or kept to small numbers for other missions, a covertly reconstituted system would have limited capabilities and would not endanger the most critical U.S. satellites in geosynchronous orbits. In the absence of further tests, Soviet confidence in the system would decline and retained payloads could not be used with other boosters.

With regard to new systems employing directed energy (high-energy lasers or particle beams) as kill mechanisms, supporters argue that deployment of such systems would be a major undertaking and would require an extended test program that would be easily identifiable. While ground-based high-energy lasers might present a more difficult verification problem, supporters argue that use of such a system against U.S. targets would certainly be identified. Testing it with cooperative Soviet satellites could also be detected by monitoring the illumination and heating of the target, they contend.

With regard to the potential future threat posed by "space mines," supporters argue that this development could be monitored with confidence, since all satellites are tracked and a Soviet satellite following a critical U.S. satellite closely in the same orbit would be immediately apparent. This threat could be contained by including in the agreement "rules of the road" that prohibit such trailing activities.

Finally, supporters argue that various indirect ASAT capabilities that are inherent in other military and civil space activities are not in fact serious threats at present. Moreover, efforts to upgrade these systems for an ASAT role would be easily verified, they assert. Specifically, while the existing Soviet ABM system deployed at Moscow and intercontinental ballistic missiles armed with nuclear warheads have an inherent capability against low-altitude satellites, it is extremely unlikely that they would be used for this purpose in peacetime or in a

conventional conflict due to the risk of escalation and the danger to
Soviet satellites. Supporters argue that any attempts to arm these systems with nonnuclear homing warheads would require extensive testing that could be easily detected. With regard to civil systems such as
the U.S. Space Shuttle or the Soviet Progress resupply vehicles that
have a rendezvous capability, supporters argue that the capabilities of
these systems are very limited in an ASAT role and that efforts to
operate them in this manner would be easily detected.

Breakout. Supporters of ASAT arms control also disagree with the
assertion that under an arms control regime the Soviet Union would be
in a unilateral position to constrain a new U.S. ASAT system because
the Soviet system is tested whereas the U.S. system is not. It is argued
that the United States has already conducted enough engineering tests
of its own ASAT interceptor to be ready for immediate space tests, if a
moratorium or ASAT agreement were terminated. It is also argued that
incorporation of reasonable survivability measures in U.S. satellites
would erode Soviet confidence in its ASAT system in the absence of
tests. Supporters also disagree with the government's assessment that
since U.S. satellites are so few in number, Soviet possession of only a few
ASAT interceptors after a breakout would pose a prohibitive risk. They
note that most of the important U.S. satellites are out of range of the
present Soviet ASAT system, which would be the only Soviet capability
available after a breakout from an agreement.

Breakout using an entirely new ASAT weapon that had been tested
only on the ground or covertly would involve prohibitive technical risk.
Breaking the ASAT system down into component parts for covert testing entails a high risk that the whole system may not work. Without
tests the Soviet Union, which has experienced many failures in its space
technology plans and performance over the years (particularly in the
ASAT field), could not confidently predict how soon a new device could
be made to work after breakout. Soviet attempts to conduct a series of
full-system ASAT tests in space would almost certainly be detected even
if deceptive tactics were attempted.

ASAT Link with ABM. Another reason for ASAT arms control, according to some supporters, is to assure that the SALT I ABM Treaty
limiting anti-ballistic missile systems is not undermined. These critics
are concerned that the lack of restrictions on ASAT development potentially provides a way to circumvent the ABM Treaty. Either side can
claim that a weapon system under development is intended to be deployed as an ASAT when the longer-term objective is really ballistic

missile defense. Conversely, a program that really is intended only to achieve an ASAT capability may be perceived by the other side as a nascent ABM system, sparking fears of ABM breakout. In this context some supporters charge that the real reason the United States does not want to negotiate ASAT controls is because they could interfere with the government's Strategic Defense Initiative.

Comprehensive Versus Partial Limitations. Many supporters of ASAT arms control emphasize the criticality of achieving comprehensive limitations on ASAT development as opposed to partial limitations designed to avoid problems raised by a comprehensive proposal. For example, one of the partial limitations that is reportedly under consideration by the Reagan Administration is a ban on ASAT systems capable of attacking satellites in high (geosynchronous) orbits. This would allow both sides to develop their low-altitude ASATs while protecting the most vital U.S. satellites, which are in high orbits. An argument for this approach is that it will avoid some of the verification difficulties of a comprehensive ban, including the elimination of existing interceptors, while giving the United States a chance to match the Soviet Union with an operational ASAT system. However, supporters of ASAT arms control oppose this limited approach on the grounds that it would not prevent the development of dangerous and destabilizing ASAT capabilities that would threaten some critical satellites on each side. They argue that this limited approach would undercut the ABM Treaty in the same manner as unrestricted competition. Finally, they question whether it would in fact resolve the verification problem, since it might be difficult to distinguish permitted activities from developments leading to a high-altitude capability.

The Soviet Draft Treaty. Some supporters of ASAT arms control disagree with the administration's negative assessment of the Soviet 1983 draft treaty. They argue that it is a significant improvement over the much less comprehensive 1981 Soviet draft and a sign that the Soviets may have a serious interest in negotiating a ban on space weapons, including ASAT systems. While cautioning that the Soviet draft treaty should not be taken as the final word and that deficiencies, such as the apparently discriminatory handling of manned space vehicles (the Space Shuttle), need to be resolved satisfactorily in negotiations, these supporters believe that the U.S. government should respond positively to this proposal. They note with approval that the Soviet Union has announced an ASAT moratorium, and that the new draft treaty is broader in scope and includes more precise definitions of the types of activities and systems that are limited than the 1981 draft. For the first time the Soviet Union has indicated a willingness to agree to dismantle its existing ASAT system. The new draft appears to have eliminated language from the earlier ASAT negotiations that some interpreted as legalizing the use of ASATs against "hostile" satellites and third parties. In short, these supporters see the Soviet draft treaty as providing a basis for serious negotiations.

7 Nuclear Test Bans

INTRODUCTION

The banning of nuclear testing has been a central and continuing objective of arms control since the mid-1950s. At the end of the Eisenhower and beginning of the Kennedy administrations, the United States and the United Kingdom made a major effort to negotiate a comprehensive test ban (CTB) treaty with the Soviet Union. Although these trilateral negotiations failed to produce a comprehensive test ban, agreement was finally reached in 1963 on the Limited Test Ban (LTB) Treaty (Appendix D), which banned all nuclear tests except those conducted underground. In 1974 the Nixon Administration negotiated the Threshold Test Ban (TTB) Treaty (Appendix E), which banned underground tests above 150 kt; and in 1976 the Ford Administration negotiated the companion Peaceful Nuclear Explosions (PNE) Treaty (Appendix F), which provided for the special handling of peaceful explosions under the threshold. The Carter Administration renewed the effort to negotiate a comprehensive test ban treaty but failed to produce an agreement. The Reagan Administration has taken the position that, while a comprehensive test ban remains a long-term U.S. goal, such a treaty would not be in the security interests of the United States at the present time.

BACKGROUND

The Eisenhower Administration

By the mid-1950s, public opposition to nuclear testing had become a significant domestic and international political force. The recurring

187

U.S. and Soviet nuclear test series, involving growing numbers of explosions with rapidly increasing yields, were a constant reminder of the threat and consequences of nuclear war. With the unexpected discovery of the extent of the danger of fallout during the U.S. test series in 1954, nuclear testing was also widely seen as a direct threat to public health and safety. Early proposals to stop testing were opposed within the U.S. government by both military and civilian officials on the grounds that the requirements for more advanced nuclear weapons were so urgent as to far outweigh any immediate health dangers that might be associated with nuclear tests. Questions were also raised about the ability to verify a ban on nuclear tests. In early 1958, following a major Soviet test series, the Soviet Union seized the political initiative by announcing that it would stop testing unilaterally if the United States would do likewise.

In a major policy shift in the spring of 1958, President Dwight Eisenhower proposed to Soviet Secretary Nikita Khrushchev that scientists from the two sides meet to assess the verifiability of a ban on nuclear tests and to recommend a possible control system. In addition to political concern about the mounting international opposition to testing, President Eisenhower's decision reflected the advice of the newly formed President's Science Advisory Committee under James Killian. Challenging the positions of the Department of Defense and the Atomic Energy Commission, the committee advised the President that a test ban could be monitored and would be in the security interests of the United States given the relative status of the nuclear weapons programs of the two sides.

The Conference of Experts, which was held in Geneva, Switzerland, during the summer of 1958, brought together a remarkable group of outstanding scientists and specialists on nuclear test detection from the West (the United States, the United Kingdom, France, and Canada) and the East (the Soviet Union, Poland, Czechoslovakia, and Romania). The conference, which was conducted as a technical study and not as a political negotiation, examined the technical problems of monitoring nuclear tests in the atmosphere, in the oceans, and underground. The report of the conference found that an international control system, using available techniques and on-site inspection, would "make it possible to detect and identify nuclear explosions, including low-yield explosions (1-5 kt)." The proposed system would have been a worldwide network made up of some 160 to 170 land-based manned control posts and ten ships with appropriate instrumentation. Subsequently, it was agreed in a separate technical working group that the control system could also be applied to tests in space if satellite-borne detectors were incorporated into it.

On the basis of the findings of the Conference of Experts, President Eisenhower called for formal negotiations on a comprehensive test ban. At the same time he announced a one-year moratorium on all testing provided the Soviet Union did the same. This moratorium was subsequently extended to the end of 1959 and testing was not resumed until 1961. On October 31, 1958, the United States, the Soviet Union, and the United Kingdom, then the only nuclear powers, opened the Conference on the Discontinuance of Nuclear Tests in Geneva, Switzerland. Despite the technical agreement at the Conference of Experts, the political negotiators quickly found that the sides were far apart in defining how the control system would actually operate and how on-site inspections, which were supposed to resolve questions regarding unidentified events, would be conducted. The United States and the United Kingdom envisaged a system administered by international personnel and operating by a majority vote, while the Soviet Union insisted on a system that it could control within its own borders.

The United States soon complicated the negotiations further by introducing new technical data and new technical problems that brought into question the findings of the Conference of Experts. The U.S. delegation first reported that analyses of new data from U.S. underground tests conducted after the Conference of Experts indicated that the lowest seismic yield that could be identified as an earthquake was about twice as high as that originally estimated. More significantly, the U.S. delegation then reported that new studies revealed a number of techniques that could permit a violator to conduct relatively large-yield underground tests so that they would not be identified or even detected by the proposed control system.

The most striking of the clandestine testing techniques was the concept of testing in huge underground cavities. Such cavities were calculated to be capable of decoupling the seismic signal from a nuclear explosion by a factor of 100 or more. The United States also suggested the possibility of conducting tests during very large earthquakes to bury the seismic signal from the test in the much greater signal from the earthquake. The Soviet delegation rejected these technical developments as simply efforts to prevent agreement. In the United States, opponents of the test ban in the executive branch and Congress seized upon the technical developments as conclusive proof that a comprehensive ban could not be verified. The U.S. government initiated an extensive research and development program (Project Vela) directed at improving seismic monitoring and other verification capabilities.

In an attempt to bypass the increasing controversy over the verification of underground tests, the Eisenhower Administration proposed to ban only those tests that could be verified by the control system devised

by the Conference of Experts. Early in 1960 the United States introduced a draft threshold treaty that would have banned all atmospheric
and underwater tests, underground tests above magnitude 4.75 on the
Richter scale, and tests in space to a distance (unspecified) at which
detection was feasible. By defining the threshold in terms of seismic
magnitude rather than yield, the proposal sought to avoid the problem
of the substantially different coupling factors of explosions in different
types of rock and in large cavities. This proposal established criteria
that would have called for an estimated average of some 20 on-site
inspections per year instead of the open-ended number of the previous
proposal. The United States also proposed that a joint U.S.-Soviet seismic research program develop techniques to lower the threshold. In
response, the Soviet Union called for a ban on all space tests, a five-year
moratorium on underground tests below magnitude 4.75 while the joint
seismic research program was under way, and a political decision on a
specific number of on-site inspections.

After meeting with British Prime Minister Harold Macmillan, President Eisenhower agreed to the concept of a moratorium on tests below
the magnitude 4.75 threshold, but only after a threshold treaty with an
agreed quota of on-site inspections had been signed and a joint research
program agreed upon. (The 1958-59 moratorium was no longer formally
in effect, though neither side had conducted any tests since that time.)
The Soviet Union accepted this approach, and arrangements were made
for a Seismic Research Program Advisory Group to meet in Geneva to
develop the joint program. The questions of the length of the moratorium and the quota of on-site inspections remained. There were also
unresolved political problems relating to the organization and operation of the control system.

Whatever prospects the threshold approach might have had ended
when a U.S. U-2 reconnaissance aircraft was shot down near Sverd-
lovsk on May 2, 1960. This led to a crisis in U.S.-Soviet relations and the
cancellation of the Paris summit at which it had been planned to seek
agreement on the duration of the moratorium and the quota of on-site
inspections. The meetings in Geneva on the joint seismic research program adjourned at the end of May without filing a report when the
Soviet delegation indicated there was no point in continuing. The formal treaty negotiations in Geneva continued but made no further progress during the remaining months of the Eisenhower Administration.

The Kennedy Administration

The new Kennedy Administration moved promptly to revive the
threshold approach that had appeared to be within reach before the U-2

incident. On April 18, 1961, after intensive internal reviews, the United States presented a revised draft treaty banning all nuclear tests, including those in space, except for underground tests below magnitude 4.75. The treaty was to be coupled with a three-year moratorium on underground tests below magnitude 4.75. The moratorium could be reviewed annually while the joint seismic research program continued.

Despite a number of compromises with the Soviet position, the new positions of the United States and United Kingdom and the Soviet Union were still far apart in many respects. The annual quota of on-site inspections became the symbol of these differences. The U.S. draft permitted 12 to 20 annual inspections in the Soviet Union, depending on the number of unidentified seismic events in the Soviet Union (the same formula would apply independently to events in the United States and the United Kingdom); the Soviet Union would only accept three inspections. There also remained fundamental organizational differences in the two sides' approaches. For example, the United States wanted a single neutral administrator for the control system; the Soviet Union wanted a three-member administrative council (one Soviet, one Western, and one neutral member) that could only operate by unanimous consent. The United States wanted the detection stations in each country to be manned by personnel from other countries; the Soviet Union wanted the stations to be manned almost entirely by personnel of the host country. The negotiations were stalemated, and neither side was prepared to make further significant concessions.

This phase of the test ban negotiations abruptly ended on August 30, 1961, when the Soviet Union announced its intention to resume nuclear testing, which began the next day. Although there was actually no moratorium in effect at the time, the Soviet action, which came as a complete surprise, generated concern and outrage in official circles and among the public at large. When President Eisenhower had originally proclaimed a one-year moratorium in August 1958, the Soviet Union announced that it would abide by the moratorium as long as the West did. After extending the moratorium through the end of 1959, President Eisenhower, who was concerned that the United States might have to resume testing, terminated the moratorium but stated that the United States would announce any resumption in advance. When the French began testing in February 1960, the Soviet Union denounced the French action as a cover for Western testing.

The Soviet Union proceeded to carry out an unprecedentedly intensive test series. Within 60 days the Soviet Union conducted 30 atmospheric tests, with greater total megatonnage than the total of all previous tests. The series included a gigantic 57-Mt test that was judged in the United States to be a reduced-yield version of the previously claimed Soviet 100-Mt bomb. As soon as Soviet testing began, President Kennedy ordered the immediate resumption of U.S. testing, and the first test was conducted within two weeks. The initial U.S. tests, which were essentially a political reaction, were conducted underground at small yield. By the spring of 1962 the United States was fully prepared and conducted Operation Dominic, a series of some 40 atmospheric tests in the Pacific that lasted over six months. Among the tests was STAR-FISH, a megaton-yield explosion at an altitude of 400 km that produced unexpected and severe high-altitude effects, including damage to satellites at great distances. By the end of Operation Dominic the Soviet Union was engaged in yet another major test series, including a 30-Mt explosion in early August. The cumulative effect of the massive Soviet and U.S. test series was to increase domestic and international concern about both the immediate health effects and longer-range military implications of what appeared to be a completely unbridled competition in atmospheric nuclear tests.

Although the trilateral negotiations on the test ban were adjourned indefinitely in January 1962, world opinion would not permit the negotiations to die. Negotiations were resumed in the spring of 1962 in the Eighteen Nation Disarmament Conference (ENDC), the multilateral forum for arms control negotiations. The United States began to relax its verification demands, but this did not narrow the gap with the Soviet position because the Soviet Union hardened its position, proposing a test ban verified only by national means of detection. In the late summer of 1962 the United States and the United Kingdom proposed two alternative approaches. One was a treaty banning all nuclear tests without a threshold on underground tests. The provisions of this comprehensive test ban were essentially those of the previous threshold test ban, although it was suggested that the number of inspections would be reduced. The other approach was a treaty banning tests in or above the atmosphere and in the sea. The Soviet delegation rejected both approaches, the first because it required inspections and the second because it permitted testing to continue.

In mid-October 1962 the Cuban missile crisis suddenly brought home to leaders and ordinary citizens everywhere the stark realization that nuclear war could happen. President Kennedy and his advisors were clearly deeply moved by their close involvement in the events. Secretary Khrushchev and his advisors also appeared to be sobered by the experience. Following the intense and continuing U.S. and Soviet atmospheric test series, the missile crisis intensified world pressure for progress in the nuclear test negotiations, which were then the only serious, well-advanced arms control negotiations in progress. Significantly, the

UN General Assembly passed two resolutions in the immediate aftermath of the Cuban missile crisis, one calling for a cessation of nuclear testing and another calling for either a comprehensive test ban or a limited ban coupled with a moratorium on underground testing.

Despite these strong pressures for an early agreement and intensive efforts over the next six months to negotiate formally at the ENDC and informally on a personal basis at various levels, the two sides were unable to resolve the remaining differences in their positions. The quota on inspections remained the major, but not the only, issue. Khrushchev reinstated his earlier offer of two or three annual inspections, reportedly in the mistaken belief that this would be acceptable to the United States. Kennedy eventually agreed to reduce the quota to seven annual inspections. Neither Kennedy nor Khrushchev apparently considered themselves sufficiently secure politically to propose a final compromise of five inspections, which appeared to some participants to be a logical outcome of the negotiating process. Kennedy was concerned over the strong opposition to further compromise from the military, the weapons laboratories, and influential members of Congress. Khrushchev told Western visitors that he had used up his political credit with his colleagues by agreeing to permit three inspections.

The number of on-site inspections was not the only difference. There was a similar impasse over the number of unmanned automatic seismic stations, or "black boxes," to be located in each country. The United States had accepted the Soviet proposal that these black boxes, which could be safeguarded to ensure the authenticity of their seismic data, should be used in place of manned control posts to eliminate the issue of the nationality of the staff at the posts. The Soviet Union had offered to locate three black boxes in the Soviet Union, and the United States had insisted on eight to ten. The gap was not narrowed. In addition to these quantitative differences that dominated the negotiations on the test ban, the two sides were far from agreement on the so-called modalities governing the conduct of individual on-site inspections and the installation and operation of the black boxes. Whether these detailed procedural issues, which were critical to the satisfactory operation of the control system whatever the quotas might be, could have been resolved if a political decision had been reached is difficult to judge. Certainly, the United States would have had to back off from the very elaborate inspection procedures it envisaged, and the Soviet Union would have had to grant considerably more access than it had yet shown signs of accepting.

The treatment of peaceful nuclear explosions was an issue that had not been resolved within the U.S. government and would eventually have to be faced with the Soviet Union. Within the Atomic Energy Commission and among influential members of Congress, there was strong support for a program of peaceful nuclear explosions, called Project Plowshare, for which great economic claims were being made. But it was also recognized within the government that continuation of Project Plowshare was inherently incompatible with a comprehensive test ban. The two sides had earlier tried to finesse the issue by permitting explosions for peaceful purposes provided the other side could inspect the internal design of the device to assure that it was not a weapon development test. Advocates of Project Plowshare, who recognized that such a provision was tantamount to stopping the program since it was most unlikely that either side would agree to it in practice, proposed instead that each side be given a quota for peaceful tests or projects. Such a proposal was recognized as being inherently contradictory to the goal of a comprehensive test ban.

At the urging of Prime Minister Macmillan, President Kennedy decided in the spring of 1963 to attempt to break out of the deadlocked ENDC negotiating framework by sending Averell Harriman to Moscow as a special personal representative to see if some resolution of the test ban issue was possible. In an exchange of personal letters, Khrushchev agreed to receive the Harriman mission. On June 10, 1963, Kennedy announced in his famous American University speech that agreement had been reached to hold high-level discussions in Moscow on the test ban. In the speech, which examined the issues of war and peace and U.S.-Soviet relations in a nuclear world, Kennedy also declared a unilateral moratorium on atmospheric nuclear tests for as long as other states did likewise.

Averell Harriman's instructions were to seek a comprehensive treaty and, if this appeared unattainable, a limited agreement along the lines of the draft treaty the United States had originally submitted to the ENDC the previous year. The impasse on the comprehensive treaty and developments immediately prior to the meeting made it clear that a limited agreement was the hoped-for outcome on both sides. On July 2, Khrushchev announced that the Soviet Union was withdrawing its offer of three on-site inspections, claiming that the West would exploit them for espionage. He also stated that the Soviet Union was prepared to conclude an agreement banning testing in the atmosphere, in outer space, and underwater. The Soviet Union had previously rejected the possibility of such a limited treaty. In the United States there was growing support in Congress for this approach, which the military strongly preferred over a comprehensive test ban.

The negotiations began on July 15, and ten days later the Treaty Banning Nuclear Weapons Tests in the Atmosphere, in Outer Space and

Under Water, or simply the Limited Test Ban Treaty (Appendix D), was initialed. There was essentially no discussion of a comprehensive ban, which was clearly out of reach for quick resolution, and the negotiations proceeded directly to the text of the limited treaty. Both sides clearly wanted an agreement, and the few matters of substance and drafting problems were quickly resolved and cleared directly with President Kennedy and Secretary Khrushchev. The Soviet delegation objected to a proposed U.S. provision permitting atmospheric tests for peaceful purposes if unanimously approved. The U.S. delegation withdrew this proposal when agreement was reached on a provision permitting treaty amendment by a majority of the parties, including the three original nuclear weapon parties, and on a U.S. provision explicitly permitting withdrawal from the treaty.

The Limited Test Ban Treaty, which was of unlimited duration, banned nuclear tests in all environments except for underground tests that contained the resulting radioactive debris so that it would not be present outside the territory of the country conducting the test. The treaty was to enter into force when ratified by the United States, the United Kingdom, and the Soviet Union and was open to signature by all countries. The treaty, which was considered verifiable by the National Technical Means (NTM) of the two sides, contained no special verification provisions.

After extensive hearings the Senate advised ratification of the treaty by a vote of 80 to 19. Support for the treaty in the hearings was not universal, with representatives of the weapons laboratories emphasizing the technological limits imposed by confining testing to underground shots. An important factor was the support of the Joint Chiefs of Staff, whose position was uncertain until the administration formally agreed to four safeguards that the chiefs proposed. These safeguards involved presidential commitments to conduct a comprehensive and continuing underground test program, to maintain the vitality of the weapons laboratories, to maintain the resources necessary for the prompt resumption of atmospheric testing, and to improve verification capabilities. The treaty was ratified by President Kennedy on October 7, 1963, and entered into force three days later.

In general, the treaty was very well received in the United States and throughout the world despite its failure to stop all testing. After the extreme tensions of the Cuban missile crisis, the first major arms control agreement between the United States and the Soviet Union came as a welcome relief. The termination of atmospheric testing also relieved widespread anxiety about immediate health effects. A large number of countries moved promptly to sign the treaty, and others have joined over the years. As of September 1984, 111 countries had signed the treaty and all but 15 had ratified it. France and the People's Republic of China have not signed the treaty. Initially, both countries continued to test in the atmosphere, but since 1974 France has not conducted any atmospheric tests.

Although the preamble to the Limited Test Ban Treaty proclaimed the objective of "the discontinuance of all test explosions of nuclear weapons for all times," the treaty, by stopping atmospheric testing by the United States and the Soviet Union, had the effect of reducing domestic and international pressure for a comprehensive test ban. As a result, there was little serious effort to achieve a comprehensive test ban until trilateral negotiations were resumed 14 years later in the Carter Administration. Arms control activities shifted to other fields.

During the Johnson Administration the focus of arms control was on the negotiation of the Non-Proliferation Treaty (NPT) (Chapter 8). The Non-Proliferation Treaty was inherently discriminatory, since it divided the world into nuclear weapon states and non-nuclear weapon states. To balance the commitment of the non-nuclear weapon states not to obtain nuclear weapons or any other nuclear explosive device, the nuclear weapon states agreed to share the benefits of the peaceful uses of atomic energy and to negotiate an end to the nuclear arms race. Article VI of the NPT specifically committed all parties to the treaty "to pursue negotiations in good faith on effective measures relating to cessation of the nuclear arms race at an early date." Moreover, the preamble to the treaty recalled the determination expressed in the preamble of the Limited Test Ban Treaty "to achieve the discontinuance of all test explosions of nuclear weapons for all time and to continue negotiations to this end."

In the eyes of most non-nuclear weapon states, nuclear testing, even though it was underground, remained the symbol of a continuing policy of active discrimination under the Non-Proliferation Treaty. Many states considered the failure to pursue serious efforts to achieve a comprehensive test ban to be a violation of the obligation to pursue this agreement "in good faith." This dissatisfaction continued even after the United States and the Soviet Union began the SALT process and achieved significant agreements. At the NPT review conferences in 1975 and 1980, key non-nuclear weapon states strongly criticized the United States and the Soviet Union for failing to make further progress on a comprehensive test ban.

The Nixon and Ford Administrations

Under President Nixon, arms control focused on the SALT process, which became the centerpiece of his foreign policy with the Soviet Union. These negotiations produced the ABM Treaty and the SALT I Interim Agreement in 1972. Although both sides gave little attention initially to the nuclear test ban issue, in the second Nixon Administration interest was suddenly rekindled in the threshold approach to a nuclear test ban. One must review the political situation existing at the time to appreciate this unexpected turn of events. With a summit meeting long scheduled for mid-1974, both sides shared a common interest in achieving in advance an agreement that would maintain the momentum of arms control and the détente process. It was clear that the SALT II Treaty, which was proving more difficult to negotiate than anticipated, could not be completed by then. As the shadow of the Watergate scandal grew, President Nixon had an additional motivation to demonstrate that he was in control of a dynamic foreign policy. In these circumstances, a threshold treaty with the threshold set sufficiently high to eliminate all verification problems and permit a significant level of testing provided an opportunity for a quick, noncontroversial agreement. It bypassed the problem of establishing quotas and procedures for on-site inspections and the persistent opposition of the military and the weapons laboratories to a ban on all testing. The treaty was rapidly negotiated and signed at the summit meeting in Moscow on July 3, 1974. One month later, President Nixon resigned.

The Treaty on the Limitation of Underground Nuclear Weapon Tests, or more simply the Threshold Test Ban Treaty (Appendix E), prohibited any underground nuclear weapon test having a yield exceeding 150 kt. At the same time it was agreed that negotiations should continue on a comprehensive test ban. The treaty was a bilateral undertaking between the United States and the Soviet Union and did not even contain provisions for other nuclear weapon states to join. The treaty included a protocol in which the parties agreed to designate the geographic boundaries of their test areas and to exchange other technical data. These technical data, which were to be made available at the time the instruments of ratification were exchanged, included detailed information on the geology of the test sites and data from two calibration shots at each test site. These cooperative measures were designed to assist the other side in translating the seismic magnitude measured by its own seismic monitoring system into an equivalent yield. The seismic signal from an explosion of a given yield depends on both the local and regional geology of a nuclear test site.

The two sides were concerned that tests with design yields near the threshold might accidentally produce yields above it, since the weapon laboratories wanted to test as close to the threshold as possible. To cover this contingency a separate understanding was subsequently reached that "one or two slight unintended breaches per year would not be considered a violation of the treaty" but would be cause for concern and on request would be the subject of consultations. This unusual understanding was submitted to the Senate as part of the negotiating record.

The Threshold Test Ban Treaty was directed specifically at weapon tests and did not deal with peaceful nuclear explosions (PNEs). In the period since the Limited Test Ban Treaty, the Soviet Union had taken over the earlier U.S. enthusiasm for this activity, and at the time of the negotiations it was apparently seriously considering some ambitious earth moving projects to divert rivers from a northern to a southern course. Since nuclear devices really intended for use in nuclear weapons could be tested as part of a legitimate PNE experiment or project, any relaxation of the provisions of the Threshold Test Ban Treaty for PNEs had to be very carefully worked out to maintain the integrity of the treaty. The time pressure of the upcoming summit did not permit this, and it was decided to deal with the PNE problem in a separate companion treaty. The Threshold Test Ban contained a provision obligating the signatories to negotiate such a treaty at "the earliest possible time."

After 18 months of intense negotiations, the Treaty on Underground Nuclear Explosions for Peaceful Purposes, or simply the Peaceful Nuclear Explosions (PNE) Treaty (Appendix F), was signed by Presidents Ford and Brezhnev on May 26, 1976. The treaty extended the threshold of 150 kt on individual nuclear explosions to PNEs but permitted "group explosions" or salvos with yields up to 1.5 Mt, provided that the individual explosions in the salvo did not exceed 150 kt. (Such salvos could be used to excavate large ditches suitable for the Soviet river diversion projects.) The treaty contained an elaborate protocol that spelled out in great technical detail the information that would be exchanged before any peaceful nuclear explosion. The protocol also provided for detailed on-site observations for salvos whose total yield would exceed the 150-kt threshold. The observers would be allowed to use specified instrumentation to confirm that the yields of the individual shots in the salvo were below the 150-kt threshold. The procedures for making these observations possible were also spelled out in great detail.

Reaction within the United States to the Threshold Test Ban and Peaceful Nuclear Explosions treaties was mixed. Supporters argued that they were a significant step toward a comprehensive ban that could be achieved by lowering and finally eliminating the threshold as confidence in the agreement was established. They emphasized that the PNE treaty established a sound precedent for serious on-site inspection. The treaties were criticized from different directions. On the one hand, some critics saw the treaties as unnecessarily restricting U.S. testing without accomplishing a serious arms control objective. These critics also saw them as being difficult to verify because of the uncertainty in the coupling of yield to seismic signal. On the other hand, some advocates of a comprehensive test ban argued that the 150-kt threshold was so high as to be meaningless and that this approach would in fact delay the achievement of a comprehensive ban since the concept of a threshold would be difficult to eliminate once it had been formalized in a treaty. In addition, they argued that institutionalizing PNEs in a formal treaty created yet another barrier to a comprehensive test ban and stimulated interest in this field to the detriment of the Non-Proliferation Treaty.

For a variety of political and tactical reasons, the Threshold Test Ban and Peaceful Nuclear Explosions treaties have never been ratified by the United States. Due to the domestic political crisis in mid-1974 and the logical coupling of the two treaties, the Threshold Test Ban Treaty was not submitted for ratification immediately after being signed. By the time the Peaceful Nuclear Explosions Treaty was signed in the spring of 1976, President Ford decided that he did not want to risk a Senate debate in the midst of his political campaign. President Carter did not want to complicate the prospects for ratification of the hoped-for SALT II or CTB treaties for a marginal agreement that both sides were already honoring. During this period both sides made clear at various times their intention not to act inconsistently with these unratified treaties as long as the other side did likewise.

The Carter Administration

President Carter came into office determined to make rapid progress in arms control, including the achievement of a comprehensive test ban. When Secretary of State Cyrus Vance and Soviet Foreign Minister Andrei Gromyko met in Moscow in March 1977 to establish the arms control agenda for the next four years, it was agreed to resume trilateral negotiations on a comprehensive test ban agreement. The negotiations, which began in the fall of 1977 in Geneva, made significant early progress but slowed as the SALT II negotiations began to dominate the bureaucratic and political processes in the United States. The domination became complete when the SALT II ratification process began.

Faced with a difficult and uncertain SALT II ratification debate, the Carter Administration was anxious to avoid reducing the prospects for success by presenting a controversial test ban agreement too soon. The weapons laboratories were strongly opposed to a comprehensive test ban, and senior military officers, whose support was critical to the ratification of SALT II, were known to be concerned about the consequences of a ban, particularly as it might affect confidence in the reliability of an aging stockpile of nuclear weapons. In this atmosphere, difficult policy decisions on the test ban were deferred, and the pace of the trilateral negotiations slowed.

Despite the lack of priority afforded the trilateral negotiations, considerable progress was made on a draft text for a comprehensive test ban treaty. Although there are differences of opinion as to how close the two sides actually were to final agreement, the overall framework of the approach had been agreed upon and most of the provisions had been drafted. The perennial problems of verification and the peaceful uses of nuclear explosives appeared to have been handled to the general satisfaction of all sides.

The form of the agreement was to be a multilateral treaty prohibiting the testing of all nuclear weapons in all environments. The treaty, which would be of short duration, would be open to signature by all countries. A protocol to the treaty established a moratorium on all explosions for peaceful purposes for the duration of the treaty unless an agreed way could be found sooner to preclude the acquisition of military information from the conduct of such tests. The verification problem was to be handled at two levels: international arrangements spelled out in the treaty, and special arrangements among the United States, the United Kingdom, and the Soviet Union.

The treaty itself was to be verified by the National Technical Means of the individual signatories, supplemented by an international exchange of seismic data. Any participant could request an on-site inspection to determine whether a suspicious event had been a nuclear explosion. After reviewing the reasons for suspecting the nature of the event, the challenged state would either grant the request or explain why this was not necessary. The treaty also made specific provision for any two or more parties to agree to additional measures to facilitate verification of the treaty. Detailed provisions were developed in the negotiations defining how this would be done in the case of the United States, the United Kingdom, and the Soviet Union. It was agreed that in the case of these parties, National Technical Means would be supplemented by a system of unmanned seismic stations, new versions of the old black boxes, using agreed-upon high-quality seismic equipment with sophisticated encryption devices to ensure authenticity of data. It was agreed that ten such stations would be located at specified locations in both the United States and the Soviet Union; the number of stations for the United Kingdom was unresolved. Using data from this system as well as the full resources of their respective National Technical Means, the countries could request inspections that would have to be granted unless explanations were provided. Detailed provisions were agreed upon to govern the rights and procedures to be followed in carrying out invitational inspections under the special arrangements involving the United States, the United Kingdom, and the Soviet Union.

The perennial issue of on-site inspections was resolved by accepting the Soviet approach of an unlimited number of challenge or invitational inspections, as opposed to the Western approach of a specific quota of mandatory inspections based on seismic events meeting agreed-upon seismic criteria. The United States had concluded that the value of mandatory on-site inspections had been exaggerated, since it was most unlikely that an on-site inspection would ever be allowed in the case of an actual violation. Similarly, an invitational inspection would not be offered in the case of an actual clandestine test. In both cases this denial would have to be taken into account in assessing the probability that an alleged test had actually occurred.

By agreeing on a very short duration for the treaty, reportedly only three years, both sides essentially finessed several underlying problems and differences. This had the effect of deferring but not finally resolving these issues. Thus, continuing Soviet interest in peaceful uses of nuclear explosions was accommodated by a three-year moratorium, after which the subject would have to be reopened. Similarly, the Soviet Union was willing to defer its insistence that China and France join the treaty provided the subject would automatically be reopened after a short period. From the U.S. perspective, the short duration responded to the concern of the military that tests for reliability or other critical purposes might be required after a few years. The process to be followed upon the expiration of the treaty was not agreed upon. The United States wanted to renegotiate and reratify the treaty, while the Soviet Union wanted simply to review and renew the treaty. Within the U.S. government there were differences as to whether the United States should only retain the option to resume testing after the three-year period or actually commit itself in advance to such action.

Among the troublesome minor issues that remained was the number of unmanned seismic stations to be located in the United Kingdom. Although the United States and the Soviet Union had each agreed to accept ten such stations, the United Kingdom rejected the Soviet demand that it accept an equal number and argued for a much smaller number related to the size of its territory. Both sides considered this a matter of principle and refused to modify their positions.

A potentially serious issue that remained to be addressed was the definition of a "nuclear weapons test explosion." This issue, which had always been postponed in earlier attempts at a comprehensive test ban, is important, since tests of at least some significance for weapons development can be conducted in the laboratory at yields equivalent to tons or even a few pounds of high explosives. As a further complication, a continuing series of very low-yield nuclear explosions are the basis for the inertial confinement fusion programs that both countries are pursuing in the attempt to develop thermonuclear power for peaceful purposes. The negotiations did not address the question as to how these low-yield military and peaceful experiments, which are far below any verification threshold, would be treated under the treaty.

The successful ratification of SALT II might have generated the political will to permit rapid resolution of the few remaining issues that divided the two sides. Nevertheless, however these issues might have been resolved, the ratification of a comprehensive test ban treaty would certainly have been strongly opposed by influential groups, including the weapons laboratories and many retired military leaders. With the Soviet invasion of Afghanistan and the suspension of the SALT II ratification process, the prospects for agreement evaporated. The trilateral negotiations dragged on through the remainder of the Carter Administration, but nothing further of significance was accomplished. The negotiations recessed immediately after President Reagan's election and have not resumed.

The Reagan Administration

Reversing the position of all U.S. administrations since Eisenhower, the Reagan Administration has formally opposed as contrary to U.S. security interests further efforts at this time to achieve a comprehensive nuclear test ban. Speaking before the multilateral Committee on Disarmament in Geneva on February 9, 1982, Eugene Rostow, director of the U.S. Arms Control and Disarmament Agency (ACDA), stated that "while a comprehensive ban on nuclear testing remains an element of the full range of long-term U.S. arms control objectives, we do not believe that, under present circumstances, a comprehensive test ban could help reduce the threat of nuclear weapons or to maintain the stability of the nuclear balance." In March 1983, James George, acting director of ACDA, stated in a written response to questions at appropriation hearings that the United States requires continued nuclear testing for "the development, modernization, and certification of warheads, the maintenance of stockpile reliability and the evaluation of nuclear weapons effects." Other spokesmen have emphasized the problems with verifying a comprehensive test ban.

The Soviet Union has continued to advocate a comprehensive test ban. When it became clear that the trilateral negotiations would not be resumed by the Reagan Administration, the Soviet Union presented at the 37th UN General Assembly a document entitled "The Basic Provisions of a Treaty on the Complete and General Prohibition of Nuclear-Weapon Tests." This document was essentially an outline of the draft treaty that had been negotiated in the trilateral negotiations during the Carter Administration. The Soviet draft treaty banned all nuclear weapon test explosions in all environments. It was coupled with a coterminous moratorium on nuclear explosions for peaceful purposes unless an acceptable technique for conducting such peaceful explosions could be found. Verification was by National Technical Means supplemented by the international exchange of seismic data. In the case of suspicious events, any party could request an on-site inspection and would either be granted an "invitational" inspection or be given an explanation. Provision was made for special arrangements, such as those that had been previously worked out at the trilateral negotiations, to be made separately between parties concerning arrangements for on-site inspections. The treaty was to be of unlimited duration once ratified by all permanent members of the UN Security Council, but it could enter into force for a limited period if ratified by only the United States, the United Kingdom, and the Soviet Union.

On December 9, 1982, the United States was the only country to vote against a resolution calling for the Committee on Disarmament to continue the consideration of the issues and "to take the necessary steps to initiate substantive negotiations." The vote on this resolution was 111 to 1 with 35 abstentions. The United States continues to refuse to negotiate a treaty in the Committee on Disarmament, although it is participating in a committee working group to discuss verification and compliance issues relating to a comprehensive test ban.

In the absence of progress on a comprehensive test ban or other arms control treaties, the Reagan Administration came under strong pressure from Congress, particularly the Senate Foreign Relations Committee, to proceed with the ratification of the Threshold Test Ban and Peaceful Nuclear Explosions treaties, which had been on the calendar of the Foreign Relations Committee since 1976. In July 1982 it was reported that President Reagan had decided that the two treaties would have to be renegotiated to seek unspecified improvements in their verification measures before they could be ratified. The administration argued that the Soviet Union appeared to be testing over the threshold and that this problem could not be resolved satisfactorily even when the verification provisions in the two treaties were fully implemented upon ratification. When the United States sought to reopen the negotiations to modify the verification procedures, the Soviet Union refused to renegotiate the treaties on the grounds that they were adequate as signed by the heads of state of the two countries.

On January 23, 1984, President Reagan declared in a report to Congress entitled "Soviet Non-Compliance with Arms Control Agreements" that the Soviet Union "is likely to have violated the nuclear testing yield limit of the Threshold Test Ban Treaty." This charge was denied by the Soviet Union and was publicly questioned by a number of U.S. scientific authorities in the field. In September 1984, President Reagan proposed in his speech to the UN General Assembly that the United States and Soviet Union exchange visits of observers to each other's nuclear test sites as a way to resolve questions about the calibration of the yields of underground tests. Both sides still maintain their intention to continue to observe the threshold in the unratified treaties.

In the 1984 presidential campaign, Democratic candidate Walter Mondale attacked the Reagan Administration's repudiation of a comprehensive test ban and called for prompt negotiation of a treaty as part of his proposed package of arms control measures. To demonstrate the seriousness of the United States in the matter and to facilitate the negotiations, candidate Mondale pledged that upon taking office he would immediately initiate a temporary moratorium on all nuclear weapon tests. The proposed moratorium would last for a fixed period during the negotiations provided the Soviet Union also refrained from testing.

THE MAIN ISSUES SURROUNDING A COMPREHENSIVE TEST BAN

Overall U.S. Security

The specifics of the debate on a comprehensive nuclear test ban have varied over the years, but the underlying issue has been whether such a ban would be in the overall security interest of the United States.

The Reagan Administration and other critics of a CTB argue that such a treaty would not be in the U.S. security interest at this time in view of high-priority military requirements that can only be met by continued testing. They assert that these tests are needed to maintain the reliability of nuclear stockpiles, to develop state-of-the-art warheads for new systems, to develop a new third generation of highly sophisticated weapons for possible application in the Strategic Defense Initiative and other programs, to understand the effects of nuclear explosions on complex weapon systems, and to improve the safety of nuclear weapons. Critics of a CTB also argue that it would not be adequately verifiable because the Soviet Union could continue to test clandestinely at low yields below the threshold of seismic detection and even at higher yields using elaborate concealment techniques. Since they assume the United States would honor the ban while the Soviet Union would not, they assert that the Soviet Union could gradually improve its nuclear weapons capabilities relative to the United States by preventing the deterioration of the reliability of its stockpile. Moreover, some critics assert that with such clandestine testing, the Soviet Union might make significant improvements in its nuclear weapons that could affect the strategic balance.

Supporters of a CTB argue that achievement of such an agreement would be in the net interest of the United States because it would help slow down the qualitative arms race that threatens strategic stability. Moreover, they emphasize the contribution that such an agreement, which has been the worldwide symbol of progress in arms control for more than 25 years, would make toward creating an international environment conducive to non-proliferation, since it would eliminate the inherently discriminatory character of the present situation, in which nuclear weapon states continue to test while non-nuclear weapon states are forbidden to do so. These supporters also argue that verification capabilities are now clearly adequate to ensure that clandestine tests could only be conducted at such low yields that they would not contribute to Soviet nuclear weapon capabilities.

In response to opponents, these supporters of a CTB assert that such a ban would not endanger the effectiveness of any component of the U.S. strategic deterrent because nuclear testing is not in fact necessary to maintain stockpile reliability. They also assert that new warheads are not necessary for future delivery systems and that a case has not been made for a new or third generation of nuclear weapons except as a component of an accelerated arms race. Similarly, nuclear tests to understand the effects of nuclear explosions are not necessary except as part of a program that would accelerate the arms race.

Importance of Testing to U.S. Security

The debate over the importance of nuclear testing to national security has shifted over time, as the technology of nuclear weapons matured and the United States lost the substantial technical lead it initially had in the field. Opponents of a CTB have always emphasized the value to U.S. military capabilities of continued testing; supporters of a ban have emphasized the net advantage to the United States of a freeze on both sides' nuclear weapon technology as compared with Soviet and U.S.

gains from continued testing. In the 1960s the technical debate focused primarily on the extension of existing technology to develop a range of strategic and tactical weapons with improved yield-to-weight ratios and to help understand sophisticated weapon effects relating to the anti-ballistic missile programs then under discussion. Opponents of a CTB also emphasized the importance of developing weapons with enhanced neutron emission for use as antitank and antipersonnel weapons in Europe. By the time of the Carter Administration the focus of the debate had shifted to the need for testing to assure the reliability of stockpiled weapons. Today the focus of the debate has expanded from the issue of weapon reliability to include the requirement for a third generation of highly sophisticated weapons. In addition, more emphasis has been placed on requirements for continued tests to determine the effects of nuclear weapons, to develop new state-of-the-art warheads optimized for new weapon systems, and to improve weapon safety.

Reliability of Stockpiled Weapons

Opponents of a CTB have particularly emphasized the problem of maintaining the reliability of the aging stockpile of nuclear weapons. With the passage of time, corrosion and other effects of aging can reduce the yield of a weapon or even cause it to fail completely. Reliability is normally monitored by nondestructive checks or by testing without a nuclear detonation. However, CTB opponents from the weapons laboratories and the military argue that circumstances could arise where these techniques would not reveal the serious impacts of certain aging effects. Thus, a nuclear test would be required to confirm the performance of an existing weapon or to certify a modification or replacement to correct the problem. Moreover, these opponents point out that these defects, which are not random and may occur after a certain period of time in weapons of a particular design, could over a relatively short period render a significant fraction of one leg of the deterrent force inoperative, since the same type of nuclear weapon is often used on all delivery vehicles of a particular type.

Opponents frequently cite two historical incidents to illustrate their point: the malfunction of the Polaris A1 warhead, and a problem relating to the primaries of thermonuclear weapons. The Polaris A1 warhead as originally designed contained a mechanical safety device to prevent a low-order nuclear explosion in case the high-explosive component was accidentally detonated by fire or shock. After a number of years the safety devices began to jam, so that the weapons would have given a greatly reduced yield if they had actually been fired. Opponents recall that modification and replacement of the safety device failed to resolve the problem satisfactorily, and the warhead was eventually replaced with a new improved warhead that required testing. In the case of the thermonuclear primaries, it was discovered through a test that the natural decay of tritium had reduced the yield of the primary (the initiator of the explosion) in a thermonuclear device more than had been calculated and that as a result after a period of time the device failed to operate correctly and gave a greatly reduced yield. Further tests were needed to determine the acceptable shelf life of such weapons. Opponents argue that in both cases nuclear tests were necessary either to resolve the problem or to obtain promptly the most desirable solution to the problem.

In the long term, opponents argue, the aging of components will inevitably require substitution of parts or refabrication of the weapons. However, due to changing manufacturing procedures and available materials, slight changes would gradually be introduced into the designs over time, gradually reducing confidence in the weapons performance. In these circumstances, opponents argue, confidence in the reliability and effectiveness of the deterrent force would also gradually decline.

Supporters of a CTB argue that, while stockpile reliability is obviously a very important problem, nuclear testing is not required to deal with it satisfactorily. Weapons undergo frequent nondestructive tests to ensure proper operation of all components. If there is a cause for concern, the entire weapon except for the active nuclear component can be detonated and sensitive instrumentation can determine whether the physical conditions necessary to initiate a nuclear detonation had been created. It is asserted that this type of quality control will identify any deterioration in materials or components or any malfunction in the operation of the weapon. If problems are discovered, components can be replaced or, if necessary, the entire weapon can be refabricated according to the precise specifications that were originally judged satisfactory for the certification of the weapon's reliability.

Historically, nuclear testing has been used very rarely to resolve reliability problems in proven designs, supporters of a CTB point out. In the few cases where nuclear tests were conducted for reliability purposes, they were not in fact necessary to ensure reliability of the weapons but allowed a quick or optimized resolution of the problem. With regard to the Polaris A1 warhead, supporters observe that the mechanical problem could have been solved by enforcing production standards and periodic replacement. The need for testing arose when a new design that was inherently safe and did not require a safety device was substituted to eliminate the source of the problem. With regard to the problem of thermonuclear primaries, supporters point out that this unique situation arose from the fact that the primaries contained tritium, which decays relatively rapidly (it has a 12-year half-life). This decay leads to a reduced concentration of tritium, a critical component in the weapon design. When a question of shelf life arose because of this well-known tritium decay, it could always have been resolved conservatively at some cost and inconvenience by maintaining a reasonably fresh supply of tritium in the warheads. In any event, supporters argue, this unique situation of rapid and predictable aging is now understood and compensated for, and there is no reason to think a comparable situation will arise with existing weapons. Looking 20 or 30 years into the future, supporters assert that stockpiled weapons can be periodically refabricated according to original specifications on a carefully preplanned schedule. If any genuine concerns exist as to the inability of obtaining identical materials or components in the future, this problem can be solved by long-range planning of procurement and stockpiling.

Supporters of a CTB also point out that the reliability currently required of nuclear weapons far exceeds the reliability of delivery systems. The nuclear component of a weapon system is expected to have almost perfect reliability, while the reliability of a major missile system is often estimated in the 80 to 90 percent range. Supporters also point out that the hypothetical problem of having all the warheads on one leg of the strategic triad become inoperable due to a common failure would be eliminated by having some mix of existing warheads on each major delivery system.

Some supporters of a CTB would acknowledge that after a generation without any nuclear testing, political and military leaders might well have less personal confidence in the stockpile regardless of its intrinsic reliability. They argue, however, that this would in fact be a stabilizing factor, since a preemptive counterforce strike demands the highest confidence in the reliability of the strike force while the deterrent value of the retaliatory force will always remain even if there is some putative uncertainty as to the reliability of individual weapons, or even types of weapons. Moreover, the deterrent value of the U.S. retaliatory force would be reinforced by the fact that an adversary could not possibly have detailed information on the reliability of U.S. nuclear weapons and would have to assume that they would function as designed.

Development of New Weapons

Opponents of a CTB argue that future weapon systems will require specially designed nuclear warheads that will have to be tested even if the warheads use state-of-the-art technology. These new warheads will be necessary to optimize the effectiveness of the particular delivery system, they point out, since existing warheads were designed to optimize other systems. New or modified designs will also be necessary to meet specific physical constraints of new systems, such as weight, volume, center of gravity, or ability to withstand high accelerations. Consequently, the use of existing warheads could significantly reduce the potential effectiveness of future delivery systems. In some cases, the use of existing warheads may not even be technically possible.

Opponents also argue that nuclear weapons technology is far from being mature and that a new or third generation of nuclear weapons could be of tremendous military significance. They foresee nuclear weapons that would use special effects to extend the power of nuclear explosions to great distances. In connection with President Reagan's Strategic Defense Initiative, some opponents of the CTB have focused particular attention on the potential of an X-ray laser, pumped by the explosion of a thermonuclear weapon, as a kill mechanism against attacking ICBM boosters and RV busses. In this concept, the X-ray radiation from an exploding thermonuclear weapon would power an external laser device that could project a narrow cone of intense soft X-ray radiation to distances of thousands of kilometers in space. Other exotic applications that use nuclear explosions as a source of power in space are also under consideration as part of the Strategic Defense Initiative. These opponents argue that a CTB would clearly preclude these and other undiscovered concepts that could prove to be critical components of a truly effective ballistic missile defense system. They maintain that these new concepts illustrate the potential of further nuclear testing to contribute to the security of the United States in fundamental and revolutionary ways.

Supporters of a CTB agree that it would indeed impede the continued qualitative improvement of nuclear delivery systems and would prevent some developments entirely. But they argue that this is precisely why a CTB is relevant to controlling the U.S.-Soviet arms race. Moreover, the constraints imposed by a test ban would tend to stabilize the strategic positions of the two sides, since they would discourage the optimization of counterforce systems without altering the inherent deterrent capabilities of current or future retaliatory forces.

With regard to the specific problem of providing warheads for planned or foreseeable delivery systems, supporters point out that for better or worse these systems could use existing tested designs without significantly changing the basic missions involved. This might involve some adjustments in asserted requirements or even in system design to accommodate existing warheads. In the past, they point out, requirements were stated and systems optimized on the premise that warheads could be treated by the military as a free good, separately funded and ordered to specification. In the future, assuming that the arms control regime permitted new systems, the availability of acceptable warheads would have to be treated as one of the fixed parameters of the system. Supporters argue that this inconvenience would discourage unnecessary improvements but would not prevent any high-priority developments that were really needed to protect the survivability of the deterrent. Supporters also argue that most deterrent systems could now be diversified by equipping them with more than one type of nuclear weapon. Thus, in the unlikely case that a defect would remain uncorrected, only a fraction of that delivery system would be affected.

Finally, these supporters argue that a CTB would make a major contribution to stopping the arms race by effectively blocking a third generation of radically new nuclear weapons, since such weapons would clearly require extensive and prolonged testing. Although some of these supporters of the CTB strongly question the technical status, practical prospects, and military significance of the concepts advanced so far, they emphasize that the initiation of such a program with grandiose claims would open a major new front in the nuclear arms race. Consequently, the ability of a CTB to contain this development, according to supporters, again illustrates its importance to containing a U.S.-Soviet arms race in space. Moreover, attempts to develop such systems, whether successful or not, would lead not only to the termination of the SALT I ABM Treaty but to the abrogation of the Limited Test Ban Treaty as well, since such applications would eventually require testing in space.

In short, supporters of a CTB argue that it would preclude a range of weapons developments that could contribute to the acceleration of the nuclear arms race without precluding future weapons systems that might contribute to the survivability of the deterrent.

Nuclear Effects

Opponents of a CTB argue that there is an important military requirement for better understanding of the effects of nuclear explosions on various critical U.S. military systems. Although a great deal of experimental data and calculations exist on this subject, opponents point out that it can be extremely difficult to assess the ability of particular components or overall systems to function in a nuclear environment without actual experiments. Components such as transistor circuits, sensors, or guidance systems can be affected in ways that would cause an entire system to malfunction. For this reason, they argue, tests of nuclear effects must be continued to protect the reliability of the deterrent force.

Some opponents of a CTB go further and emphasize the potentially critical importance of the complex electromagnetic effects of nuclear explosions in space on the atmosphere and on systems in space and on earth. These effects could have a major impact on communications, on radars, and on any satellite systems in space. Until these effects are better understood, opponents argue, there will be serious questions about the survivability of the command and control system for the strategic forces on which deterrence depends. Opponents also emphasize that knowledge of these effects may play a critical role in developing an effective ballistic missile defense system. On the one hand, it may be possible to exploit such effects to the advantage of the defense. On the other hand, if not understood and successfully countered, these effects could cause the complete collapse of the system, because radars, other space- and ground-based sensors, and communications on which the system depended could fail catastrophically in a nuclear environment even though they were not directly attacked. This argument brings into question the Limited Test Ban Treaty, since such tests would generally have to be conducted in space.

Some opponents of a CTB also emphasize the need for full-scale tests to determine the actual hardness of missile silos to blast and other effects in order to assess the survivability of the land-based ICBM force. This argument again brings into question the Limited Test Ban and the Threshold Test Ban treaties, which would ban most tests of this type.

Supporters of a CTB agree that it would limit the information on nuclear effects that could be obtained and reinforce the already severe constraints on assessing the potential effects of high-altitude explosions imposed by the Limited Test Ban Treaty. However, they point out that the tests permitted under the Limited Test Ban would add little additional useful effects information that cannot be calculated or determined by nonnuclear sources, such as X-ray generators, particle accelerators, high-explosive impulses to simulate X-ray shock, and other techniques. They assert that this nonnuclear approach would provide sufficient information to assess the vulnerability of existing warheads and delivery systems to nuclear radiation and to ensure adequate communications to release retaliatory strategic forces.

Supporters of a CTB argue that the limitations on high-altitude tests significantly constrain the development of ballistic missile defense systems, which they believe are essentially destabilizing and a major stimulant to the nuclear arms race. As long as these high-altitude effects are very uncertain, they argue, it would be difficult to justify proceeding with a ballistic missile defense system in which little confidence could be placed. Even though some supporters are confident that more information on this subject would not improve the prospects for an effective defense system, they argue that pursuit of such a test program would prove a provocative stimulant to the arms race. Each side would fear that the other side had developed new information that might make its defense system more effective.

With regard to the determination of the precise hardness of missile silos to overpressure and other nuclear effects, supporters argue that this is not a critical issue in the survivability of the U.S. strategic deterrent. Moreover, full-scale atmospheric tests or even underground tests with very large yields, which would be required to advance substantially the current state of knowledge, are already banned by the Limited Test Ban and Threshold Test Ban treaties. The underground tests permitted under the Threshold Test Ban would not yield significant data that could not be obtained by sophisticated high-explosive tests of the type already conducted by the United States, according to supporters. In any event, the exact level of hardening is not a critical factor in assessing the credibility of the deterrent, since the extremely high missile accuracies that have already been achieved and the further improvements that can be expected will make all silos vulnerable to attack. Moreover, with the demise of the "dense pack" concept of MX deployment, there is no current need for superhardening of silos.

Finally, supporters emphasize that while some tests of nuclear effects on components can be carried out in the laboratory, a CTB has an equal impact on such tests for both the United States and the Soviet Union.

Safety

Some opponents of a CTB argue that it would interfere with efforts to improve the safety of nuclear weapons in the event of an accident involving nuclear weapons. Although such an accident or terrorist attack would not produce a nuclear explosion, it could cause the high-explosive component of a nuclear weapon to go off, dispersing several kilograms of dangerously radioactive plutonium in the general vicinity. Such accidents can present a local health problem with potentially serious political implications. There have been several such accidents, including one at Palomares, Spain, where a bomber loaded with nuclear weapons crashed and contaminated a small community. Opponents point out that this problem can be eliminated by substituting special insensitive high-explosive components that will not detonate from the shocks and heat experienced in accidents or terrorist attacks. The substitution of the new insensitive high explosives will require retesting of the improved weapons, since differences in the burning characteristics of these high explosives require minor modifications in design.

Supporters of a CTB argue that, while improved safety is certainly desirable, the dispersal of plutonium is a relatively minor problem compared with the contribution a CTB would make to controlling nuclear weapons. Moreover, they observe that this problem has been recognized for some time and that any modifications that have not already been made could be incorporated and tested before a ban on nuclear weapon tests entered into effect.

Peaceful Nuclear Explosions

Some opponents of a CTB argue that such a ban would preclude the development and use of peaceful nuclear explosives, which they claim hold great economic promise. This was a major issue during the 1960s, when the U.S. weapons laboratories were widely proclaiming the tremendous contributions PNEs would make in such diverse fields as gas and oil production, mining, electric power generation, and large-scale earth moving. Earth moving on a grand scale appeared the most immediate and dramatic application, and serious study was given to such proposals as constructing an alternate sea level canal to the Panama Canal and creating a new harbor in Australia. A major development effort, Project Plowshare, was undertaken to explore these applications. But with a more realistic assessment of the technical and political problems associated with these projects as well as their economic prospects, the early enthusiasm in the United States for PNEs waned.

By the 1970s the Soviet Union had become the principal advocate of PNEs. It expressed serious interest in large earth moving projects to divert rivers so that they would flow south to the Caspian Sea instead of north into the Arctic Ocean, and a number of tests of various other applications were conducted. As a consequence of this interest, the Soviet Union, in the trilateral negotiations during the Carter Administration and in its recent treaty outline, called for a ban on all military nuclear explosions but only a moratorium on PNEs until a satisfactory method could be found for preventing such activities from being used for military purposes. It is not clear whether the Soviet move was simply an effort to finesse internal pressure for PNEs or a serious effort to hold open a PNE option.

Supporters of a CTB generally agree that such a treaty would prevent further progress in PNEs, since it does not appear possible as a practical matter to distinguish PNE developments from activities of potential military significance. (A few supporters of a CTB suggest that a mutually satisfactory technique for continuing PNE activities might be developed in the future.) All supporters of a CTB contend that PNEs are not sufficiently important to economic development to be allowed to interfere with a CTB treaty that would contribute significantly to the prospects for peace. Moreover, many supporters challenge the economic claims of PNE advocates. They also argue that many PNE applications, such as major earth moving projects, would be politically unacceptable in today's world because of the associated nuclear fallout and other effects, such as blast and earth shock.

Finally, supporters of a CTB point out that efforts to accommodate a future PNE option in a CTB treaty undercut the non-proliferation value of the treaty. Some potential nuclear weapon states seized on advocacy of PNEs by the United States during the 1960s as a reason for seeking an independent capability to produce nuclear explosives. For example, the Indian government claimed that its first nuclear explosion in 1974 was actually part of a PNE program. With this history in mind, supporters of a CTB argue that continued efforts on the part of the nuclear powers to retain a PNE option will provide potential nuclear weapon states with a rationale for keeping open a nuclear explosives option on the grounds that it might be needed for a PNE program.

The Weapons Laboratories

Opponents of a CTB argue that such an agreement would seriously weaken U.S. weapons laboratories without having a comparable effect on Soviet laboratories. They assert that the agreement would deny U.S. weapons designers the opportunity to test their ideas and products. As a result, morale at the weapons laboratories would suffer, gifted weapons designers would leave, and new ones could not be recruited or trained. This would seriously impair the ability to resume vigorous weapons programs in case the agreement ended. While granting that such factors would to some extent also affect Soviet laboratories, opponents assert that the Soviet Union would find ways to ensure that key personnel remain at the weapons laboratories and continue to make significant contributions. Moreover, the potential for cheating through low-yield clandestine tests could give Soviet laboratories much greater opportunities to test new ideas and maintain a vigorous research and development program. Some opponents of a CTB argue that leadership in nuclear weapon developments has been a major element in U.S. military strength, offsetting other Soviet military advantages, and that any weakening of the weapons laboratories undermines the source of this U.S. technological advantage.

Supporters of a CTB argue that it is the purpose of the treaty to slow down to the extent possible all nuclear weapon developments. On balance, they assert that it is not obvious that Soviet weapons laboratories would have a significant advantage in such an environment. While weapons research would clearly be curtailed, U.S. laboratories have a major advantage in their greatly superior computational equipment and their extensive experience with simulated nonnuclear effects tests. Moreover, in contrast to the Soviet laboratories, the two U.S. weapons laboratories have highly diversified research programs, with roughly one half of their present activities outside the nuclear weapons area. In addition, many of the technical and analytical problems in the peaceful inertial confinement fusion programs of the two laboratories overlap problems encountered in weapons design. Such programs should help maintain the skills of weapons designers while engaging them in interesting and important work. Finally, scientific productivity cannot be coerced, supporters of a CTB argue, even though the Soviet Union may be able to inhibit the departure of key scientific personnel.

Verification

General

Verification has been a central issue in the CTB debate since the mid-1950s. The underlying question has been, What would be the military significance of testing that could be conducted clandestinely beneath the threshold of the monitoring system? Most CTB opponents have argued that activities of military significance could be clandestinely conducted by the Soviet Union under a CTB. In contrast, most CTB supporters have argued that there would be adequate verification to ensure that the Soviet Union or other countries were not conducting nuclear tests that could have any real military significance.

Nuclear explosions are unique events. The large amount of energy and forms of radiation they generate produce a variety of physical phenomena that can be detected at great distances. From the beginning the debate on verification focused primarily on underground tests, since tests in the atmosphere, oceans, and space appeared to be adequately verifiable. Early tests were all in the atmosphere, and by the late 1940s techniques were developed to detect even low-yield tests by their acoustic signals and unique radioactive debris. Today these tests can be monitored very effectively from satellites through the visible and near-infrared light emitted from the explosion. Underwater nuclear explosions can be monitored to very low yields, far below the threshold of underground tests, by existing acoustic sensors associated with anti-submarine warfare systems. Nuclear explosions in space at vast distances can be monitored from satellites by their characteristic X-ray emissions.

Underground tests present a more serious technical challenge. The seismic signals they produce must be not only detected but distinguished from a large background of seismic signals from natural earthquakes whose numbers rapidly increase at lower magnitudes. The identification process benefits from the fact that the seismic signals of explosions differ in a number of significant ways from those of earthquakes. Explosions are a point source of energy in space and time, while earthquakes result from the slipping of faults over a considerable distance. Over the last 25 years there has been a major effort to understand this problem and to improve the capabilities to monitor underground tests.

A number of seismic techniques that were available from the beginning of the debate have been refined. These include seismic determination of the location and depth of the event and the "first motion" of the initial compression wave transmitted through the earth. Over the years a number of additional techniques have been developed that depend on the differing spectrums of the seismic signals from explosions and earthquakes. Specific criteria are sometimes difficult to formulate, but many seismologists believe that the overall spectrum and complexity of seismic signals clearly differentiate the two types of events.

Experts, including those opposed to a test ban, generally agree that these collective criteria can separate explosions from earthquakes and identify suspicious events, but assessments of the threshold of identification and level of confidence differ. Location can normally be determined with confidence to within 25 km. Depths can be determined to within 15 km, which is greater than current drilling capabilities. Most earthquakes (more than 90 percent) are either located in the deep ocean or more than 30 km underground, which automatically eliminates them from concern. Of the earthquakes in the Soviet Union and its coastal waters, 75 percent are in or near the Kamchatka Peninsula and the Kurile Islands and tend to have very deep focuses and to be offshore. With these events eliminated, an estimated average of around 100 earthquakes per year occur in the Soviet Union with body wave magnitudes greater than 3.8 on the Richter scale (equivalent to less than a 1-kt explosion in hard rock).

Originally it was proposed to identify these residual earthquakes by the so-called first-motion criteria. An explosion sends an initial compression wave through the earth in all directions, since it compresses the surrounding medium symmetrically. An earthquake, which is generated by a slipfault, sends initial compression waves in some directions and rarefactions in others. This provides a very powerful technique for differentiation provided there are enough stations in the monitoring system and the signal is sufficiently strong to be recorded with confidence.

The discriminants that compare various segments of the seismic spectrum take advantage of the fact that earthquakes put a substantially larger fraction of their energy into modes other than compression waves because of the complex nature of their sources. In exploiting this well-established phenomenon, particular attention has been given to comparing the magnitudes of a given event as measured by surface waves and by body waves. An explosion with the same body wave magnitude as an earthquake has a much smaller surface wave magnitude, since much less of its energy goes into surface waves. This same general phenomenon leads to a number of more qualitative criteria that are very persuasive to most seismologists.

The coupling of energy from a nuclear explosion to its surroundings creates another problem. The seismic monitoring system measures the seismic magnitude, not the yield, of an explosion. The yield equivalent to a seismic signal must be estimated by calibration shots or calculations. The coupling factor depends not only on the immediate medium in which the explosion occurs but also on the general geologic location of the event. In general, equivalent yields are referred to hard rock, which gives the best coupling of energy between an explosion and the surrounding medium. The poorest coupling occurs in deep dry alluvium deposits, which may have a coupling factor only one-tenth that of hard rock. If a nuclear test occurs in a sufficiently large cavity that does not collapse during the explosion, it is theoretically possible to reduce the coupling by a factor of as much as 100. In addition, there are regional biases in the coupling of seismic signals. For example, most seismologists believe that seismic body waves from the shots at the U.S. test site in Nevada, where the United States obtains its calibration data to relate magnitude and yield, are by virtue of the regional geology less well coupled than are Soviet explosions in the Semipalatinsk test area. If correct, this would cause a systematic overestimation of the test yields at Semipalatinsk. However, in the absence of reliable calibration data, the precise magnitude of such an effect remains uncertain.

By visiting the site of a suspicious seismic event, it is possible to obtain direct or indirect physical evidence of a nuclear explosion. From seismic data alone the location of the event is uncertain to within 10 to 25 km. However, information from other National Technical Means may focus attention on a specific location because of surface collapse, scarring, or evidence of unusual human activity. On the ground, investigators might find telltale traces of escaping radioactive gases or circumstantial evidence of surface fracturing or human activity that would lead to further efforts to obtain definitive samples of radioactive debris.

Threshold of Detection and Identification

Opponents of a CTB argue that tests with significant yields could be tested under that threshold of identification or even the threshold of detection. Even without sophisticated evasion techniques, according to some opponents, the Soviet Union could be confident of successfully conducting explosions with yields in the 10- to 20-kt range by testing in dry alluvium, a light sandy material. They point out that thresholds of a kiloton or so assume hard rock coupling. Some opponents also argue that by firing shots in large underground cavities, the Soviet Union could clandestinely conduct tests with yields up to 100 kt.

Supporters of a CTB argue that the system that the Soviet Union appeared to be prepared to accept would have an effective threshold of identification of around 1 kt. They assert that in hard rock the threshold would actually be less than 1 kt. While agreeing that equivalent yields would be greater in other media, they assert that the extreme case of dry alluvium is misleading since this material does not exist in the Soviet Union at depths necessary for clandestine testing above a kiloton. Moreover, they point out that even when fully contained, shots in alluvium and other media that couple less well than hard rock at the Nevada test site leave distinctive subsidence craters that can easily be identified from the air. By properly distributing 10 or 15 unmanned seismic stations within the Soviet Union, according to supporters, the threshold of identification for explosions in any medium, including alluvium, would be a kiloton or less.

Supporters of a CTB argue that the threat of evasion posed by "big hole" decoupling has been greatly exaggerated by opponents of a CTB treaty. They point out that the cavity required to decouple a 100-kt shot would have to be 150 m in diameter at a depth 2 km below the surface, an unprecedented engineering project in hard rock. A more practical approach would be to create a cavity in a salt dome by solution mining, although such cavities would be restricted to much smaller sizes. However, these supporters assert that the only suitable salt domes in the Soviet Union are located in the Caspian Sea area and could be specially monitored by a few properly located unmanned seismic stations. These stations, which because of their location could record close-in seismic signals, would be extremely effective since any seismic disturbance in this aseismic area would be a suspicious event. With such a monitoring system, some of these supporters argue, clandestine tests could not as a practical matter be successfully concealed in these salt domes at yields much above a kiloton.

Hiding in Earthquakes

Some opponents of a CTB contend that it would be possible to test clandestinely by carrying out the test during a major earthquake so that the seismic signal from the explosion would be lost in the earthquake's extended seismic disturbance. They assert that it would be possible to design a test so that a device with a yield far above the threshold of the monitoring system could be held in readiness for an extended period and fired at the proper moment to avoid separate detection.

Supporters of a CTB dismiss this as a serious clandestine testing technique. A test program conducted by holding devices in readiness for months or years to be fired on one or two minutes notice after the initial detection of an entirely unpredictable event is not credible, they maintain. Moreover, in the very unlikely event that such a clandestine test were attempted, it would be extremely difficult to hide from the proposed monitoring system if it were significantly above the threshold. Supporters point out that it would be virtually impossible for the Soviet Union to conduct such a test near the origin of a large earthquake. The only area of the Soviet Union that has such earthquakes with reasonable frequency is the Kamchatka Peninsula and Kurile Islands region, where seismic events can be carefully monitored by nearby seismic equipment in Japan and Alaska, by very sensitive underwater seismic arrays, and by any unmanned seismic stations located in the immediate area as part of the monitoring system in the Soviet Union. Since the test would have to be hidden in the signal from a distant earthquake, it would be very difficult to match the exact timing and magnitude of the test with the arriving earthquake signal. Even if this extremely difficult task were successfully accomplished, supporters assert, there is a good chance that the test would still be identified as a separate, and therefore very suspicious, event.

Nonseismic Information

Some opponents of a CTB argue that nonseismic sources of information cannot be counted on to help verify compliance. They point out that if the Soviet Union attempted to evade the agreement, it would be extremely careful to avoid obvious activities that might arouse suspicion in advance or help identify a test after the fact.

Supporters of a CTB argue that nonseismic sources of information would contribute significantly to verification capabilities. Satellite photography would contribute in many ways to determining whether an unidentified seismic event was in an area of other suspicious activities. Such reconnaissance could narrow, or even pinpoint, areas of concern for an on-site inspection. For example, efforts to construct a big hole for decoupling purposes or the subsidence crater from a shot in alluvium or other media would be easily identified by such observations. These supporters emphasize that the deterrent value of the entire intelligence operation should not be underestimated since the Soviet Union could not be certain that information suggesting the evasion of a test ban would not come to the attention of U.S. or allied intelligence from technical or human sources.

On-Site Inspection

Opponents of a CTB tend to argue that the right to a substantial number of on-site inspections is essential to a verifiable agreement. Some opponents take the position that this is the only way to prove a violation. Others assert that individual on-site inspections would be so ineffective that even substantial numbers would have little chance of discovering a clandestine test program. The seismic monitoring system would only locate an event to within 10 to 25 km, they point out, and a deeply buried shot would produce little, if any, surface disturbance.

Opponents argue that on-site inspections must be mandatory to be useful. Invitational inspections of the type envisaged in the trilateral negotiations during the Carter Administration would be of little or no value, they assert. It is unrealistic to imagine that a country would invite an inspection if there had actually been a test unless it was certain that the test could not be discovered.

Supporters of a CTB argue that on-site inspections, even if they require an invitation after a challenge, are a useful deterrent to clandestine testing. However, they disagree that an on-site inspection is needed to make the case that a nuclear test has in fact occurred since seismic criteria are capable of identifying most explosions as explosions above the threshold with high confidence. While agreeing that an invitation would be unlikely in the case of an actual test, these supporters argue that if serious suspicions existed, the denial of an invitation without a very persuasive explanation would indicate that a violation had in fact occurred. Consequently, there would be a strong motivation to grant an invitational inspection to clear the record if, in fact, there had not been a clandestine test.

Supporters contend that individual inspections could be very effective in resolving specific suspicious events. While seismic means alone could only locate the event to within 10 to 25 km, satellite reconnaissance could identify surface subsidence or suspicious human activities that could focus the inspection process. During the inspection, the presence or absence of traces of unique radioactive gas and of characteristic surface disturbances could adequately resolve suspicious cases.

Significance of Testing Below the Threshold

Opponents of a CTB argue that even below a kiloton—the threshold claimed by many CTB supporters—important weapons work could still continue on tactical weapons, weapons effects, and the physics of weapons design. They also contend that at a threshold of 10 to 20 kt, which many opponents believe is a more realistic estimate, it would also be possible to carry out some important reliability tests that could help maintain confidence in the stockpile. Opponents would also argue that if the threshold can be pushed up to 25 to 100 kt with big hole decoupling or by hiding tests in earthquakes, it would be possible to make major developments in certain types of nuclear weapons and test the reliability of any nuclear weapon regardless of its design yield. In this connection, some opponents would note that, while a major development program would require many tests, only a few tests would be required to maintain confidence in stockpile reliability. Opponents also emphasize the importance of any of these test programs to sustaining the effectiveness and responsiveness of Soviet weapons laboratories. In this way, the Soviet Union could have an option to resume testing openly with an asymmetric advantage.

Supporters of a CTB argue that clandestine testing below a kiloton or even several kilotons would not contribute to existing Soviet weapons capabilities. It would also not give the Soviet Union an asymmetric advantage in reliability testing, they continue, since the operation of most warheads, including all of those on strategic systems, could not be confirmed at these low yields. Moreover, supporters question whether such testing would give the Soviet laboratories any real advantage in maintaining interest and morale, since U.S. laboratories would have opportunities to continue work with computers and simulated nonnuclear effects and to carry out related peaceful research activities. Supporters dismiss as unrealistic the possibility that testing at yields much above a few kilotons could be carried out and note that many tests would be required to conduct a sustained weapons development program. Consequently, these supporters conclude that a clandestine program would not have any effect on the military balance.

Significance of Very Low Yield Tests

Some opponents of a CTB argue that nuclear tests of military interest can be conducted at such low yields, a few tons or less, that there is no possibility of detection. They assert that these experiments, which can be conducted at a laboratory, are useful in studying the physics of nuclear explosions, in advancing safety, and in maintaining laboratory competence and interest. They also point out that, if taken literally, a complete ban on all nuclear explosions would also ban the peaceful inertial confinement fusion program, which involves a continuous process of tiny explosions and possibly other approaches to controlled thermonuclear power as well. It is a fundamental error, they contend, to include in a ban a category of activities that is clearly unverifiable.

Supporters of a CTB argue that this is not a real problem affecting U.S. security interests since these tests would not significantly advance either side's nuclear weapons program. The problem can be handled either by a formal definition of nuclear explosions at a very low level or by an unstated *de minimis* interpretation that would effectively exclude very low level laboratory activities from the ban. Supporters add that the history of the CTB negotiations makes it clear that neither side had any intention of banning the inertial confinement fusion program or other efforts to develop fusion or fission reactors.

Impact on Nuclear Proliferation

Opponents of a CTB argue that it would not be an important factor in non-proliferation. The potential nuclear weapon states of real concern would not sign such a treaty, they assert, since these states wish to maintain a nuclear weapons option. These countries have their own security concerns that have little or nothing to do with whether the United States and Soviet Union are continuing to improve their nuclear weapons. The pressure to join a CTB is further reduced by the fact that France and China most likely will not join such a treaty in the foreseeable future.

Opponents also argue that potential nuclear weapon states will be able to develop a nuclear weapons option and even stockpile simple nuclear weapons without testing. In support of this contention, they point out that Israel has apparently been able to develop a first generation nuclear weapon without a test. Other countries may follow this example even if they join a CTB or are indirectly constrained by it from testing.

Supporters of a CTB argue that it would be a key factor in creating a nuclear regime conducive to non-proliferation. Many non-nuclear weapon states have consistently and bitterly complained about the discriminatory nature of the present international nuclear regime. These states particularly object to the continuation of nuclear testing, which they see as a symbol both of the threat of nuclear war and of the inequitable nature of the Non-Proliferation Treaty. Supporters argue that a CTB treaty would go a long way in the eyes of the non-nuclear weapon states to meeting the obligation that the United States and the Soviet Union undertook in Article VI of the NPT "to pursue negotiations in good faith on effective measures relating to cessation of the nuclear arms race at an early date."

Many potential nuclear weapon states would sign a CTB treaty, according to supporters. Even those that chose not to sign would be under greatly increased international, and in some cases domestic, pressure not to undertake nuclear testing. Thus France, a nonsignatory to the Limited Test Ban Treaty, discontinued atmospheric testing in 1974. Despite the apparent example of Israel, these supporters contend that it would be much more difficult technically and politically for most countries to develop a nuclear weapons capability without testing. This would be particularly true in countries without the technical expertise of Israel or in which the military did not have confidence in a relatively inexperienced scientific community. In any event, it would be extremely difficult for present non-nuclear weapon states to go beyond relatively primitive first generation nuclear fission weapons without testing, and it would be impossible for them to develop thermonuclear weapons.

8 Non-Proliferation of Nuclear Weapons

INTRODUCTION

There is general agreement in the United States that the acquisition of nuclear weapons by additional states would be contrary to the security interests of the United States and the world at large. Most other states share this view. Despite this broad international consensus, the development of an effective regime to prevent the spread of nuclear weapons presents many difficult practical problems. This chapter discusses the specific measures that presently contribute to the non-proliferation regime and the issues associated with them.

The underlying issues associated with the proliferation of nuclear weapons have both political and technical dimensions, which are discussed in the first part of this chapter. The motivation to acquire nuclear weapons reflects a state's deepest fears and ambitions. Given sufficient political incentives, most states could in time produce a nuclear weapon, if they obtain the necessary technical talent, know-how, and materials. A peaceful nuclear power program can provide many of the technical assets that could contribute to a nuclear weapons program. The technical component of the non-proliferation regime seeks either to deny critical technical assets to potential nuclear weapon states or to assure that these assets are only made available under safeguarded conditions. Technical controls alone, however, can delay but not prevent a state from obtaining a nuclear weapons capability if it judges this to be in its overriding political interest. The political component of the non-proliferation regime seeks to create both an international environment and specifically targeted incentives and disincentives to discourage states from making this decision.

At the end of World War II the United States had a monopoly on nuclear weapons and held most of the related know-how and technology. Today there are five nuclear weapon states and many states with extensive nuclear power capabilities. The United States acting alone obviously cannot prevent nuclear proliferation, since there are now many suppliers of nuclear equipment and materials and importers with many different political, economic, and military perspectives. Efforts to establish an international non-proliferation regime therefore involve a complex of unilateral, bilateral, and multilateral undertakings. The practical problems associated with these efforts reflect the underlying issues and the priority that participating states are prepared to give to non-proliferation. The second part of this chapter briefly reviews the history of the proliferation problem and efforts to deal with it.

The overlap of the technologies of peaceful nuclear power and nuclear weapons has created a potential conflict between worldwide interest in extending the benefits of nuclear power and concern over further proliferation of nuclear weapons. This dilemma underlies the continuing debate as to whether U.S. export policy should be based on "denial" of nuclear capabilities that might be used in a nuclear weapons program or on "constructive engagement" with other states developing nuclear power to encourage them to accept an international non-proliferation regime. The third section in this chapter reviews the development of these approaches and the rationales and issues involved with them.

The Non-Proliferation Treaty and the Treaty for the Prohibition of Nuclear Weapons in Latin America, which underlie the international non-proliferation regime, are efforts to catalyze and codify national decisions not to acquire nuclear weapons or assist others in doing so. The fourth part of this chapter addresses the role and future prospects of these treaties and identifies the principal issues that have arisen among the nuclear weapon states and the non-nuclear weapon states in defining and developing this international regime.

International safeguards are the principal means of verifying compliance with the two major multinational non-proliferation treaties. Safeguards are also the most common condition imposed on nuclear exports. They are therefore a particularly important tool for dealing with states that have not joined the international non-proliferation treaties. The fifth section of this chapter reviews the questions that have arisen over the application and effectiveness of these safeguards.

A few states have insisted on remaining outside the international non-proliferation regime. The final section of this chapter addresses the problems of constructively influencing these holdout states and of coordinating the efforts of other states that share the concerns of the United States about this problem.

An examination of current approaches to non-proliferation demonstrates not only the difficulty and complexity of the problem but also the opportunities that do exist to slow, if not stop, the spread of nuclear weapons. Although several states appear to be seriously considering the development of nuclear weapons, a range of possible actions exists to deter this decision in each case. In 1964, when China joined the United States, the Soviet Union, the United Kingdom, and France as the fifth nuclear weapons state, few observers would have predicted that in the next 20 years only one other state would test a nuclear device and that, after the test, India would emphasize its "peaceful" purpose and apparently discontinue an active weapons program.

THE NATURE OF THE RISK

The Relationship of Proliferation to U.S. and International Security and Stability

There is a broadly based international consensus that worldwide security and stability would be best served by limiting the number of states with an independent nuclear explosive capability. From the perspective of the United States, this would have several advantages. Not only would it reduce the threat of nuclear weapons being used directly against the United States and its allies, but, more significantly, it would prevent the use of nuclear weapons in circumstances that might provoke nuclear retaliation against the United States or in local hostilities that might escalate into a broader nuclear war threatening the United States.

An increase in the number of states with nuclear weapons, particularly those with unstable governments or limited capabilities for technical controls, would increase the probability that nuclear weapons might be used by accident or miscalculation or that nuclear weapons might come under the control of irresponsible leaders. Further proliferation would also increase the probability that weapons might be seized by dissident groups or stolen by terrorist organizations. In view of the complex of potential threats, further proliferation would also greatly complicate the military planning of the United States.

These concerns about nuclear proliferation are not confined to the United States or even to states that already have nuclear weapons. Indeed, all states should have an interest in preventing the creation of these potential threats to their security. States directly threatened by regional hostilities should have a particular interest in keeping nuclear weapons from being introduced into these conflicts. States without nuclear weapons generally see the danger of the potential proliferation of nuclear weapons to their adversaries as outweighing the advantages of acquiring nuclear weapons or even the future option to produce them. The adherence of more than 120 states to treaties designed to support non-proliferation and the willingness of these states to pay the price of forswearing the future option of acquiring nuclear explosives demonstrates the widespread recognition of the international security advantages of a non-proliferation regime. Significantly, these states include most of the non-nuclear weapon states that could easily develop nuclear weapons. In each case, the state independently concluded that it would be against its overall security interests to undertake such a program or even maintain an option to do so.

A small minority of states have been unwilling to accept this non-proliferation regime. However, none of these states has openly proclaimed that it intends to develop an independent nuclear weapon capability. While some of these states have argued that the regime is blatantly discriminatory and offensive to their concept of national sovereignty, they have at most only implied that they were maintaining a future option to obtain nuclear weapons. Those few states that appear to be actively developing this option are clearly weighing its consequences very carefully. These consequences include the acceleration of nuclear weapon programs by their adversaries, the possibility of preemptive attacks, and competition with their own conventional military forces for scarce technical resources. Above all, these states must decide whether nuclear weapons are relevant to their actual defense needs. They must ask whether the nuclear option would not only fail to deter their adversaries but endanger their own survival.

There have been a few proponents of the concept that the general proliferation of nuclear weapons would serve a useful purpose by extending the concept of mutual nuclear deterrence to potential adversaries other than the present nuclear weapon states. They have suggested that the possession of nuclear weapons would engender a deeper sense of responsibility among leaders of those states. This notion has received very little support in the United States and has not become a major official argument even among the non-nuclear weapon states that are apparently developing a nuclear weapons option.

In addition to the proliferation of nuclear weapon capabilities to more states, there is the frightening possibility of proliferation to subnational groups within states. This form of proliferation would most likely occur by the direct seizure of existing nuclear weapons, but the construction of crude weapons from stolen weapons-grade fissionable material and available components cannot be completely ruled out. In the present nuclear weapon states, this threat may exist if weapons are deployed or stored at poorly protected sites and are inadequately secured against unauthorized use or where large quantities of weapons-grade material exist. However, it would present a particular danger in potential nuclear weapon states that are unstable and contain subnational groups prepared to engage in terrorism.

There are differences of opinion as to the actual extent of the nuclear threat from subnational groups. The ability of a subnational group to detonate a stolen weapon would depend on the state's technical precautions to secure its weapons against unauthorized use, precautions that in principle can be very effective. Presumably, these precautions would be least developed in a state that had not had time to develop technical sophistication in the control of its weapons.

The seizure of fissionable material or production facilities by subnational groups is certainly a real possibility, but the fabrication of this material into a successful explosive, while conceivable, would in practice be an extremely difficult technical task for such a group. However, even the claimed existence of a primitive device in credible circumstances could be a powerful tool for blackmail.

The Technical Problem

The problem of controlling nuclear proliferation is greatly complicated by the overlap of underlying technologies needed for developing and producing nuclear weapons and peaceful nuclear power. At the outset of the nuclear age, some scientists believed that the dangers inherent in the proliferation of nuclear weapons were so great and the technologies so closely related that nuclear power should not be developed at all. With the massive worldwide investment in nuclear power in the postwar period, the option of prohibiting the peaceful uses of nuclear energy has long been dead, if in fact it ever existed.

The critical and indispensable ingredient of a nuclear explosive device is either highly enriched uranium or separated plutonium. Highly enriched uranium is produced by separating the fissionable isotope uranium-235 from uranium-238 in enrichment plants by a variety of different processes. Plutonium, which is produced in a reactor when uranium-238 is irradiated with neutrons, becomes available in a form suitable for weapons when it is separated in a chemical reprocessing facility from the residual uranium and fission products. The underlying technical problem arises because these materials and "sensitive" facilities with the capability to produce them can, and do, exist in nuclear power programs devoted entirely to peaceful purposes.

Most current power reactors use low-enriched uranium, which is not suitable for weapons. However, some high-enriched uranium suitable for weapons can exist in peaceful programs as fuel for certain advanced power and research reactors. In any event, facilities that produce low-enriched uranium could in general be modified to produce high-enriched uranium or be used to feed other facilities to produce a highly enriched product.

Plutonium is produced in all nuclear reactors fueled with natural or low-enriched uranium. Although it is not now economically advantageous to do so, this spent fuel can be reprocessed as part of a peaceful power program to separate the plutonium for reuse as fuel either in present generation power reactors or future fast breeder reactors. Reprocessing can also facilitate the disposal process by separating radioactive waste products into a more manageable form. Although plutonium produced in a nuclear power plant optimized for power production would be much less desirable for weapons than plutonium produced expressly for that purpose, it is now clear, contrary to earlier hopes, that the reprocessed plutonium could be used in weapons.

Stimulated by unrealistically low estimates of the cost of nuclear power and by inflated projections of worldwide electric power consumption, a massive buildup and proliferation in uranium enrichment capacity occurred during the 1960s and 1970s. Moreover, the demand for nuclear power was confidently expected to be so great that the price of uranium, which was thought to be in short supply, would quickly rise to the point where reprocessed plutonium would be competitive with enriched uranium for use in ordinary reactors. The anticipated need for reprocessed plutonium stimulated and legitimized worldwide interest in reprocessing facilities that could be built on a more modest scale than an enrichment facility.

In this economic environment, the plutonium breeder reactor, which could efficiently exploit the full energy potential of uranium, seemed the next logical technical step despite its very high capital cost. The resulting "plutonium economy" based on plutonium reprocessing and plutonium breeder reactors would have resulted in vast quantities of plutonium at power reactors, reprocessing plants, and fuel fabrication plants and in transit domestically and internationally between these facilities. For example, the plutonium fuel load for a single plutonium breeder could contain enough plutonium to make 50 nuclear weapons. With plutonium recycled in existing thermal reactors, every power re-

actor would become a potential recipient of plutonium fuels. A plutonium economy would increase the opportunities for theft, seizure, or
diversion of plutonium and make it more difficult to achieve safeguards
in which a high degree of confidence could be placed.

The drastic slowdown in the growth of nuclear power in recent years
has caused a glut of natural and enriched uranium on the international
market, and the projected plutonium economy has receded into the
future. Domestically, the changed prospects in this area are dramatically illustrated by the termination of the Barnwell reprocessing plant
and the Clinch River breeder reactor and by the increasingly uncertain
future of the large new centrifuge uranium separation plant at Ports-
mouth, Ohio. Nevertheless, in the longer term the basic proliferation
problems inherent in the plutonium economy will have to be faced.
Moreover, despite today's unfavorable economic prospects, many states,
for a variety of reasons discussed later in this chapter, are still pursuing
their earlier interest in sensitive facilities.

In the International Nuclear Fuel Cycle Evaluation (INFCE) study,
the Carter Administration undertook a major effort to engage the international nuclear community in a critical review of the problems and
prospects of the nuclear fuel cycle. The goal was to develop a consensus
on the advisability of deferring dangerous developments in the nuclear
fuel cycle that presupposed the early attainment of an international
plutonium economy. Many of the INFCE participants, confident of the
basic economic viability of the plutonium fuel cycle, opposed this position and refused to consider restructuring their approaches to nuclear
power. However, the continued decline in the fortunes of domestic and
international nuclear power since the 1980 INFCE report has raised
further questions about arguments in favor of the current need for a
plutonium fuel cycle. Consequently, despite the long history of international opposition to such efforts, the issue of whether the nuclear fuel
cycle should be limited or restructured to build a more secure non-
proliferation regime will probably remain active.

The current situation, in which uranium enrichment services are in
ample supply at reasonable prices and the electric power industry is
eager to avoid taking further risks, holds out some prospect of avoiding
or at least substantially deferring a worldwide plutonium economy.
Plutonium recycled in existing reactors is now clearly uneconomical,
and commercial fast breeder reactors are at best a future prospect in a
few highly industrialized states. This economic reality should reduce
the pressure for additional commercial reprocessing facilities. However, it will not necessarily eliminate interest in small reprocessing
plants for the declared purpose of gaining technical experience as a

hedge against another change in the economics of nuclear power. Nor will it eliminate the current risks posed by the problem states described later in this chapter, which have unsafeguarded enrichment and reprocessing facilities already in existence or under construction.

Although not an immediate problem, technological advances now under development, such as laser isotope separation, could complicate non-proliferation efforts in the future. Laser isotope separation might be used to produce highly enriched uranium in a single step, as opposed to the current multistage processes that can be more reliably dedicated to the production of only low-enriched uranium. It could also be used to remove the isotope plutonium-240, which has an undesirably high spontaneous fission rate, from the plutonium in spent fuel from power reactors, improving the plutonium's utility for weapons purposes. If applied by non-nuclear weapon states to the accumulated spent fuel from their reactors, it could greatly enlarge the quantity and quality of weapons-usable material accessible to them. Laser isotope separation plants may also present a special monitoring problem, since they could be much smaller and less distinctive than the large gaseous diffusion plants or even production-size centrifuge plants.

Once adequate quantities of enriched uranium or plutonium are available, the problem of fabricating a simple fission weapon should not prove too difficult for any state that has developed even a modest level of competence in the nuclear field. The basic design features of first generation fission weapons are now widely known. A small number of scientists and engineers whose experience was derived from a peaceful nuclear power program could develop a workable design. The actual fabrication of a device would require a small team of fairly qualified experts in a number of fields with access to laboratory and fabrication facilities using easily obtainable equipment.

Opinions vary as to how necessary a proof test would be for an initial weapon. Some argue on the basis of hindsight that such a test would not be necessary since there would be reasonably high confidence that a weapon produced by a technically competent group would give a significant yield. However, the weapon designers would probably have little confidence in the actual yield. More significantly, military or political authorities might not share the confidence of their scientists and technicians, particularly in a state with little history of technical sophistication. Moreover, a test is the only way a state could demonstrate to others that it had actually acquired a weapons capability. In any event, the development of more advanced fission weapons, and certainly thermonuclear weapons, would require testing.

At the same time, the ease of making a fission weapon should not be exaggerated. It cannot be done by a clever university student in a garage with tools from the local hardware store. As a practical matter, it would probably be beyond the capabilities of almost any subnational group operating outside of an organized nuclear program. Besides obtaining an adequate supply of fissionable material, such a group would need qualified experts and technicians in a number of fields with access to equipment and considerable advanced planning to develop a design. Such a group could conceivably construct a primitive device with an uncertain yield of a few hundred tons, although it might have no yield at all. Nonetheless, a primitive device of this type could be extremely destructive in an urban area. Consequently, as noted above, even the claim of such a capability in credible circumstances could be a powerful tool for blackmail.

Capabilities Versus Intentions

In developing a strategy to prevent the further proliferation of nuclear weapons, a fundamental issue has been how much weight to give to "capabilities" as compared with "intentions." This issue reflects the inevitable interaction of the technical capabilities associated with nuclear power with the technology required to fabricate nuclear weapons.

To date, non-proliferation efforts have primarily sought to build legal and political barriers against decisions to fabricate nuclear weapons. International agreements have drawn the line at the actual manufacture of a nuclear explosive, rather than at the acquisition of underlying technical capabilities that might make this possible. States such as Japan and the Federal Republic of Germany clearly have the technical capability to produce weapons on short notice. The barriers to their doing so are their conviction that it would not be in their net interest and the international legal obligations they have undertaken on the basis of this conviction. These convictions and commitments are very substantial barriers for these countries. The risk that they will override these constraints in the foreseeable future appears correspondingly small.

Critics of this approach argue that intentions change and that international obligations can easily be violated or abrogated. Only by limiting the underlying technical capabilities can proliferation be prevented in the longer term, they argue. While some half a dozen states may now intend either to build nuclear weapons or to consciously develop an option to do so quickly, probably 20 to 30 countries already have most of the technical capabilities needed to make such a decision.

The acquisition of the nuclear materials and facilities needed for a nuclear power program, the associated industrial support, and the related infrastructure of trained nuclear scientists and technicians go a long way toward giving a state the technical capability to make weapons, especially if weapons-usable material or facilities for its production are available. But there is considerable difference of opinion over the extent to which the prior acquisition of these capabilities makes a nuclear weapon decision more likely. The French and the Indians built up such capabilities some years before deciding to manufacture nuclear weapons, though at least one of their objectives in undertaking this buildup was to establish the option to make such a decision. On the other hand, Sweden built up such capabilities and seriously debated initiating a weapons program in the 1960s but finally decided that this would be counter to its security interests.

Since the mid-1970s, concern over nuclear weapon capabilities has focused on the most sensitive nuclear materials (highly enriched uranium and separated plutonium) and on enrichment and reprocessing facilities for their production. Some critics of the present U.S. approach have essentially redefined proliferation as access to these sensitive materials. They have largely discounted the offsetting effect of legal and political commitments against proliferation.

THE HISTORY OF NON-PROLIFERATION

Shortly after World War II, while it still had an absolute monopoly on nuclear weapons, the United States offered in the Baruch Plan to work out arrangements for the international ownership and control of all nuclear materials and facilities. Pending agreement on such an arrangement with appropriate international safeguards, the United States prohibited through the MacMahon Act of 1946 the export of nuclear materials, equipment, or technology. Nevertheless, without direct assistance from the United States, four additional nuclear weapon states emerged in the next 18 years: the Soviet Union (1949), the United Kingdom (1952), France (1960), and China (1964). By 1953, with the reported test of a Soviet thermonuclear device and growing international interest in peaceful nuclear power, it became apparent that the United States could not dictate a policy of denial simply by unilaterally prohibiting exports.

In response to this changing situation, President Dwight Eisenhower reversed the policy of denial in December 1953 and inaugurated a policy of constructive engagement with the Atoms for Peace program. This policy was designed to promote internationally the peaceful applications of nuclear energy, provided the recipient state guaranteed that there would be no diversions to military use and agreed to accept safeguards. The U.S. Atomic Energy Act was revised to reflect this new approach of constructive engagement, and the first of several international Atoms for Peace conferences, at which much previously classified nuclear technology was released, was held in Geneva. The U.S. efforts to implement the new policy included agreements for cooperation with over 30 states, liberal grants to facilitate the purchase of research reactors, training programs, and disclosures of technology. The International Atomic Energy Agency (IAEA), established in 1957, was based on this approach of open international cooperation in the peaceful uses of nuclear energy in return for pledges against diversion to military use and the acceptance of international safeguards.

Following the Cuban missile crisis in 1962, interest in preventing the spread of nuclear weapons intensified not only in the United States and the Soviet Union but in many non-nuclear weapon states as well. One result was the negotiation in 1963 of the Limited Test Ban Treaty (Appendix D), which had non-proliferation as one of its main objectives. Nearly every state having a potential capability to make nuclear weapons eventually joined the treaty. Another consequence was the Latin American initiative that led to the 1967 Treaty for the Prohibition of Nuclear Weapons in Latin America, often referred to as the Treaty of Tlatelolco.

Following the Limited Test Ban Treaty, the United States and the Soviet Union turned their attention directly to the proliferation problem, embarking on the complex four-year multinational negotiation of the Treaty on the Non-Proliferation of Nuclear Weapons (Appendix G) or simply the Non-Proliferation Treaty (NPT). The treaty, which was signed on July 1, 1968, and entered into force in 1970, has so far been joined by 124 states, including most of the advanced industrial nuclear states.

Both the Non-Proliferation Treaty and the Treaty for the Prohibition of Nuclear Weapons in Latin America were remarkable achievements. They involved commitments by sovereign states that had not yet acquired nuclear weapons to forswear that option and to accept international safeguards to verify their compliance with this self-denying obligation. At the same time, both treaties embodied commitments to facilitate international cooperation in the peaceful uses of nuclear energy under safeguards. They thus reinforced the policy of constructive engagement.

At the time the NPT was negotiated, the United States was the free world's principal supplier of nuclear power reactors and its only exporter of enriched uranium to fuel these reactors. This situation changed radically in the 1970s, when reactor vendors in other states (most notably France, Germany, and Canada) won a substantial proportion of new orders, several other exporters of enriched uranium emerged (EURODIF, controlled by France; URENCO, a German-British-Dutch consortium; and the Soviet Union), and the bottom fell out of the reactor market as a result of reduced demand for electric power and a rapid rise in nuclear power costs. With the anticipated loss of the U.S. monopoly now a fact, the need for collective action by the nuclear supplier states on nuclear export conditions became urgent.

In the mid-1970s a series of proliferation problems, including the detonation by India of a nuclear device made with materials pledged to peaceful use, the purchase of French reprocessing plants by Pakistan and South Korea, and the purchase of German uranium enrichment and plutonium reprocessing plants by Brazil, convinced many that the liberality of nuclear exports had been excessive and had created potential proliferation threats that would not otherwise have existed. This led to new pressures for denial of sensitive nuclear materials and technology, enrichment and reprocessing facilities, and heavy water production technology.

In response to a U.S. request, the principal nuclear supplier states met in 1975-76 to work out common guidelines for nuclear exports.* The resulting Nuclear Suppliers' Guidelines, which do not constitute a formal international agreement, include a clarification of the safeguard requirements for exports to nonparties to the NPT, the first multilateral attempts to deal with physical security against subnational threats, and constraints on the export of sensitive nuclear materials and facilities. A major accomplishment was to persuade France, which is not a party to the NPT, to require safeguards on its exports. While the guidelines did not categorically forbid exports of sensitive materials or facilities, they called for restraint in such exports. France, the Federal Republic of Germany, and the United States also announced that no new commitments to export reprocessing technology or equipment would be made until further notice.

In the United States, President Gerald Ford's nuclear policy statement of October 28, 1976, and President Jimmy Carter's policy statements of April 1977 called for a reconsideration of plans to proceed with the commercial reprocessing of reactor fuel and the widespread use of plutonium-based fuels in existing reactors. These statements coupled

---

*The following states have stated that their nuclear export policies follow the guidelines: the United States, the United Kingdom, the Soviet Union, France, the Federal Republic of Germany, Japan, Canada, Belgium, Italy, the Netherlands, Sweden, Switzerland, Czechoslovakia, the German Democratic Republic, Poland, Australia, and Finland.

concern over the increased risk of proliferation from these activities with serious questions about the economic basis for moving in this direction. The call by the United States for an international reexamination of the technical side of these questions met considerable resistance in the International Nuclear Fuel Cycle Evaluation held from 1977 to 1980. Nevertheless, pressures to move in the direction of commercial reprocessing and plutonium fuel recycling have markedly decreased since then in light of the dramatic reduction in the number of nuclear power plants on order or still under construction and the greatly improved supply picture for uranium-based nuclear fuels.

The U.S. Nuclear Non-Proliferation Act of 1978 reflected the growing concern over the spread of sensitive nuclear capabilities. The target of the act was not only the proliferation of nuclear explosive devices (the objective of the NPT) but also the proliferation of "the direct capability to manufacture or otherwise acquire such devices." The act was even more restrictive about the export of sensitive nuclear technology than the Nuclear Suppliers' Guidelines. It included provisions designed to discourage any further international transfer of reprocessing technology, the reprocessing abroad of fuel originating in the United States, and the international transfer of plutonium. The act also provided for termination of nuclear cooperation with any nation or group of nations found by the President to have entered into an agreement, after the date of enactment, for the transfer of reprocessing equipment, materials, or technology to a non-nuclear weapon state. In addition, the Symington and Glenn amendments to the Foreign Assistance Act provided for a cutoff of economic aid and military grants and credits to states that delivered or received reprocessing or uranium enrichment equipment or technology. An amendment to the Export-Import Bank Act forbade that institution to give credit, guarantees, or insurance for the purchase of any liquid metal fast breeder reactor or any nuclear fuel reprocessing facility.

These U.S. statutes were criticized abroad as discriminatory and inconsistent with the obligations of the United States under the NPT to help non-nuclear weapon states develop peaceful nuclear energy. The most vociferous critics of this new emphasis were major allies and trading partners of the United States such as Japan, Korea, and the Western European states, all of which were parties to the NPT, as well as various developing states such as India, Pakistan, and Brazil, which were not. The latter group posed a qualitatively different problem, since the risks inherent in their access to weapons-usable material were compounded by the absence of any international commitment not to use the material to make nuclear weapons.

Although the Reagan Administration initially asserted that its predecessors had overemphasized capabilities as opposed to motivations and intentions in approaching non-proliferation, the United States has in fact continued the policy of withholding the transfer of sensitive nuclear materials, equipment, and technology to states considered to present a risk of proliferation. At the same time, the fortunes of nuclear power have continued to decline domestically and internationally, with the plutonium economy and breeder reactors becoming increasingly less economical. These trends were underscored by the final demise in 1983 of the Barnwell reprocessing plant and the Clinch River breeder reactor.

NUCLEAR EXPORT POLICY

Denial Versus Constructive Engagement

The historical record outlined above points toward the earliest and most persistent issue that has shaped nuclear export policies: the choice between "denial" and "constructive engagement." Under the first Atomic Energy Act, U.S. export policy was one of total denial. The Atoms for Peace program and the NPT shifted this policy to one of constructive engagement, with liberal international cooperation under safeguards and conditions of supply. These safeguards and supply conditions were tightened up somewhat after the Indian nuclear explosion. As interest in commercial reprocessing and plutonium began to emerge in the mid-1970s, the policy of constructive engagement was qualified by selective denial of the most sensitive nuclear exports. These exports included plutonium, highly enriched uranium, and facilities for separating plutonium and enriching uranium. This mix of constructive engagement and selective denial remains U.S. policy today, although the Reagan Administration has declared that the standards of selective denial will be less rigorously applied to the Euratom countries (Belgium, Denmark, France, the Federal Republic of Germany, Greece, Ireland, Italy, Luxembourg, the Netherlands, and the United Kingdom) and Japan.

Thus, current U.S. policy, like that of most states, accepts the general policy of constructive engagement. But there remain controversies over the nature and severity of the conditions imposed on nuclear cooperation and the extent to which the selective denial of sensitive nuclear exports is applied. Both issues were raised by the Nuclear Suppliers' Guidelines and by the U.S. legislation of the late 1970s.

The controversy over supply conditions related to both the extent of supplier control and the method by which such conditions were imposed. The principal complaint about the Nuclear Suppliers' Guidelines was that they unilaterally imposed new conditions without consulting the recipients. The complaints about the U.S. legislation were both that it made unilateral changes in previously agreed supply arrangements and that it went too far. For example, the legislation required prior U.S. consent for the handling of materials derived from U.S. exports and created tight new criteria for the granting of such consents. Considerable friction with Japan, Sweden, and Switzerland, as well as India, resulted from delays and uncertainties caused when these new criteria were applied to the granting of consents for the reprocessing, or the transfer to another country for reprocessing, of spent fuel produced through the use of U.S. exports. The legislation also required a recipient country to accept full-scope safeguards on wholly indigenous activities (as the NPT requires signatory non-nuclear weapon states to do) and to refrain from transferring reprocessing technology or equipment. In the United States, however, and especially in Congress, the legislation has been criticized as not being tight enough.

Selective denial defines the effort to prevent non-nuclear weapon states from gaining direct access to weapons-usable material. It is based on the premises that these materials are the pacing items for the fabrication of nuclear weapons and that international safeguards would not provide "timely warning" of their diversion (safeguards are discussed later in this chapter). These premises bring into focus the basic controversy over the relative emphasis that should be given to capabilities versus intentions. Some believe that this policy should be more rigorously applied than it is at present, that exceptions should not be made for the Euratom countries and Japan, and that efforts should be renewed to ensure that other states do not undercut the policy. Others point out that even the current policy has caused a number of unfavorable reactions, including resentment by the nuclear industrial states over what they see as unwarranted interference with their nuclear energy programs, resistance by other states to what they characterize as a violation of their "inalienable right" to make their own policy choices, complaints about discriminatory treatment, and charges that such selective denial is incompatible with the basic intent of the NPT. Opponents of selective denial also argue that it is counterproductive in several ways. It alienates other states whose cooperation is necessary to an effective non-proliferation regime, they contend, and it could increase the determination of states denied access to this technology to develop independent nuclear capabilities free of international controls or stimulate a nuclear black market among the states affected.

A few of the most extreme critics of the present non-proliferation regime advocate a return to a policy of total denial, a cessation of all international nuclear commerce. They argue that the Atoms for Peace program was a fundamental mistake, since it created the basic stepping stones to a nuclear weapons option. They also contend that the risk of proliferation far outweighs the value of nuclear power for world development, since the economic advantages of nuclear power have now essentially vanished.

Supporters of the present approach to exports argue that a policy of total denial is impractical and would be counterproductive since the United States is no longer the sole source of nuclear exports. Thus, even if the United States stopped all of its nuclear exports, the gap would be filled by other suppliers and by indigenous programs under much less rigorous non-proliferation controls than those that now exist, which are largely a result of U.S. influence. Moreover, a policy of total denial would clearly be inconsistent with Article IV of the NPT and would undermine support for that treaty, which is indispensable to an effective non-proliferation regime. Finally, whatever the validity of the economic and other arguments against nuclear power, many states are not yet convinced by them and genuinely believe that their need for this added energy option outweighs other considerations.

International controversy over the proper balance between denial and constructive engagement is bound to continue. It will be an important issue at the 1985 NPT review conference, and it is expected to be the central focus of the next UN Conference on the Peaceful Uses of Nuclear Energy, which is scheduled for 1986.

Uniform Versus Discriminatory Policies

The Reagan Administration has declared that the policy of selective denial of sensitive nuclear exports will be applied less rigorously to Euratom countries and Japan. This decision highlights the issue of the extent to which it is desirable and feasible to establish uniform non-proliferation policies, applicable to all potential trading partners without discrimination. This question has arisen in legislative approaches to non-proliferation,* in attempts to coordinate the policies of nuclear

---

*Most such legislation has included a compromise that grants the President authority to waive the application of the statutory requirements in particular cases, though Congress has attempted to reserve a right to veto such waivers. Congressional veto provisions (except those requiring a joint resolution passed by two thirds of both houses of Congress) were declared unconstitutional in 1983 by the U.S. Supreme Court in *U.S.* v. *Chadha.*

supplier states,* and in relations between supplier and recipient states.†

Supporters of a uniform export policy argue that it underscores the seriousness of the U.S. approach to non-proliferation and prevents the government from subordinating this objective to other policy objectives that appear to have temporary priority. By prohibiting preferential treatment of certain states, it obviates charges of "discrimination" by which problem states seek to discredit the entire non-proliferation regime.

Critics of this approach point out that non-nuclear weapon states present a wide variety of very different situations. Some states (such as Japan and the Federal Republic of Germany) have extensive commercial nuclear programs, current access to substantial quantities of weapons-usable material, and strong national commitments not to develop nuclear weapons. Other states may lack all three of these characteristics. Critics assert that it does not serve national policy, or in the long run non-proliferation policy, to ignore these differences and apply exactly the same restrictions to all states. This would be the case if the shipment of an advanced research reactor to Japan were subject to the same restrictions as its shipment to Libya or South Africa. They assert that the principle of nondiscrimination should be construed, as it is under present domestic U.S. law, to treat as unfair only discrimination among "parties similarly situated."

The Carter Administration's overall non-proliferation policy and the Nuclear Non-Proliferation Act of 1978 have been criticized as being too inflexible, since their heaviest impact was on the United States' closest allies and most substantial trading partners, who posed no near-term threat of proliferation, rather than on the states of greatest proliferation concern. While these policies did cause friction with U.S. allies and trading partners, the underlying issue is whether this friction was too high a price to pay to achieve a uniform and consistent approach to the problem of nuclear export control.

---

*For example, the Nuclear Suppliers' Guidelines provide that "suppliers should consult, as each deems appropriate, with other Governments concerned on specific sensitive cases, to ensure that any transfer does not contribute to risks of conflict or instability" and that "in considering transfers, each supplier should exercise prudence having regard to all the circumstances of each case, including any risk that technology transfers not covered by paragraph 5, or subsequent retransfers, might result in unsafeguarded nuclear materials."

†For example, Japan and the Federal Republic of Germany have complained that they were being treated no differently from Bangladesh, Libya, or Pakistan; the late Shah of Iran demanded a guarantee that his state would be given "most favored nation treatment" in the nuclear field; and Pakistan has complained that it was being unfairly discriminated against.

Some participants in the debate have gone so far as to advocate a completely ad hoc case-by-case approach to export decisions. This would allow the government to focus on the few serious potential proliferation cases without adversely affecting nuclear trade or relations with other states, they say. Critics of this approach argue that, in addition to the uncertainties that such a case-by-case approach would create for potential customers, it would inevitably sacrifice non-proliferation objectives to immediate political expediency and would result in endless charges of discrimination.

Relationship to NPT Obligations

The Non-Proliferation Treaty places some inhibition on policies that deny nuclear exports to states that have adhered to the treaty. Article IV of the NPT recognizes "the inalienable right of all the Parties to the Treaty to develop research, production and use of nuclear energy for peaceful purposes without discrimination" when in conformity with those articles of the treaty aimed at preventing proliferation. In addition, Article IV provides for the fullest possible exchange among the treaty's parties of equipment, materials, and scientific and technological information for the peaceful uses of nuclear energy.

Some NPT parties have complained that those provisions of the Nuclear Suppliers' Guidelines and the 1978 Nuclear Non-Proliferation Act that restrict the transfer of sensitive nuclear technology are clearly inconsistent with the provisions of Article IV. In response to this criticism, it is argued that while this article was designed to ensure that the parties would not be deprived of the benefits of peaceful nuclear energy, it did not remove all discretion as to how this was to be accomplished consistent with the treaty's underlying objective of preventing further proliferation. The article provides for the "fullest possible" exchange, and what is possible must be judged in the light of that underlying objective and common sense. For example, even though Libya is a party to the NPT, if it applied to another party to purchase 500 kg of separated plutonium "for peaceful purposes," that party would clearly not be obligated to provide the plutonium since there would be a substantial risk that such a transaction would defeat the central objective of the treaty. Moreover, Article IV clearly was not intended to deprive a party of its discretion in selecting trading partners. In addition, a party is obviously not obligated to export facilities or technology that it has determined not to license domestically or has discontinued developing. The United States would not be expected to export commercial reprocessing facilities when its own program has been suspended.

As one could expect, not all parties to the NPT have accepted these arguments, and the tension between Article IV and the policy of denial must be taken seriously. More broadly, it is argued that this policy, if carried too far, could undercut non-proliferation objectives by weakening support for the NPT. This aspect of the problem, however, relates only to denials of exports to parties to the NPT and not to states that have chosen to remain outside the NPT.

Extreme Denial

Beyond the withholding of cooperation, a policy of denial might *in extremis* be extended to the destruction of existing nuclear facilities. Examples of such a policy were Israel's actions against an Iraqi reactor whose nuclear potential Israel considered an imminent threat even though Iraq was a party to the Non-Proliferation Treaty with a full-scope safeguards agreement. The actions initially involved sabotage of the French plant that was producing equipment for Iraq's reactor and later the aerial bombing of the reactor shortly before it was scheduled to commence operation. The Israeli attack was strongly condemned by the international community. It was seen as damaging to the non-proliferation regime since, by totally dismissing the assurances provided in the NPT and full-scope safeguards, it lessened the incentive to adhere to those measures. The international community rejected Israel's attempt to justify its attack as an act of anticipatory self-defense based on its assessment that the Iraqi program presented an imminent proliferation threat.

It has been argued that such preemptive actions cannot, in the long run, prevent a state determined to develop nuclear explosives from doing so, and that physical attacks will only increase the attacked state's determination to develop a nuclear weapons program. Nevertheless, there is no question that the Israeli attack set back the Iraqi nuclear program by several years and that the French and other suppliers will probably be considerably more circumspect in how they replace and fuel the damaged facility.

THE ROLE OF INTERNATIONAL NON-PROLIFERATION AGREEMENTS

A major focus of non-proliferation efforts has been to secure treaty commitments from non-nuclear weapon states not to develop or acquire nuclear explosives. It is sometimes argued that such treaties are only paper promises that can always be violated or abrogated. There is a strong international consensus, however, that this approach helps catalyze national decisions to forego nuclear weapons and erects legal and political obstacles to changing these decisions. There are appreciable costs in international relations involved in breaking a treaty commitment, as well as internal pressures against doing so. Above all, as long as nuclear proliferation can be held to a minimum, most states continue to believe that their security is better served by preventing their neighbors and adversaries from developing nuclear weapons than by undertaking a weapons program of their own.

The Non-Proliferation Treaty

The Treaty on the Non-Proliferation of Nuclear Weapons (Appendix G), which was originally signed on July 1, 1968, and has been in force since March 5, 1970, was designed to help stop the spread of nuclear weapons to additional states. The treaty, which divided the world into nuclear weapon states (the United States, the Soviet Union, the United Kingdom, France, and China) and non-nuclear weapon states (all other countries) places on all signatories the following basic obligations:

• Article I: "Each nuclear-weapon State Party to the Treaty undertakes not to transfer to any recipient whatsoever nuclear weapons or other nuclear explosive devices or control over such weapons or explosive devices directly or indirectly; and not in any way to assist, encourage, or induce any non-nuclear-weapon State to manufacture or otherwise acquire nuclear weapons or other nuclear explosive devices, or control over such weapons or explosive devices."

• Article II: "Each non-nuclear-weapon State Party to the Treaty undertakes not to receive the transfer from any transferor whatsoever of nuclear weapons or other nuclear explosive devices or of control over such weapons or explosive devices directly, or indirectly; not to manufacture or otherwise acquire nuclear weapons or other nuclear explosive devices; and not to seek or receive any assistance in the manufacture of nuclear weapons or other nuclear explosive devices."

• Article III calls on all parties not to export nuclear materials or equipment to any non-nuclear weapon state without international safeguards designed to verify that such exports are not diverted to nuclear explosive programs.

Despite ominous predictions that most states would not chose to adhere to the treaty, the NPT currently has 124 parties, including all members of NATO and the Euratom countries other than France, all members of the Warsaw Pact, and most of the other states in the world having nuclear programs. The parties to the NPT, together with France, which has made clear that it will act as if it were a party to the treaty, account for 98 percent of the world's installed nuclear power capacity, 95 percent of the nuclear power capacity under construction, and all of the world's exporters of enriched uranium. This means not only that the overwhelming majority of the world's civil nuclear activities are directly under the treaty's regime, but also that any nonparty seriously interested in developing nuclear power must rely at least in part on cooperation with parties to the treaty, whose exports are covered by international safeguards.

Despite these impressive statistics, not all relevant states have joined the treaty. In addition to France and China, which already have nuclear weapons, the most notable holdouts are Israel, India, Pakistan, South Africa, Argentina, Brazil, and Cuba. Each of these non-nuclear weapon states presents a special case (these cases are discussed in the final part of this chapter). The primary effect of the treaty for the nuclear weapon states is on their export policies. It is therefore important to note that France has voluntarily conformed its nuclear export policies to those called for by the treaty and that China has not yet become a significant nuclear exporter.

To make adherence to the treaty more attractive and to minimize charges of discrimination, three significant provisions (Articles IV, V, and VI) were added during the negotiation process. Article IV declares that nothing in the treaty should be interpreted as affecting "the inalienable right" of all parties to develop nuclear energy for peaceful purposes without discrimination. It also provides that all parties should facilitate, and have a right to participate in, the fullest possible exchange of equipment, materials, and scientific and technological information for the peaceful uses of nuclear energy.

Article V guarantees the sharing of any peaceful benefits from nuclear explosives. This was designed to prevent the nuclear weapon states from gaining special economic advantage from these developments, which they alone could pursue. During the 1960s the United States had promoted the potential of nuclear explosives for a variety of peaceful purposes. This led some non-nuclear weapon states to charge that the prohibition on the development of nuclear explosives in the treaty would inhibit their economic development. This was cited as a further example of the discriminatory nature of the NPT. Although some holdout states (including India, Argentina, and Brazil) still refer to this issue, the increasingly pessimistic assessments of the prospects for such peaceful applications have greatly reduced its appeal. Article VI contains a very important provision for all parties to "pursue negotiations in good faith on effective measures relating to cessation of the nuclear arms race at an early date and to nuclear disarmament."

Most of the controversy among treaty parties has been over the implementation of Articles IV and VI, which a number of the most significant parties declared to be of central importance to them when they ratified the treaty. Some parties have charged that the nuclear export policies of some supplier states, in particular the United States, have been inconsistent with Article IV, and that the benefits they envisaged have not been realized. The lack of progress by the superpowers in curbing their own nuclear arms race, which is sometimes referred to as "vertical proliferation," has been even more heavily criticized. There have been many protests that, contrary to the obligation of Article VI, the nuclear arsenals of the superpowers have been growing in size and sophistication. Many of the non-nuclear weapon states contend that the "basic bargain" consisted of a balance of obligations between the non-nuclear weapon states, who were forswearing nuclear weapons, and the undertakings of the suppliers and the superpowers under Articles IV and VI, and that the latter are not honoring their end of the bargain. The counterarguments are that there has been extensive cooperation in the peaceful uses of nuclear energy and some progress in nuclear arms control, and that in any event the basic non-proliferation provisions of the treaty benefit the security interests of all its parties.

Perceptions as to the military value of nuclear weapons obviously affect national decisions about whether to acquire them. These perceptions are influenced by what the nuclear weapon states do and say about the role of nuclear weapons. Opinions vary as to how the willingness of the nuclear powers to accept arms control measures on their own forces influence those perceptions. Some believe this is a critical consideration, that the failure to establish a more effective arms control regime between the nuclear powers undercuts efforts to establish an international environment conducive to nuclear self-restraint and accentuates the double standard between the nuclear and non-nuclear weapon states. Others have discounted this argument and have suggested that any perceived weakening of the U.S. nuclear umbrella could increase the incentive of states now covered by it to develop their own nuclear deterrents. Still others argue that this is not a central issue one way or the other since any state that decides to develop a nuclear option will do so for its own reasons, which will be largely unrelated to the arms control actions of the nuclear weapon states.

These issues dominated the NPT five-year review conferences held in 1975 and 1980 and are likely to be even more heated at the review conference scheduled for 1985. An important question is whether these issues will cause any withdrawals from the treaty and how seriously they may affect the prospects for extending the treaty beyond its stated expiration date of 1995. While the treaty does contain a standard provision permitting withdrawal on three months' advance notice, it is significant that no party has ever seriously threatened to withdraw.

Some critics of the NPT note its lack of security assurances to its parties (other than those inherent in the prevention of proliferation to fellow parties). In this regard, it should be noted that the treaty was accompanied by parallel declarations by the United States, the Soviet Union, and the United Kingdom that each would "seek immediate Security Council Action to provide assistance to any non-nuclear weapon state party to the Treaty that is a victim of an act of aggression or an object of a threat of aggression in which nuclear weapons are used." Thereafter, all five nuclear weapon states entered into a treaty commitment not to use or threaten to use nuclear weapons against the Latin American parties to the Treaty of Tlatelolco (discussed below). The United States, the Soviet Union, and the United Kingdom also declared at the 1978 UN Special Session on Disarmament that they would not use nuclear weapons against non-nuclear weapon states that are parties to the NPT. The formulations of these so-called negative security assurances differed in important respects, since the United States did not give up the option of first use of nuclear weapons against non-nuclear weapon states allied with a nuclear power (e.g., Warsaw Pact countries) engaged in an attack on the United States or its allies.

The Treaty of Tlatelolco and Other Possibilities for Nuclear Weapon-Free Zones

The Treaty for the Prohibition of Nuclear Weapons in Latin America (the Treaty of Tlatelolco), which entered into force before the NPT in 1968, establishes Latin America as a nuclear weapon-free zone. The Treaty of Tlatelolco, which was open for signature only by Latin American states, differs from the NPT in that it prohibits not only the acquisition of nuclear weapons by its parties but also the stationing of nuclear weapons within their territory. The NPT, in contrast, does not prevent the United States from stationing its weapons in NATO states or the Soviet Union from stationing its weapons in the Warsaw Pact area. The Treaty of Tlatelolco contains a fundamental ambiguity with respect to "peaceful" nuclear explosions, since it provides for the conduct of such explosions by or for the benefit of its parties under international supervision but also contains a definition of "nuclear weapon" that appears to preclude the possession or development of peaceful nuclear explosives by the parties to the treaty. The NPT unambiguously bans the transfer to non-nuclear weapon states, or the acquisition by non-nuclear weapon states, of all nuclear explosives.

The verification provisions of the Treaty of Tlatelolco include not only IAEA safeguards on all nuclear activities of the states concerned but also inspection and investigative rights by the regional treaty organization.

The treaty was accompanied by two important protocols. Protocol I subjects all states outside Latin America that exercise *de jure* or *de facto* jurisdiction over territories within that region to the obligations of the treaty. Protocol II requires nuclear weapon states to respect the treaty regime, not to contribute to its violation, and not to use or threaten to use nuclear weapons against its Latin American parties.

The treaty has been signed by all Latin American states except Cuba and ratified by all of its signatories except Argentina. In two cases, Brazil and Chile, the instrument of ratification deferred the actual entry into force of the treaty for these particular states until all other eligible parties have ratified the treaty and its protocols. Protocol I has been ratified by all eligible states except France, which has signed it and whose ratification is pending. Protocol II has been ratified by all five nuclear weapon states, including China, which gives it the unique distinction of being the only nuclear arms control treaty joined by all five of these states. The treaty has been signed by several parties and signatories that have not joined the NPT (Argentina, Brazil, and Chile). In defense of its failure to sign the treaty, Cuba has cited the continued existence of the Guantanamo naval base even though Protocol I would preclude the United States from stationing nuclear weapons there. Argentina officially announced in 1978 its decision to ratify the treaty but has deferred doing so pending the resolution of a dispute with the IAEA over the terms of the safeguards agreement required to implement it.

The treaty and its protocols are relevant to the current situation in Central America in several respects. The United States, the Soviet Union, and the United Kingdom are required by Protocol II to respect the treaty requirements that no nuclear weapons be stationed in El Salvador, Grenada, Guatemala, Honduras, or Nicaragua and are obligated not to use or threaten to use nuclear weapons against any of these states. However, the transit or transport of nuclear weapons by states other than the Latin American parties through the territorial waters or air space of these states (or through the Panama Canal) is not affected by the treaty. The negotiating history of the treaty makes this clear, and the United States ratified the two protocols on the basis of this understanding. The treaty obligations do not apply to Cuba, which has not signed or ratified the treaty.

The principal dispute among signatories of the treaty is whether it precludes the acquisition of peaceful nuclear explosives by its parties. Most of its parties acknowledge that it has that effect unless and until a clear way of distinguishing peaceful explosives from weapons can be found, but Argentina and Brazil maintain that it does not foreclose their right to acquire such explosives. In the case of Argentina this issue has been a major stumbling block in negotiating the safeguards agreement with the IAEA that is required under the treaty, because the IAEA maintains that a safeguards agreement must preclude the use of safeguarded materials for any nuclear explosive, whatever its stated purpose.

In addition to the importance of the treaty and its protocols to controlling proliferation in Latin America, its use as a model for other regions has been much discussed. There was some interest in an African nuclear weapon-free zone, but this has languished because of a lack of leadership and the problem of how to deal with South Africa and Israel and with the emergence in a few African states such as Nigeria of some advocacy for a nuclear option. UN resolutions calling for the negotiation of a Middle East nuclear weapon-free zone have been supported by Egypt, Israel, and others, but political obstacles to direct negotiations with Israel, a fear that such a treaty might freeze Israel's current advantage, and Arab insistence that Israel join the NPT before negotiating such a treaty have contributed to the lack of progress. Pakistan has called for the creation of a South Asian nuclear weapon-free zone, but India has opposed it on the ground that to be meaningful it would have to include China, which India considers the principal nuclear threat it faces. Other such zones, some involving members of U.S. alliances, have been proposed, including the Scandinavian states, the Balkans, and the Pacific Ocean. The nuclear weapon-free zone proposed by the Palme Commission for central Europe is different from other proposals, in that this proposal is primarily concerned with disengagement by the nuclear weapon states rather than with preventing further nuclear proliferation among non-nuclear weapon states.

In none of the other proposed nuclear weapon-free zones is there the fortunate combination of circumstances found in Latin America: there were clearly no nuclear weapon states in the region; the regional initiative was originally supported by all of the region's principal states (Argentina and Brazil were among the early proponents); and there was no serious impact on existing mutual security arrangements.

Agreements Limiting Nuclear Testing

Agreements to limit or ban nuclear testing may contribute significantly to non-proliferation. The broader ramifications of the nuclear testing issue are discussed in Chapter 7, but it is important to appreciate how the different agreements and proposals limiting nuclear testing specifically relate to the non-proliferation problem. The existing Limited Test Ban Treaty (Appendix D), to which all states of current proliferation concern except Argentina, Pakistan, and Cuba are parties, contributes to the difficulty faced by non-nuclear weapon states in developing nuclear weapons. By limiting signatory states to underground testing, it somewhat complicates the initial test and demonstration of a nuclear weapon. However, the extensive underground testing programs of the nuclear-weapon states since the treaty entered into force and India's underground test of its first nuclear explosion suggest that it is not a stringent constraint. Technically, any of the potential nuclear weapon states could conduct their first test within the constraints of the Limited Test Ban Treaty.

The unratified Threshold Test Ban Treaty (Appendix E) and the accompanying Treaty on Underground Nuclear Explosions for Peaceful Purposes (Appendix F) would appear to have little direct relevance to non-proliferation since the threshold is so high (150 kt) that it would not prohibit the initial testing of fission weapons. In fact, it is not even open for signature by non-nuclear weapon states. Moreover, any tests under the associated Treaty on Underground Nuclear Explosions for Peaceful Purposes could blunt the increasingly successful efforts to establish internationally that peaceful nuclear explosions are unpromising and cannot be excepted from non-proliferation constraints because they are indistinguishable from explosions of nuclear weapons.

In contrast, a comprehensive test ban, which is not now being negotiated, is widely held to be important to non-proliferation. Without any testing, states would not be able to demonstrate a nuclear explosive capability, and their confidence that they had in fact achieved such a capability would be greatly reduced. A comprehensive test ban is specifically mentioned as a goal in the preamble to the Non-Proliferation Treaty, and it has been the most persistently demanded step to demonstrate compliance with Article VI of the treaty. Proponents argue that such a ban would contribute significantly to reducing the discriminatory impact of the NPT by forbidding all parties to conduct nuclear tests. Its direct impact on proliferation, however, would depend in large part on whether states of proliferation concern adhere to it. Some argue that the existence of these restrictions would in many cases discourage the initiation of a program. A ban on testing would certainly preclude the development of more advanced fission weapons and thermonuclear weapons. Proponents argue that even states that do not join a comprehensive test ban would be inhibited from conducting tests in an international environment in which most of the states in the world, including the nuclear weapon states, did observe such a ban. However, some critics of a comprehensive test ban argue that testing is not indispensable to the development of first generation nuclear weapons and that, in any event, the states of primary proliferation concern would not be likely to join such a treaty or be influenced by it.

The Physical Security Convention and Related Measures

Until relatively recently, the physical security of civil nuclear facilities as well as military nuclear facilities was considered exclusively the responsibility of the states within which they were located and not an appropriate subject for international concern. However, the rise in international terrorism has raised questions as to the adequacy of existing security procedures and the international consequences of their failure. In the 1970s a number of studies were generally critical of security practices at U.S. civil nuclear facilities. These studies stimulated considerable debate as to the extent of the problem in the United States and in other states where security procedures were uncertain and the threat presumably greater.

Following a declaration on the urgency of the problem at the 1975 NPT review conference, the IAEA published recommendations on levels of physical protection for nuclear materials and facilities based on their relative sensitivity. The United States and other states took steps to bring their regulatory requirements into line with these recommendations. The Nuclear Suppliers' Guidelines included agreed standards for the levels of physical protection that were to be maintained for nuclear exports.

The U.S. Nuclear Non-Proliferation Act of 1978 made adequate physical protection a condition of nuclear export licenses and of new or amended agreements for cooperation. The act also called for the negotiation of an international agreement that would establish international procedures to be followed in the event of diversion, theft, or sabotage of nuclear materials or sabotage of nuclear facilities. In addition, the act required that adequate physical security be established for any international shipment of significant quantities of uranium or special nuclear materials.

Recognizing the problem of physical security, the IAEA negotiated the Convention on the Physical Protection of Nuclear Material. This convention now stands as the principal international instrument dealing with the threat of subnational proliferation. It was opened for signature on March 3, 1980, and now has been signed by 37 states and Euratom and ratified by 10 states, including the United States and the Soviet Union. It will enter into force 30 days after the deposit of the twenty-first instrument of ratification or accession.

The convention seeks to ensure adequate physical protection of nuclear materials during international transport by requiring each party to assure that the levels of physical protection in the Nuclear Suppliers' Guidelines will be applied to any nuclear materials that it imports or exports or that move through its territory. The convention also provides for international cooperation in protecting threatened nuclear material and in ensuring the recovery and return of material that has been seized or stolen. Finally, the convention obligates its parties to enact criminal penalties for such activities as the theft of nuclear materials; unauthorized possession, use, transfer, alteration, or disposal of such materials; or the threat to use nuclear materials for purposes of blackmail. While most states would probably welcome the existence of the convention, the underlying problem is how far the security procedures for civil nuclear facilities can and should be internationalized given the inherently national character of securing domestic activities.

THE ADEQUACY AND SUFFICIENCY OF INTERNATIONAL SAFEGUARDS

Essential Characteristics

There appears to be general agreement that international safeguards are *necessary* for an effective non-proliferation regime. A wide range of views exists, however, as to whether the existing and anticipated safeguards are *sufficient* to ensure the regime's integrity.

International safeguards provide an agreed mechanism for demonstrating and verifying compliance with commitments not to divert safeguarded material or equipment to any military or explosive use and not to transfer it to any non-nuclear weapon state without safeguards. They also provide detailed knowledge of the location, status, and use of safeguarded material or equipment. In addition, they act as separate legal commitments to the IAEA not to engage in such diversion or transfer.

These commitments, contained in every safeguards agreement, reinforce commitments made to the supplier or contained in multilateral treaties such as the NPT. In some cases they can even fill gaps in these other commitments.

To illustrate the difference that safeguards make, it is useful to compare two existing facilities in Israel: the unsafeguarded Dimona reactor and the safeguarded research reactor at Soreq. It is not known how the Dimona reactor has been operated, how much plutonium it has produced, whether the plutonium has been reprocessed, or what may have been done with the reprocessed plutonium. Israel has given no legal or political assurances not to use the plutonium from this reactor to make nuclear explosives. In the case of the safeguarded Soreq reactor, on the other hand, the IAEA has firsthand and continuing knowledge of how it has been operated, how much nuclear material has been supplied to it, and the form, location, and use of that material. Israel has a legal commitment not to divert this plutonium to any military or explosive use and to permit the IAEA to verify compliance with that commitment.

In judging whether existing safeguards are sufficient to prevent proliferation through international nuclear commerce, it is essential to recognize the substantial contribution that safeguards can make to reduce the risk of proliferation. While it would undoubtedly be possible to improve the existing system, it is extremely unlikely that a system anywhere near as effective as the existing one would be accepted internationally if it had to be negotiated from scratch at the present time. This reflects both the loss by the United States of its monopolistic position as a nuclear supplier and the increasing resistance to nuclear export controls by the developing states.

In considering these problems that relate to the international character of the IAEA, it should be recognized that safeguards administered by individual states, rather than the IAEA, are in general probably no longer a practical alternative. Consider, for example, a state that imported reactors from the Federal Republic of Germany and Canadian uranium enriched in the United States and fabricated into fuel in Belgium. The NPT requires each of those suppliers to insist that its export be covered by international safeguards, and each supplier would presumably insist on some form of safeguards even if it were not bound by the treaty. If each chose to apply bilateral safeguards, the customer state would have four sets of inspectors to deal with and four groups to whom it was obligated to furnish records and reports. Each supplier would also have to staff and fund its own complete safeguards effort. The result would probably be viewed as an unacceptable burden by the recipient state and a costly and unnecessary duplication of effort by each of the supplier states.

Technical Considerations

Safeguards are designed with several objectives: to keep track of nuclear materials; to make sure that they continue to be used for known, nonmilitary, nonexplosive purposes; and to detect any diversion of the materials to military, explosive, or unknown purposes. In general, they are intended to deter diversion by providing a high risk of detection and timely notice of its occurrence, and to provide assurance that such diversion has not occurred.

Safeguards consist of a complex of interrelated measures: a comprehensive system of checking and cross-checking the records of relevant facilities and reports of each significant change in location of nuclear material; verification by such means as surveillance of key locations by monitoring cameras and other instrumentation; physical inspections, measurements, and sampling and observation by inspection personnel; and the use of seals and other techniques to help ensure that no unreported movement of materials has occurred. Approximately 5 to 10 kg of separated plutonium or 15 to 20 kg of highly enriched uranium are needed to make a nuclear explosive. The IAEA considers 25 kg of the latter the significant benchmark with regard to losses in the fabrication process.

To account for all of the fissionable material in a large peaceful nuclear program, safeguards must be applied to a wide range of materials and facilities. Most of the material under safeguards is not in a suitable form for use in weapons and would have to be further processed in other facilities before it could be used to make a nuclear explosive. Fissionable materials that could be diverted for use in weapons are in very different forms. This affects both their availability and the practical problems of applying safeguards. Where the fissionable material is contained in discrete, countable items such as reactor fuel rods, accountability is relatively straightforward. Where it is in undifferentiated bulk form, such as a liquid solution or gas flowing through a facility's pipes, the technical problem can be very complex.

The largest quantity of safeguarded nuclear material is located in light-water power reactors. These are among the simplest facilities to safeguard with a high degree of confidence. The fresh fuel is slightly enriched uranium, which could not be used directly to make a weapon. While this fuel is in the operating reactor core, it is totally inaccessible.

Spent fuel, which contains significant quantities of plutonium produced during power production, is discharged from the reactor about once a year directly into a storage pool. There it remains in discrete, highly visible, countable items that are heavy, hot, and very radioactive, emitting a characteristic glow from the Cherenkov effect. These irradiated fuel elements can be removed from the pool only by heavy-duty remote handling techniques and must be put directly into heavy, specially designed shipping casks because of their high residual radioactivity.

Fast breeder reactors present major additional safeguards problems. They contain a very large inventory of plutonium in their fresh fuel (typically enough for several tens of weapons and in a high enough concentration to make direct use for that purpose at least theoretically possible). Additional plutonium is also produced in blanket material around the reactor.

Research reactors range from very small reactors, whose inventories and annual throughputs are insignificant in terms of the requirements for a nuclear weapon, to large reactors that use or produce significant quantities of weapons-usable material. There are a few research reactors fueled by highly enriched uranium or capable of producing significant quantities of plutonium. The most sensitive type of research facility is the so-called fast-critical facility, where highly enriched uranium or plutonium in large enough quantities for a weapon are easily accessible in metallic form. There are only a few known facilities of this type, all located in states considered "safe" from a proliferation standpoint.

Facilities for the fabrication of fuel from natural or low-enriched uranium are the least sensitive of the bulk-handling facilities. These facilities contain large quantities of uranium in bulk form and in scrap, which increases the difficulty of precisely accounting for materials, but these materials are a number of steps removed from being usable in weapons. Facilities for the fabrication of fuel from highly enriched uranium or plutonium, where accountability problems are combined with the greater sensitivity of the materials, present much greater risks.

Reprocessing facilities, where spent reactor fuel is dissolved to recover plutonium and enriched uranium, present the problem of large quantities of weapons-usable material coupled with a difficult problem of materials accounting. These materials are in liquid form in a complex facility where it is difficult for even the operators to know precisely how much material may be tied up within the system. Since possible discrepancies may amount to 1 to 2 percent of the throughput, this may be a serious problem in large facilities, where the material unaccounted for can constitute an appreciable quantity of weapons-usable material.

Most uranium enrichment facilities are in practice operated to produce low-enriched uranium, but many could be operated or modified to produce highly enriched uranium suitable for weapons. Most operating enrichment plants use either the gaseous diffusion process or the centrifuge process. Both processes present the problem of an unusually complex facility with large inventories of material, particularly in the case of gaseous diffusion, tied up in the system. The problem of safeguarding these facilities is further compounded by concern that intrusive procedures might disclose details of the technology that could help other states build or improve their own enrichment facilities. Effective ways of dealing with these facilities have been studied, but this remains on the frontier of safeguards development.

Timely Warning

One objective of safeguards is to provide "timely warning" of the diversion of nuclear materials to military, explosive, or unknown uses. Timely warning means that the monitoring organization is alerted in time to allow other interested parties to prevent the offending state from converting the diverted material to weapons. How much weight should be given to this objective in judging the efficacy of safeguards has been perhaps the most hotly debated safeguards issue in the United States in recent years.

The proponents of a strict interpretation of the timely warning test argue that it cannot be satisfied in situations where critical quantities of weapons-grade fissionable material are readily available to a state that has already acquired the other capabilities and made the necessary preparations to manufacture a nuclear explosive. This argument simply underscores the fact that access to critical quantities of weapons-usable material by a non-nuclear weapon state could be the final step in its acquisition of the technical capability to make a nuclear weapon. In light of this potential risk, proponents of a strict interpretation of the timely warning test believe that weapons-usable materials should not be available to any non-nuclear weapon state.

Those who disagree with this viewpoint do not deny the theoretical risk but believe that considerable weight should also be given to other factors, such as whether the state in question already has access to substantial quantities of weapons-usable material and whether there is any evidence that it might have decided to develop nuclear explosives. Thus, they question whether this argument should be used to prevent the acquisition of additional quantities of weapons-usable material by Japan or the Federal Republic of Germany, each of which already has substantial quantities of such material. In neither case is there any evidence of interest in violating or abrogating the present safeguards agreements, and both states have substantial political disincentives to their taking such actions.

The 1978 Nuclear Non-Proliferation Act deals with this issue by requiring a formal judgment before the United States consents to the reprocessing abroad of nuclear materials originating in the United States or to international transfers of plutonium that the risk of proliferation would not significantly increase. "Foremost consideration" must be given to whether the reprocessing or retransfer would take place "under conditions that will ensure timely warning to the United States well in advance of the time at which the non-nuclear weapon state could transform the diverted material into a nuclear explosive device." This legislative compromise did not dispose of the underlying issue as to what mix of capabilities and intentions constitutes proliferation.

Coverage

A set of important but complicated issues relates to what activities safeguards must cover to achieve the intent of the safeguards regime.

Safeguards on Exports Versus Full-Scope Safeguards. The non-nuclear weapon state parties to the NPT are required to accept safeguards on *all* of their peaceful nuclear activities, a provision defined as "full scope" safeguards. In contrast, the export provisions of the NPT only require safeguards on the nuclear material or equipment actually exported by any of its parties to a non-nuclear weapon state and on any special nuclear material produced with these exports. In other words, the state receiving these exports is not required to have full-scope safeguards on all of its own nuclear activities unless it also is a party to the NPT. Historically, the reason for this formulation was that the United States was not sure when the treaty was being negotiated that all of its European allies and major trading partners would join the treaty. In these circumstances the United States did not want to create a treaty commitment that might require it to violate its existing nuclear supply agreements. This reason no longer carries much weight, since most U.S. allies that are non-nuclear weapon states and with which the United States has nuclear cooperation agreements (including all of those in NATO, Japan, the Republic of Korea, and Taiwan) have joined the treaty. This arrangement provides assurance that the exporter is not adding to the stock of fissionable material that the recipient could legally use for nuclear explosives. However, the recipient state can use safeguarded nuclear imports to build a peaceful nuclear power program and technical base while maintaining separate unsafeguarded facilities to make nuclear explosives.

Short of amending the NPT, exporters can make it a unilateral condition of export that importers accept full-scope safeguards on all of their peaceful nuclear activities. The importer then has to choose between obtaining the benefits of nuclear imports from any supplier having this requirement or foregoing these benefits to escape international nonproliferation controls on all of its nuclear activities. For at least the next decade (and probably considerably longer), it will be virtually impossible for any non-nuclear weapon state other than the major suppliers themselves to develop a nuclear power industry without importing major components or materials. Thus, if all major nuclear supplier countries required full-scope safeguards as a condition for exports, it would exert powerful leverage on most states with a serious interest in nuclear power.

U.S. law now requires full-scope safeguards as a condition of most nuclear exports. This has not solved the problem but has made it a triangular issue among the United States (together with Canada, Australia, and Sweden, which have comparable policies), supplier countries (such as France and the Federal Republic of Germany, which have resisted this export condition), and importing non-nuclear weapon states without full-scope safeguards. Even in the United States, debate continues as to whether the full-scope safeguards condition should be extended to cover exports of nonnuclear components of nuclear facilities and related materials such as heavy water. In fact, pending congressional legislation that is opposed by the Reagan Administration proposes such an extension.

The stated French argument against a full-scope safeguards requirement rests on the doctrinal point that it is improper to use supply conditions to deprive a recipient country of its "sovereign right" to decide what it does with domestic materials and equipment that are not supplied by others. In response, the United States has maintained that the effectiveness of the safeguards on exports is affected by whether or not the recipient country has safeguards on its other nuclear facilities. The United States has also pointed out that the requirement does not in fact deprive a country of its sovereign right of decision. It simply makes the country face a harder decision, since it could still go ahead with its internal program if it were prepared to forego foreign imports. Some members of the nuclear industry have argued against the full-scope safeguards requirement on the grounds that it is unrealistic to hope that it will be made universal. Consequently, they contend, those supplier states that ignore the requirement will quickly capture the market. Some supporters of a strong non-proliferation policy have also argued against the full-scope safeguards requirement, saying that it encourages independent nuclear programs and nuclear trade among the states interested in resisting international controls. While this consideration is relevant for countries without a serious interest in peaceful nuclear power, it does not apply to countries that are deeply involved in nuclear technology, since they will have to depend at least in part on imports from the major supplier countries for some time to come.

De Jure Versus De Facto Full-Scope Safeguards. The NPT includes a commitment by its non-nuclear weapon parties to accept IAEA safeguards on all of their present and future peaceful nuclear activities. These are known as *de jure* full-scope safeguards. The Nuclear Non-Proliferation Act of 1978 introduced a variant known as *de facto* full-scope safeguards. The act makes it a condition of supply that, at the time of export, the recipient must have all of its peaceful nuclear activities under international safeguards. However, the act does not require a legal commitment to place all future peaceful nuclear activities under safeguards. This provision was included to make the full-scope safeguards condition more acceptable to non-NPT states. The principal leverage behind a full-scope safeguards requirement is the disadvantage of being cut off from nuclear imports. A *de facto* full-scope safeguards requirement preserves this leverage in practice, but at the same time leaves open the future option of developing nuclear weapons if the state should decide that this objective is sufficiently important to justify foregoing further nuclear imports.

The *de facto* approach clearly provides less protection against future proliferation than *de jure* full-scope safeguards. *De facto* safeguards would enable a state to acquire a substantial nuclear power program through imports and then, when its dependence on further imports decreased, to decide to build an unsafeguarded fuel cycle for weapons purposes without breaking any international agreement. In reality, the state would probably continue to need spare parts for existing reactors, as well as new reactors, for which further imports from the major supplier states would be necessary.

The United States is the only exporter to have adopted the *de facto* variant of full-scope safeguards. The other chief proponents of full-scope safeguards (Canada, Sweden, and Australia) have criticized this approach as an undesirable compromise.

The Retroactivity of Full-Scope Safeguards. The Nuclear Non-Proliferation Act of 1978 required the cutoff of exports under *existing contractual arrangements* if a recipient non-nuclear weapon state did not, within two years of its enactment, meet the condition of *de facto* full-scope safeguards. Thus, the law was retroactive in the sense that it required a unilateral modification of previously agreed terms after a two-year grace period.

The proponents of this provision argued that the new requirement was essential to achieve comprehensive *de facto* full-scope safeguards. Moreover, they pointed out that this modification fell within a provision in the existing contracts that exports be subject to the laws, regulations, and licensing requirements of the United States. However, it can be questioned whether that clause was originally intended to cover laws that fundamentally changed the conditions of the sale. In any event, the unilateral imposition of this new condition drew substantial international criticism.

The practical policy issue resulting from the new requirement was dramatically illustrated by the case of fuel shipments to India's Tarapur reactors. The agreement for cooperation with India signed in 1963 provided that the United States would furnish all of the fuel for the Tarapur reactors until 1993 and that the Indians would use no other fuel in the reactors. The safeguards requirements and other non-proliferation controls (such as the need for U.S. consent to the reprocessing of the spent fuel in Indian facilities) were tied to this guaranteed fuel supply, as was the clarification obtained by the United States in 1974 that India would not use this fuel in any nuclear explosive, even if labeled "peaceful."

When the Nuclear Non-Proliferation Act forced the United States to terminate fuel deliveries to India, India took the position that since this was a material breach of the agreement, India was relieved of its reciprocal obligations. Among other things, this jeopardized the safeguards on the substantial quantities of spent fuel that had accumulated at the reactor site. More broadly, it reduced U.S. influence on Indian nuclear policy and adversely affected relations between India and the United States in general. A compromise settlement was finally worked out in which the United States waived the Indian obligation to buy the fuel for the reactors exclusively from the United States, and France agreed to supply fuel under the existing limited safeguards arrangements. While this settlement failed to institute the *de facto* full-scope safeguards on the Indian program, proponents of the Nuclear Non-Proliferation Act argued that failing to apply the legislation to India would have been widely viewed as a lack of determination to enforce its full-scope safeguards requirement. Furthermore, they noted that it would have complicated efforts to obtain safeguards over Pakistan's enrichment and reprocessing facilities. Efforts by the United States to persuade other countries to adopt full-scope safeguards as a condition of their nuclear exports have attempted to avoid this retroactivity problem and the issue of breach of existing agreements by suggesting that this requirement be applied only as a condition of future supply commitments.

Safeguarding Transfers of Technology. The NPT requires safeguards on transfers of nuclear materials and equipment. It does not require safeguards as a condition of the transfer of information about nuclear technology. The problem this can create was illustrated by the Indian construction of an unsafeguarded power reactor on the basis of the designs and know-how India acquired from Canada in connection with the construction of a safeguarded Canadian reactor.

To deal with problems arising from technology transfer, the Federal Republic of Germany (as part of its agreement to sell safeguarded reprocessing and enrichment facilities to Brazil) and France (as part of its agreement to sell a safeguarded reprocessing plant to Pakistan) required that any plants of the same general type built by the recipient in the next 20 years would have to be put under IAEA safeguards since they would be presumed to have benefited from the transferred technology. The Nuclear Suppliers' Guidelines of 1977 reflected this tightening of safeguards, but only with respect to sensitive nuclear technology, i.e., technology relating to enrichment, reprocessing, and heavy water production plants. The tighter safeguards would not have covered the loophole exploited by India when it constructed an unsafeguarded power reactor based on the design of the safeguarded Canadian reactor. This problem does not arise where the recipient has a *de jure* full-scope safeguards commitment, since all of its future nuclear facilities automatically come under IAEA safeguards. Consequently, this loophole is an issue only with non-nuclear weapon states not party to the NPT.

Safeguards on Exports to Nuclear Weapon States. The NPT does not obligate a nuclear weapon state to safeguard its own nuclear programs or to require safeguards on nuclear exports to other nuclear weapon states. The rationale was that since these states were allowed to produce nuclear weapons, it did nothing for non-proliferation to demonstrate that certain other parts of their nuclear programs were peaceful. Moreover, this requirement would have diverted scarce IAEA resources from the task of verifying that the fundamental purpose of the NPT was being fulfilled.

Initially, non-nuclear weapon states often complained that the resulting discrimination could adversely affect their civil nuclear programs by revealing commercial secrets. The United States and the United Kingdom attempted to mollify those complaints by voluntarily placing some of their peaceful nuclear activities under IAEA safeguards. France subsequently took a similar, though more limited, action. In 1982 the Soviet Union also made a similar offer and is currently negotiating an agreement with the IAEA to implement it.

The Nuclear Non-Proliferation Act also does not require full-scope safeguards as a condition of exports to nuclear weapon states, but it does call for "safeguards as required in the agreement for cooperation." U.S. commercial nuclear exports to the United Kingdom and France are covered by international safeguards agreements. Current efforts to enter into an agreement for nuclear cooperation with China have reopened the issue. The question has been asked whether the requirements of the Nuclear Non-Proliferation Act would be met if such an agreement included no safeguards. It has been argued that safeguards are necessary to verify compliance with the basic statutory requirement that materials and equipment exported for civil uses not be diverted to any military use. These are significant questions, since China is not a member of the NPT and has in the past denounced the non-proliferation regime as discriminatory against the Third World.

Adequacy of Safeguards Agreements

Another set of issues relates to the adequacy of the safeguards agreements negotiated with the IAEA.

Prohibition on All Nuclear Explosives. The NPT and the Nuclear Suppliers' Guidelines both require safeguards against diversion to any nuclear explosive device, regardless of its intended use. This is to avoid a loophole by which a nation might develop nuclear explosives for purposes asserted to be "peaceful."

There is general agreement that a nuclear explosive device suitable for peaceful applications, such as nuclear excavation or the stimulation of oil or gas wells, would be usable as a weapon and could not be developed without acquiring a nuclear weapons capability. Nevertheless, some earlier safeguards agreements, including that with India, simply required guarantees of "peaceful" use and safeguards against use "to further any military purpose." India asserted that its 1974 nuclear detonation, which did not use safeguarded material, was compatible with its assurances of peaceful use, since it was part of a program directed at peaceful applications. Disagreeing with this interpretation, the United States sought clarifications from all nonparties to the NPT with which it had similar earlier agreements that they would not interpret the safeguards on U.S. nuclear exports in the same way.

The IAEA took the position that any nuclear explosive device inherently furthers a military purpose. It has insisted in all safeguards agreements since the Indian explosion on language explicitly precluding any military or explosive use, while maintaining that this was the meaning of the earlier language. Some of the principal holdouts from the NPT, notably India, Pakistan, Argentina, and Brazil, are still contesting this position. They assert that this concept stems from the NPT, to which they have not agreed, and that their option to develop nuclear explosives for peaceful purposes should not be restricted.

Duration.   In connection with the purchase of a reactor from the Federal Republic of Germany in the early 1970s, Argentina negotiated a safeguards agreement having an initial duration of only five years, extendable by agreement of the parties. In view of the expected 40-year life of the reactor, the duration of the agreement was obviously inadequate. Consequently, the IAEA established a policy that the duration of a safeguards agreement should be reasonably related to the life of the facility and that, once applied, safeguards should continue as long as the safeguarded facility or material could be misused for explosive purposes, regardless of the duration of the agreement under which they were initiated. This policy was embodied in the Nuclear Suppliers' Guidelines in 1977 and in the Nuclear Non-Proliferation Act. It has also been reflected in all safeguards agreements subsequently negotiated by the IAEA with nonparties to the NPT.

Limitation of Safeguards to Declared Facilities.   IAEA safeguarding activities are carried out only at declared facilities, and the IAEA does not have the right to look for undeclared facilities. However, safeguards agreements negotiated under the NPT obligate the state to declare all existing peaceful nuclear activities and to notify the IAEA of any future acquisitions of nuclear materials or equipment. Therefore, failure to declare a facility or nuclear import would in itself be a material breach of the safeguards agreement.

In this connection, national intelligence resources have considerable capability to discover not only undeclared facilities but illegal imports as well. As will be apparent in the discussions of problem states in the final part of this chapter, national intelligence has been very successful in providing "timely warning" of the existence of activities outside the safeguards regime. While some of the details of these programs and the ultimate intention of the governments may not be well known, the ability to identify the existence of undeclared activities, which is all that is necessary to verify compliance with the NPT obligation to declare all facilities, demonstrates the power of national intelligence to reinforce IAEA safeguards on declared facilities. National intelligence presumably has obtained, and is capable of obtaining, considerably more information on foreign nuclear programs than has been made public. The extent of this undisclosed information and the unpredictable nature of future intelligence capabilities further deter states under the safeguards regime from undertaking clandestine activities contrary to their obligations under the NPT or other international commitments.

Inspection Rights

The IAEA Statute, to which all states (including China) with serious nuclear activities are now parties, provides that IAEA safeguards inspectors shall have broad access to facilities and persons concerned with nuclear activities. Specifically, it provides that inspectors shall have "access at all times to all places and data and to any person who by reason of his occupation deals with materials, equipment or facilities which are required . . . to be safeguarded, as necessary to account for source or special nuclear materials supplied and fissionable products and to determine whether there is compliance with the undertaking against use in furtherance of any military purpose . . . and with any other conditions prescribed in the agreement between the Agency and the State or States concerned." It also declares that IAEA inspectors shall be accompanied by representatives of the authorities of the state concerned, if that state so requests, "provided that the inspectors shall not thereby be delayed or otherwise impeded in the exercise of their functions." Most safeguards agreements actually negotiated under this provision limit the frequency of routine inspections, depending on the quantity of sensitive material present at the facility. They also limit the access of normal inspection to certain agreed "strategic points" at the facility. This access is often limited even further by confidential "subsidiary arrangements," in which the IAEA and the state agree on how inspection will normally be implemented in practice. But the IAEA has the right to make special inspections, where the circumstances warrant, without regard to these limitations.

The IAEA has the theoretical right to make surprise inspections, but in practice arrangements must be made for visas and travel except where there is resident inspection. Consequently, advance notice is almost always given before an inspection takes place. The agreements also entitle the host state to reject particular inspectors as long as such rejection does not impede the exercise of the IAEA's responsibilities. There have been instances of abuse of this right, such as Iraq's refusal during an earlier period to accept any inspectors that were not nationals of the Warsaw Pact countries.

Potential Loopholes. Critics point out that there are a number of potential loopholes in the safeguards regime. Since the NPT only requires a commitment not to make nuclear explosives, NPT safeguards permit a state to withdraw materials from safeguards if it declares that those materials are to be used for a nonexplosive military purpose such as submarine propulsion. This apparently glaring loophole, which was included in the NPT in response to earlier desires of certain NATO allies to retain an option to develop nuclear-powered submarines and warships, has in fact never been exercised by any party to the NPT. The principal protection against abuse of this right is to provide by separate agreements (as the United States does in all of its agreements for civil cooperation) that special nuclear materials derived from commercial exports may not be used for any military purpose.

Safeguards are designed to ensure that nuclear materials are accounted for and that they are not being used in a nuclear explosive program. They do not require any authorization for specific uses short of those prohibited. Thus, if safeguards inspectors found declared and safeguarded highly enriched uranium or plutonium fabricated into shapes suitable for nuclear explosives, they would have no authority to object so long as the material remained accounted for and was not actually made into a weapon, although they would presumably report the finding. The proposed international plutonium storage system under discussion in the IAEA could partially remedy this defect.

Implementation of Existing Safeguards

Whatever the theoretical effectiveness of the existing safeguards regime, a general question remains as to how effectively the IAEA uses its existing safeguards rights. Opinions on this issue differ considerably and reflect a complex of practical political and bureaucratic problems involved in operating the safeguards system. The IAEA has limited funds to carry out its safeguard functions. These funds are restricted both by the total resources provided by member states and by the demands of a majority of the member states that the IAEA devote a greater proportion of its resources to assist the development of nuclear energy than to safeguards. For example, in 1984 the total budget of the IAEA was $130 million, of which $32.5 million was for safeguards. Moreover, in view of demands for nondiscriminatory treatment, the IAEA cannot necessarily allocate its limited safeguards resources on a priority basis to the states that are the most serious proliferation risks.

The IAEA also has a continuing problem of finding and retaining technically qualified safeguards personnel, who must be selected with regard to "geographical balance." More serious is the inherent question of the willingness of international civil servants to take the politically difficult or even dangerous actions involved in charging that safeguards have been materially violated. Finally, the Board of Governors of the IAEA is a very broadly based political body, representing 34 countries. This raises the question of the board's willingness to fulfill its function in finding that a material violation has occurred and to call for implementation of the sanctions contemplated by the IAEA Statute. At the same time, it can be argued that, if the safeguards system discovered a serious violation, it would soon be widely known and the international community would be fully informed about the problem.

HOLDOUTS FROM THE INTERNATIONAL REGIME

The most immediate and challenging problem in the field of non-proliferation is how to deal with the small number of states that have deliberately avoided international commitments and controls in order to preserve a nuclear weapons option that some of them appear to be pursuing. These states must be persuaded that the acquisition of nuclear weapons would not be in their net interest. This is a formidable diplomatic task.

The circumstances are so varied that each country requires separate analysis, and even this must be adjusted to account for changes of leadership and local conditions. For example, in the case of Israel the United States has considerable potential leverage, but its use is subject to severe domestic political constraints. In the cases of Korea and Taiwan the United States has great leverage and has used it effectively. In the case of Pakistan, competing security and foreign policy objectives have affected the United States' willingness to use its available leverage. With others, such as Libya, the United States has little or no direct influence and must rely on other concerned states. In every case the relative priority to be given to non-proliferation is an issue. The most effective diplomatic efforts in this area are usually bilateral, but it is obviously important that other countries not undercut these efforts and, if possible, that they be reinforced through other direct diplomatic approaches.

The Countries of Greatest Near-Term Concern

The six countries generally considered to present the greatest near-term risks of proliferation are India, Pakistan, Israel, South Africa, Argentina, and Brazil. Each of them has avoided formal commitments to the international non-proliferation regime. None is a party to the NPT; none has accepted full-scope safeguards; neither Argentina nor Brazil has brought the Treaty of Tlatelolco into effect; and Argentina and Pakistan have not ratified the Limited Test Ban Treaty.

India. The only one of this group that is known to have demonstrated a nuclear explosive capability is India, which in 1974 carried out an underground fission explosion that it insisted was for "peaceful" purposes. India has a large cadre of trained nuclear scientists, engineers, and technicians. Its unsafeguarded facilities include a large research reactor at Trombay that produced the plutonium for the 1974 explosion, a reprocessing plant at Trombay that reprocessed the plutonium, and a reprocessing plant at Tarapur that will be under safeguards only while reprocessing fuel from safeguarded reactors. It has a number of other unsafeguarded nuclear facilities under construction, including several power reactors and an experimental fast breeder reactor.

A change of administration in India in 1977 led to the apparent suspension of its nuclear explosive program. But Indira Gandhi, who originally authorized the 1974 explosion, returned to power in 1980 and there have been recurrent concerns that the program may be revived. While India has repeatedly disclaimed any intention of manufacturing nuclear weapons, it has cited China's nuclear weapons capability as a major reason against forswearing a future option to do so. India is also obviously concerned about Pakistan's nuclear developments, which were in turn spurred on by the 1974 Indian detonation.

The principal non-proliferation objectives with respect to India are to head off the resumption of its nuclear explosive program and to persuade it to continue to maintain a prudent export policy that minimizes the risk of India's becoming a source of further proliferation. Since it is no longer possible to deprive India of the capability to manufacture some nuclear weapons, the practical problem is to persuade it that nuclear weapons are not in its overall interest and that it should refrain from further developments in this direction. Failure to contain Pakistan's nuclear weapon developments would make this task extremely difficult if not impossible.

Pakistan. Pakistan has been constructing a large, unsafeguarded centrifuge enrichment plant using plans stolen from the URENCO enrichment facility in the Netherlands and parts and components procured from various Western sources. Pakistan also has an unsafeguarded fuel fabrication facility and is constructing an unsafeguarded reprocessing facility. The only known indigenous source of plutonium for the reprocessing facility is the safeguarded power reactor at Kanupp. While the IAEA earlier acknowledged some deficiencies in its safeguards arrangements at that reactor, these problems have reportedly been remedied. Over the years considerable evidence has accumulated indicating that Pakistan was headed toward a nuclear explosive capability.

At the same time, there are important disincentives to a Pakistani decision to carry its nuclear weapons option all the way to a nuclear explosion. This event would almost certainly lead to the resumption of the Indian nuclear weapons program. It would probably also result in a cutoff of U.S. economic and military aid under the present five-year $3.2 billion program, since the authorizing legislation calls for such a cutoff if the recipient detonates a nuclear explosive device or transfers such a device to a non-nuclear weapon state. The transfer provision was included in view of the concern, which Pakistan claims is unfounded, that it might be persuaded to help other Muslim states acquire a nuclear explosive capability. Pakistan is also reportedly interested in acquiring an additional power reactor from Western sources, although it is not clear how it would pay for it. The United States has attempted to persuade potential European suppliers that they should insist on a full-scope safeguards commitment from Pakistan as a condition of such a sale.

The principal non-proliferation objectives with respect to Pakistan are to head off a test of a nuclear device; to inhibit the completion of its unsafeguarded enrichment and reprocessing facilities; to obtain full-scope safeguards coverage and, if possible, an explicit dedication of the enrichment facility to low enrichment; and to persuade Pakistan to maintain a prudent nuclear export policy that minimizes the risk of its becoming a source of further proliferation.

Israel. The declared policy of Israel has long been that it would not be the first to introduce nuclear weapons into the Middle East. Nevertheless, Israel is generally believed to have the capability to manufacture a nuclear weapon on very short notice if it has not already done so. As long ago as 1976, a senior CIA official stated that Israel was estimated to have a stockpile of 10 to 20 weapons or their necessary components.

Israel has a strong base of highly trained nuclear scientists, engineers, and technicians. It has an unsafeguarded research reactor capable of producing enough plutonium for at least one bomb a year and an unsafeguarded reprocessing facility. In addition, it may have acquired enough highly enriched uranium for a number of nuclear weapons from a U.S. facility that could not account for the loss of the material under incriminating circumstances.

Israel has been careful to maintain a studied ambiguity between its declared nuclear policy and its nuclear capabilities. Apparently, it wishes to maintain an implicit nuclear deterrent while avoiding an open confrontation on the nuclear issue with its neighbors.

The principal non-proliferation objectives with respect to Israel are to head off a nuclear detonation, to discourage the deployment of any weapons that might exist, and to persuade Israel that the acquisition of nuclear weapons is not in its net interest. Above all, Israel should be dissuaded from shifting to an openly declared military policy based on a nuclear deterrent. In pursuing these objectives, the United States has much more potential leverage than in the case of any other high-risk country if it is prepared to use it.

South Africa. The principal focus of proliferation concern in South Africa is its unsafeguarded uranium enrichment plant, which is probably capable of producing highly enriched uranium from ample supplies of indigenous uranium. In 1977 a Soviet satellite reconnaissance discovered apparent preparations for an underground nuclear weapon test in the Kalahari Desert, which were privately called to the attention of the U.S. government. As a result of strong diplomatic approaches by the United States and others, plans for the test were apparently discontinued. In 1979 a U.S. satellite recorded a signal that many initially thought to be from a nuclear detonation over the ocean between South Africa and Antarctica. A prestigious group of scientists appointed by President Carter concluded that the signal was most likely caused by a micrometeor triggering the detection device and not by a nuclear explosion; nevertheless, an element of suspicion has remained.

Attempts to persuade South Africa to join the NPT and to subject its enrichment facility to safeguards have proved unavailing. While a nuclear weapon capability would single it out from its potential African adversaries, the military relevance of nuclear weapons to South Africa's security problems is not obvious.

The principal non-proliferation objectives with respect to South Africa, in addition to persuading it that the acquisition of nuclear weapons is not in its best interest, are to head off a nuclear detonation; to persuade it to accept safeguards on its enrichment facility and, if possible, on all of its nuclear activities; to dedicate its enrichment plant to the production of low-enriched uranium; and to persuade it to continue to maintain a prudent nuclear export policy that will minimize the risk of its becoming a source of further proliferation. The South African government has recently announced that it will accept safeguards on its projected commercial enrichment plant (but not on its operating pilot plant) and will require safeguards on all of its nuclear exports to non-nuclear weapon states.

Argentina. The proliferation concern in Argentina, which has indigenous uranium resources and is acquiring a large heavy water production plant, focuses on an unsafeguarded reprocessing plant, a large unsafeguarded research reactor that has long been planned, and a recently announced unsafeguarded uranium enrichment facility that is under construction.

There appears to be no evidence of a definitive decision by Argentina to acquire nuclear explosives or undertake a program with that end. There are differences of view as to whether the Falkland crisis increased the likelihood that the Argentines might move in this direction. The recent change of government in Argentina appears to present a new opportunity to pursue the principal non-proliferation objectives with respect to the country: to persuade it on economic grounds not to proceed with the enrichment plant or, if it does, to subject the plant to safeguards and dedicate it to low enrichment; to proceed with ratification of the Treaty of Tlatelolco and the negotiation of a full-scope safeguards agreement with the IAEA; and to exercise prudence in its nuclear exports so as to minimize the risk of its becoming a source of further proliferation. The last of these objectives is important because of Argentina's past aspirations to become a nuclear supplier to Third World countries.

Brazil. The focus of proliferation concern in Brazil has been its 1975 agreement with the Federal Republic of Germany to acquire a uranium enrichment facility and a reprocessing plant. Under the agreement, both facilities would be placed under IAEA safeguards, as would any of their output and any facilities of the same type built by Brazil in the next 20 years. Although the agreement was concluded in 1975, none of the facilities is yet in operation, and there has been a major cutback in the scope of the Brazilian nuclear power program. There have been reports, however, that Brazil is conducting research on advanced isotope separation techniques, which would not be covered by the safeguards provision of the German agreement.

While Brazil now has a number of minor unsafeguarded facilities, there is some prospect that it may soon be covered by *de facto* full-scope safeguards. Brazil has not yet carried out its commitment in the Treaty of Tlatelolco to negotiate a full-scope safeguards agreement. After ratifying the treaty, Brazil made its entry into force contingent on all Latin American countries joining the treaty and all eligible countries joining its protocols. This requires Argentina and Cuba to adhere to the treaty (although Brazil might waive the latter requirement) and completion of the pending French ratification of Protocol I.

Thus, non-proliferation objectives in Brazil are to achieve full-scope safeguards coverage, to complete the preconditions so that the Treaty of Tlatelolco will come into force for Brazil, to persuade Brazil that a nuclear explosive program would not be in its security or economic interests, and to persuade Brazil to exercise prudence in its nuclear exports so as to minimize the risk of its becoming a source of further proliferation.

Other Holdouts. Cuba's limited technical capabilities make its threat of proliferation less severe than for that of any of the six countries listed above, but it is acquiring a safeguarded research reactor and a safeguarded power reactor from the Soviet Union. Although there is a question in this case whether the Soviet Union will follow its usual practice of requiring the return of the spent fuel, Cuba will probably continue to have *de facto* full-scope safeguards. Moreover, the Soviet Union is committed by the NPT not to help or encourage Cuba to acquire a nuclear explosive capability, and it seems most unlikely that it would tolerate independent Cuban efforts in that direction. Future proliferation concerns about Cuba would be greatly reduced if Cuba signed and ratified the Treaty of Tlatelolco, an action that the Soviet Union should be in a position to encourage. As discussed above, Cuba's signing and ratifying the treaty would put great pressure on Argentina to ratify, which would bring the treaty into force for all Latin American countries.

Notwithstanding their acceptance of full-scope safeguards and other obligations of the NPT, there has been some concern about the actual intentions of a few parties to the NPT. The most notable of these is Libya, whose leader has reportedly made a number of unsuccessful efforts to acquire nuclear weapons directly from other states. These actions, coupled with his open support for terrorist organizations and activities, justify greater proliferation concerns about Libya than its immediate technical capabilities would indicate. Libya has little capability to produce nuclear weapons indigenously, even though it is acquiring a safeguarded research reactor and plans to acquire a safeguarded power reactor from the Soviet Union. This situation would be substantially improved if the Soviet Union required the return of the spent fuel, as it has previously done in similar situations.

The Middle Eastern parties to the NPT that have aroused some proliferation concern are Iraq and Iran. The former was building a safeguarded nuclear research center, which may have presented some proliferation risks, but this was destroyed in 1981 by Israeli bombing. Reconstruction of the facility has not yet begun. Iran's extremely ambitious civil nuclear program under the Shah was suspended by the Khomeini government, but there are reports that Iran is now negotiating with Western European countries for the completion of one or more of the power reactors involved. All potential suppliers are bound by the NPT to require safeguards on any such reactor and its fuel. But the international irresponsibility displayed by Iran provides serious grounds for concern about longer-term proliferation problems in this country.

In the Far East, two NPT parties whose potential interest in developing a nuclear weapons capability was of considerable concern in the past are Taiwan and the Republic of Korea. A major disincentive in both cases was the damage that such development would have done to their security relations with the United States and with their ability to continue to obtain vital supplies for their civil nuclear power plants, which provide a substantial portion of their electric power. The United States made use of its special relations with these two countries to convince them that the development of an independent nuclear capability would be ill advised. These two countries do not now appear to constitute serious proliferation risks.

While this completes the list of countries generally thought to be of near-term proliferation concern, it must be recognized that if the international security environment changed dramatically—if, for example, the current non-proliferation regime or certain alliance relationships disintegrated—some of the major industrial states, which clearly have the capabilities to make nuclear weapons on a large scale relatively quickly, might decide to do so. This would obviously have extremely serious implications for international security.

The Problem of Coordination with Other Concerned Countries

Improved controls on nuclear exports are an important, but clearly not sufficient, tool in seeking to achieve the non-proliferation objectives for the problem countries considered above. The United States can no longer effectively impose these export controls alone but must seek coordination among all significant potential suppliers. These suppliers remain relatively few in number and share some commitment to non-proliferation. A question for the future is how to keep new suppliers who may not share that commitment from undercutting these efforts. This could be a particular problem in the case of China, whose government has in the past opposed non-proliferation on doctrinal grounds. The non-proliferation export regime will be tightened if the Chinese government adheres to its recently announced policy of requiring safeguards on Chinese nuclear exports.

The existing non-proliferation regime provides an indispensable framework for coordinating the safeguards on all nuclear exports required by the NPT and the Nuclear Suppliers' Guidelines among the present suppliers. However, differences remain among these suppliers on several important issues. The most fundamental difference is the relative priority to be accorded to non-proliferation compared with other foreign policy objectives. Many of the major supplier countries, while willing to endorse non-proliferation as a goal, do not consider it a major concern from the point of view of their national security. NATO allies have frequently reacted defensively to U.S. non-proliferation initiatives. Accordingly, they have given it relatively low priority in cases where it conflicted with other foreign policy interests, as the United States has also done on occasion. Some of the supplier countries also disagree with the emphasis that has sometimes been placed on capabilities rather than intentions. This is understandable, since they also have the requisite capabilities for a nuclear weapons program and therefore wish to emphasize the importance of legal and political commitments to non-proliferation.

The ambivalence between a policy of constructive engagement and denial that has characterized U.S. policy is shared by other suppliers. The Federal Republic of Germany has argued that the safeguards and other commitments it obtained from Brazil in connection with the agreement to sell enrichment and reprocessing technology were a positive contribution to non-proliferation. While France did not complete its sale of a reprocessing facility to Pakistan, some French leaders have questioned whether the completion of that sale, which would at least have resulted in safeguards on the facility, would not have been preferable to the present situation where there is an unsafeguarded reprocessing facility. The Swiss have argued that the expanded safeguards coverage obtained through the sale of a heavy water production facility to Argentina offsets any proliferation risk involved.

The various attitudes toward these basic issues also affect the prospects for concerted action in using nonmilitary leverage to head off or respond to actions that clearly portend further nuclear proliferation. Such concerted action is limited by the special relationships of certain of the participants to the problem countries, such as that of the United States to Israel or that of France and Italy to Iraq (from whom those countries obtain substantial supplies of oil). Conflicting foreign policy goals, such as U.S. interest in supporting Pakistani resistance to the Soviet threat in South Asia, encouraging the new Argentine government, and not precipitating default on international loans by Argentina or Brazil, also limit such action.

Finally, the failure of the United States and the Soviet Union to make progress in nuclear arms control, and the continued emphasis of the nuclear weapon states on the need for and utility of nuclear weapons in an ever-broadening military context, greatly complicate the task of persuading the problem countries not to acquire their own nuclear weapons.

Acronyms

ABM   *Anti-Ballistic Missile.* A missile designed to defend against a ballistic missile attack by destroying incoming ballistic missile warheads.

ACDA   *Arms Control and Disarmament Agency.* The independent U.S. agency that deals with arms control matters.

AEC   *Atomic Energy Commission.* The U.S. agency for military and peaceful atomic energy development from 1946 to 1975.

ALCM   *Air-Launched Cruise Missile.* A cruise missile launched from an aircraft in flight. A cruise missile is a pilotless, aerodynamic vehicle with an air-breathing jet engine designed to operate in the atmosphere.

ASAT   *Anti-Satellite.* A weapon used to attack satellites.

ASBM   *Air-to-Surface Ballistic Missile.* A ballistic missile launched from an airborne carrier to hit a surface target.

ASW   *Antisubmarine Warfare.* The complex of activities involved in the detection, identification, tracking, and destruction of hostile submarines.

BMD   *Ballistic Missile Defense.* Measures for defending against an attack by ballistic missiles.

CD   *Committee on Disarmament.* A negotiating body of the United Nations for multilateral disarmament treaties set up in 1978. Renamed the Conference on Disarmament in February 1984.

275

CTB   *Comprehensive Test Ban.* A proposed ban on nuclear test-
            ing in all environments.

ENDC   *Eighteen Nation Disarmament Conference.* A negotiating
            organ for multilateral agreements established in 1961.

FBS   *Forward-Based Systems.* A Soviet term defining interme-
            diate-range U.S. nuclear delivery systems based in
            third countries or on aircraft carriers that can strike
            targets in the Soviet Union.

FRODs   *Functionally Related Observable Differences.* A SALT II
            term referring to differences in the observable features
            of airplanes that provide for distinguishing between
            those aircraft that are and are not capable of perform-
            ing certain functions limited by SALT.

GLCM   *Ground-Launched Cruise Missile.* A cruise missile
            launched from a land-based platform. See ALCM.

IAEA   *International Atomic Energy Agency.* An international
            organization established in 1956 to promote peaceful
            uses of atomic energy and to provide safeguards to as-
            sure that atomic installations are not used for weapons
            purposes.

ICBM   *Intercontinental Ballistic Missile.* A land-based rocket-
            propelled vehicle capable of delivering a warhead
            through space to a target at ranges in excess of 5,500
            km.

INF   *Intermediate-Range Nuclear Forces.* A U.S. term for land-
            based nuclear systems with a range capability greater
            than that of short-range nuclear forces but less than
            that of intercontinental forces (5,500 km).

INFCE   *International Nuclear Fuel Cycle Evaluation.* An interna-
            tional study initiated by the Carter Administration to
            assess the comparative economic, technical, and politi-
            cal advantages of various nuclear fuel cycles, with par-
            ticular reference to the use of plutonium for recycling
            and breeder reactors.

LTB   *Limited Test Ban.* A treaty initially signed by the United
            States, the United Kingdom, and the Soviet Union in
            1963 to prohibit the testing of nuclear weapons in the
            atmosphere, in space, and underwater.

MAD   *Mutual Assured Destruction.* The concept of recipro-
            cal deterrence that rests on the inherent ability of
            the two nuclear superpowers to inflict unacceptable
            damage on one another after surviving a nuclear first
            strike.

MARV   *Maneuverable Reentry Vehicle.* A ballistic missile reentry vehicle whose ballistic trajectory can be adjusted by internal or external mechanisms, enabling it to evade ABM defenses and/or strike its target with a high degree of accuracy.

MIRV   *Multiple Independently Targetable Reentry Vehicle.* A package of two or more reentry vehicles with nuclear warheads that can be carried by a single ballistic missile and delivered to separate targets.

MX   *Missile Experimental.* A new U.S. ICBM with ten warheads scheduled to be deployed in the late 1980s.

NPT   *Non-Proliferation Treaty.* A multilateral treaty to prevent the spread of nuclear weapons while guaranteeing the peaceful uses of nuclear energy through cooperation.

NNPA   *Nuclear Non-Proliferation Act.* Legislation providing for the current U.S. policy controlling the export of nuclear materials and equipment.

NTM   *National Technical Means.* A nation's technical intelligence assets that can monitor another country's compliance with an arms control agreement from outside of that country. NTM include satellite-based sensors such as photographic reconnaissance, aircraft-based systems such as radars and optical systems, and sea- and ground-based systems such as radars and antennas for collecting telemetry.

OST   *Outer Space Treaty.* A 1967 treaty that prohibits the placing of nuclear weapons or other weapons of mass destruction around the earth and outlaws the establishment of military bases, installations, and fortifications, the testing of any type of weapons, and the conduct of military maneuvers in outer space.

PAR   *Perimeter Acquisition Radar.* Radars on the perimeter of a nation designed to detect incoming warheads and predict their trajectories while they are still several thousand miles from their targets.

PD 59   *Presidential Directive 59.* A directive of the Carter Administration formalizing the doctrine of countervailing strategy, which emphasized the importance of flexible options and survivable command and control to assure deterrence against a wide range of threats.

PNE   *Peaceful Nuclear Explosion.* A nuclear explosion for peaceful purposes.

PNET   *Peaceful Nuclear Explosions Treaty.* A treaty that established a limit of 150 kt on individual underground nuclear explosions for peaceful purposes and a limit of 1,500 kt on any simultaneous series of nuclear explosions for peaceful purposes. It was signed by the United States and the Soviet Union in 1976.

RORSAT   *Radar Ocean Reconnaissance Satellite.* A Soviet satellite intended to provide real time location of targets such as ships at sea.

RV   *Reentry Vehicle.* The part of a ballistic missile designed to reenter the earth's atmosphere in the terminal portion of its trajectory to deliver a warhead to a target.

SALT   *Strategic Arms Limitation Talks.* Negotiations initiated in 1969 between the United States and the Soviet Union directed at limiting the strategic offensive and defensive nuclear forces of the two sides.

SAM   *Surface-to-Air Missile.* A surface-launched missile designed to operate against aircraft or other aerodynamic targets.

SCC   *Standing Consultative Commission.* A permanent U.S.-Soviet commission established in the SALT I agreements and incorporated into the SALT II agreement to deal with questions of compliance and the working out of additional procedures to implement the provisions of the SALT agreements.

SDI   *Strategic Defense Initiative.* The Reagan Administration's research and development program to investigate the possibility of developing a nationwide ballistic missile defense with the goal of ultimately eliminating the strategic role of nuclear weapons.

SLBM   *Submarine-Launched Ballistic Missile.* A ballistic missile launched from a submarine. See ICBM.

SRAM   *Short-Range Attack Missile.* A nuclear-armed short-range air-to-surface missile deployed on B-52s for defense suppression as well as target attack.

START   *Strategic Arms Reduction Talks.* Negotiations between the United States and the Soviet Union initiated in 1982 by the Reagan Administration to seek substantial reductions in strategic nuclear weapons.

TTB   *Threshold Test Ban.* A treaty signed by the United States and the Soviet Union in 1974 to prohibit underground tests of nuclear weapons with a yield greater than 150 kt.

Appendix A

Interim Agreement Between the
United States of America and the
Union of Soviet Socialist
Republics on Certain Measures
With Respect to the Limitation
of Strategic Offensive Arms

Interim Agreement Between the United States of America and the Union of Soviet Socialist Republics on Certain Measures With Respect to the Limitation of Strategic Offensive Arms

Signed at Moscow May 26, 1972
Approval authorized by U.S. Congress September 30, 1972
Approved by U.S. President September 30, 1972
Notices of acceptance exchanged October 3, 1972
Entered into force October 3, 1972

The United States of America and the Union of Soviet Socialist Republics, hereinafter referred to as the Parties,

Convinced that the Treaty on the Limitation of Anti-Ballistic Missile Systems and this Interim Agreement on Certain Measures with Respect to the Limitation of Strategic Offensive Arms will contribute to the creation of more favorable conditions for active negotiations on limiting strategic arms as well as to the relaxation of international tension and the strengthening of trust between States,

Taking into account the relationship between strategic offensive and defensive arms,

Mindful of their obligations under Article VI of the Treaty on the Non-Proliferation of Nuclear Weapons,

Have agreed as follows:

Article I

The Parties undertake not to start construction of additional fixed land-based intercontinental ballistic missile (ICBM) launchers after July 1, 1972.

Article II

The Parties undertake not to convert land-based launchers for light ICBMs, or for ICBMs of older types deployed prior to 1964, into land-based launchers for heavy ICBMs of types deployed after that time.

Article III

The Parties undertake to limit submarine-launched ballistic missile (SLBM) launchers and modern ballistic missile submarines to the numbers operational and under construction on the date of signature of this Interim Agreement, and in addition to launchers and submarines constructed under procedures established by the Parties as replacements for an equal number of ICBM launchers of older types deployed prior to 1964 or for launchers on older submarines.

Article IV

Subject to the provisions of this Interim Agreement, modernization and replacement of strategic offensive ballistic missiles and launchers covered by this Interim Agreement may be undertaken.

Article V

1. For the purpose of providing assurance of compliance with the provisions of this Interim Agreement, each Party shall use national technical means of verification at its disposal in a manner consistent with generally recognized principles of international law.

2. Each party undertakes not to interfere with the national technical means of verification of the other Party operating in accordance with paragraph 1 of this Article.

3. Each Party undertakes not to use deliberate concealment measures which impede verification by national technical means of compliance with the provisions of this Interim Agreement. This obligation shall not require changes in current construction, assembly, conversion, or overhaul practices.

Article VI

To promote the objectives and implementation of the provisions of this Interim Agreement, the Parties shall use the Standing Consultative Commission established under Article XIII of the Treaty on the Limitation of Anti-Ballistic Missile Systems in accordance with the provisions of that Article.

Article VII

The Parties undertake to continue active negotiations for limitations on strategic offensive arms. The obligations provided for in this Interim Agreement shall not prejudice the scope or terms of the limitations on strategic offensive arms which may be worked out in the course of further negotiations.

Article VIII

1. This Interim Agreement shall enter into force upon exchange of written notices of acceptance by each Party, which exchange shall take place simultaneously with the exchange of instruments of ratification of the Treaty on the Limitation of Anti-Ballistic Missile Systems.

2. This Interim Agreement shall remain in force for a period of five years unless replaced earlier by an agreement on more complete measures limiting strategic offensive arms. It is the objective of the Parties to conduct active follow-on negotiations with the aim of concluding such an agreement as soon as possible.

3. Each Party shall, in exercising its national sovereignty, have the right to withdraw from this Interim Agreement if it decides that extraordinary events related to the subject matter of this Interim Agreement have jeopardized its supreme interests. It shall give notice of its decision to the other Party six months prior to withdrawal from this Interim Agreement. Such notice shall include a statement of the extraordinary events the notifying Party regards as having jeopardized its supreme interests.

DONE at Moscow on May 26, 1972, in two copies, each in the English and Russian languages, both texts being equally authentic.

FOR THE UNITED STATES OF AMERICA

RICHARD NIXON

President of the United States of America

FOR THE UNION OF SOVIET SOCIALIST REPUBLICS

L. I. BREZHNEV

General Secretary of the Central Committee of the CPSU

Protocol to the Interim Agreement Between the United States of America and the Union of Soviet Socialist Republics on Certain Measures With Respect to the Limitation of Strategic Offensive Arms

The United States of America and the Union of Soviet Socialist Republics, hereinafter referred to as the Parties,

Having agreed on certain limitations relating to submarine-launched ballistic missile launchers and modern ballistic missile submarines, and to replacement procedures, in the Interim Agreement,

Have agreed as follows:

The Parties understand that, under Article III of the Interim Agreement, for the period during which that Agreement remains in force:

The U.S. may have no more than 710 ballistic missile launchers on submarines (SLBMs) and no more than 44 modern ballistic missile submarines. The Soviet Union may have no more than 950 ballistic missile launchers on submarines and no more than 62 modern ballistic missile submarines.

Additional ballistic missile launchers on submarines up to the above-mentioned levels, in the U.S.—over 656 ballistic missile launchers on nuclear-powered submarines, and in the U.S.S.R.—over 740 ballistic missile launchers on nuclear-powered submarines, operational and under construction, may become operational as replacements for equal numbers of ballistic missile launchers of older types deployed prior to 1964 or of ballistic missile launchers on older submarines.

The deployment of modern SLBMs on any submarine, regardless of type, will be counted against the total level of SLBMs permitted for the U.S. and the U.S.S.R.

This Protocol shall be considered an integral part of the Interim Agreement.

DONE at Moscow this 26th day of May, 1972

FOR THE UNITED STATES OF AMERICA

FOR THE UNION OF SOVIET SOCIALIST REPUBLICS

RICHARD NIXON

L. I. BREZHNEV

President of the
United States of America

General Secretary of the
Central Committee of the CPSU

Agreed Statements, Common Understandings, and Unilateral Statements Regarding the Interim Agreement Between the United States of America and the Union of Soviet Socialist Republics on Certain Measures With Respect to the Limitation of Strategic Offensive Arms

1. Agreed Statements

The document set forth below was agreed upon and initialed by the Heads of the Delegations on May 26, 1972 (letter designations added):

AGREED STATEMENTS REGARDING THE INTERIM AGREEMENT BETWEEN THE UNITED STATES OF AMERICA AND THE UNION OF SOVIET SOCIALIST REPUBLICS ON CERTAIN MEASURES WITH RESPECT TO THE LIMITATION OF STRATEGIC OFFENSIVE ARMS

[A]

The Parties understand that land-based ICBM launchers referred to in the Interim Agreement are understood to be launchers for strategic ballistic missiles capable of ranges in excess of the shortest distance between the northeastern border of the continental U.S. and the northwestern border of the continental USSR.

[B]

The Parties understand that fixed land-based ICBM launchers under active construction as of the date of signature of the Interim Agreement may be completed.

[C]

The Parties understand that in the process of modernization and replacement the dimensions of land-based ICBM silo launchers will not be significantly increased.

[D]

The Parties understand that during the period of the Interim Agreement there shall be no significant increase in the number of ICBM or SLBM test and training launchers, or in the number of such launchers for modern land-based heavy ICBMs. The Parties further understand that construction or conversion of ICBM launchers at test ranges shall be undertaken only for purposes of testing and training.

[E]

The Parties understand that dismantling or destruction of ICBM launchers of older types deployed prior to 1964 and ballistic missile launchers on older submarines being replaced by new SLBM launchers on modern submarines will be initiated at the time of the beginning of sea trials of a replacement submarine, and will be completed in the shortest possible agreed period of time. Such dismantling or destruction, and timely notification thereof, will be accomplished under procedures to be agreed in the Standing Consultative Commission.

2. Common Understandings

Common understanding of the Parties on the following matters was reached during the negotiations:

A. Increase in ICBM Silo Dimensions

Ambassador Smith made the following statement on May 26, 1972:

The Parties agree that the term "significantly increased" means that an increase will not be greater than 10–15 percent of the present dimensions of land-based ICBM silo launchers.

Minister Semenov replied that this statement corresponded to the Soviet understanding.

B. Standing Consultative Commission

Ambassador Smith made the following statement on May 22, 1972:

The United States proposes that the sides agree that, with regard to initial implementation of the ABM Treaty's Article XIII on the Standing Consultative Commission (SCC) and of the consultation Articles to the Interim Agreement on offensive arms and the Accidents Agreement,[1] agreement establishing the SCC will be worked out early in the follow-on SALT negotiations; until that is completed, the following arrangements will prevail: when SALT is in session, any consultation desired by either side under these Articles can be carried out by the two SALT Delegations; when SALT is not in session, *ad hoc* arrangements for any desired consultations under these Articles may be made through diplomatic channels.

Minister Semenov replied that, on an *ad referendum* basis, he could agree that the U.S. statement corresponded to the Soviet understanding.

C. Standstill

On May 6, 1972, Minister Semenov made the following statement:

In an effort to accommodate the wishes of the U.S. side, the Soviet Delegation is prepared to proceed on the basis that the two sides will in fact observe the obligations of both the Interim Agreement and the ABM Treaty beginning from the date of signature of these two documents.

In reply, the U.S. Delegation made the following statement on May 20, 1972:

The U.S. agrees in principle with the Soviet statement made on May 6 concerning observance of obligations beginning from date of signature but we would like to make clear our understanding that this means that, pending ratification and acceptance, neither side would take any action prohibited by the agreements after

---

[1]See Article 7 of Agreement to Reduce the Risk of Outbreak of Nuclear War Between the United States of America and the Union of Soviet Socialist Republics, signed Sept. 30, 1971.

they had entered into force. This understanding would continue to apply in the absence of notification by either signatory of its intention not to proceed with ratification or approval.

The Soviet Delegation indicated agreement with the U.S. statement.

3. Unilateral Statements

(a) The following noteworthy unilateral statements were made during the negotiations by the United States Delegation:

A. Withdrawal from the ABM Treaty

On May 9, 1972, Ambassador Smith made the following statement:

The U.S. Delegation has stressed the importance the U.S. Government attaches to achieving agreement on more complete limitations on strategic offensive arms, following agreement on an ABM Treaty and on an Interim Agreement on certain measures with respect to the limitation of strategic offensive arms. The U.S. Delegation believes that an objective of the follow-on negotiations should be to constrain and reduce on a long-term basis threats to the survivability of our respective strategic retaliatory forces. The USSR Delegation has also indicated that the objectives of SALT would remain unfulfilled without the achievement of an agreement providing for more complete limitations on strategic offensive arms. Both sides recognize that the initial agreements would be steps toward the achievement of more complete limitations on strategic arms. If an agreement providing for more complete strategic offensive arms limitations were not achieved within five years, U.S. supreme interests could be jeopardized. Should that occur, it would constitute a basis for withdrawal from the ABM Treaty. The U.S. does not wish to see such a situation occur, nor do we believe that the USSR does. It is because we wish to prevent such a situation that we emphasize the importance the U.S. Government attaches to achievement of more complete limitations on strategic offensive arms. The U.S. Executive will inform the Congress, in connection with Congressional consideration of the ABM Treaty and the Interim Agreement, of this statement of the U.S. position.

B. Land-Mobile ICBM Launchers

The U.S. Delegation made the following statement on May 20, 1972:

In connection with the important subject of land-mobile ICBM launchers, in the interest of concluding the Interim Agreement the U.S. Delegation now withdraws its proposal that Article I or an agreed statement explicitly prohibit the deployment of mobile land-based ICBM launchers. I have been instructed to inform you that, while agreeing to defer the question of limitation of operational land-mobile ICBM launchers to the subsequent negotiations on more complete limitations on strategic offensive arms, the U.S. would consider the deployment of operational land-mobile ICBM launchers during the period of the Interim Agreement as inconsistent with the objectives of that Agreement.

C. Covered Facilities

The U.S. Delegation made the following statement on May 20, 1972:

I wish to emphasize the importance that the United States attaches to the provisions of Article V, including in particular their application to fitting out or berthing submarines.

D. "Heavy" ICBM's

The U.S. Delegation made the following statement on May 26, 1972:

The U.S. Delegation regrets that the Soviet Delegation has not been willing to agree on a common definition of a heavy missile. Under these circumstances, the U.S. Delegation believes it necessary to state the following: The United States would consider any ICBM having a volume significantly greater than that of the largest light ICBM now operational on either side to be a heavy ICBM. The U.S. proceeds on the premise that the Soviet side will give due account to this consideration.

(b) The following noteworthy unilateral statement was made by the Delegation of the U.S.S.R. and is shown here with the U.S. reply:

On May 17, 1972, Minister Semenov made the following unilateral "Statement of the Soviet Side":

Taking into account that modern ballistic missile submarines are presently in the possession of not only the U.S., but also of its NATO allies, the Soviet Union agrees that for the period of effectiveness of the Interim 'Freeze' Agreement the U.S. and its NATO allies have up to 50 such submarines with a total of up to 800 ballistic missile launchers thereon (including 41 U.S. submarines with 656 ballistic missile launchers). However, if during the period of effectiveness of the Agreement U.S. allies in NATO should increase the number of their modern submarines to exceed the numbers of submarines they would have operational or under construction on the date of signature of the Agreement, the Soviet Union will have the right to a corresponding increase in the number of its submarines. In the opinion of the Soviet side, the solution of the question of modern ballistic missile submarines provided for in the Interim Agreement only partially compensates for the strategic imbalance in the deployment of the nuclear-powered missile submarines of the USSR and the U.S. Therefore, the Soviet side believes that this whole question, and above all the question of liquidating the American missile submarine bases outside the U.S., will be appropriately resolved in the course of follow-on negotiations.

On May 24, Ambassador Smith made the following reply to Minister Semenov:

The United States side has studied the "statement made by the Soviet side" of May 17 concerning compensation for submarine basing and SLBM submarines belonging to third countries. The United States does not accept the validity of the considerations in that statement.

On May 26 Minister Semenov repeated the unilateral statement made on May 17. Ambassador Smith also repeated the U.S. rejection on May 26.

Appendix B

Treaty Between the United States of America and the Union of Soviet Socialist Republics on the Limitation of Strategic Offensive Arms

Treaty Between the United States of America and the Union of Soviet Socialist Republics on the Limitation of Strategic Offensive Arms*

Signed at Vienna June 18, 1979

The United States of America and the Union of Soviet Socialist Republics, hereinafter referred to as the Parties,

Conscious that nuclear war would have devastating consequences for all mankind,

Proceeding from the Basic Principles of Relations Between the United States of America and the Union of Soviet Socialist Republics of May 29, 1972,

Attaching particular significance to the limitation of strategic arms and determined to continue their efforts begun with the Treaty on the Limitation of Anti-Ballistic Missile Systems and the Interim Agreement on Certain Measures with Respect to the Limitation of Strategic Offensive Arms, of May 26, 1972,

Convinced that the additional measures limiting strategic offensive arms provided for in this Treaty will contribute to the improvement of relations between the Parties, help to reduce the risk of outbreak of nuclear war and strengthen international peace and security,

Mindful of their obligations under Article VI of the Treaty on the Non-Proliferation of Nuclear Weapons,

Guided by the principle of equality and equal security,

Recognizing that the strengthening of strategic stability meets the interests of the Parties and the interests of international security,

Reaffirming their desire to take measures for the further limitation and for the further reduction of strategic arms, having in mind the goal of achieving general and complete disarmament,

Declaring their intention to undertake in the near future negotiations further to limit and further to reduce strategic offensive arms,

Have agreed as follows:

---

*The text of the SALT II Treaty and Protocol, as signed in Vienna, is accompanied by a set of Agreed Statements and Common Understandings, also signed by Presidents Carter and Brezhnev, which is prefaced as follows:

In connection with the Treaty Between the United States of America and the Union of Soviet Socialist Republics on the Limitation of Strategic Offensive Arms, the Parties have agreed on the following Agreed Statements and Common Understandings undertaken on behalf of the Government of the United States and the Government of the Union of Soviet Socialist Republics.

As an aid to the reader, the texts of the Agreed Statements and Common Understandings are beneath the articles of the Treaty or Protocol to which they pertain.

Article I

Each Party undertakes, in accordance with the provisions of this Treaty, to limit strategic offensive arms quantitatively and qualitatively, to exercise restraint in the development of new types of strategic offensive arms, and to adopt other measures provided for in this Treaty.

Article II

For the purposes of this Treaty:

1. Intercontinental ballistic missile (ICBM) launchers are land-based launchers of ballistic missiles capable of a range in excess of the shortest distance between the northeastern border of the continental part of the territory of the United States of America and the northwestern border of the continental part of the territory of the Union of Soviet Socialist Republics, that is, a range in excess of 5,500 kilometers.

First Agreed Statement. The term "intercontinental ballistic missile launchers," as defined in paragraph 1 of Article II of the Treaty, includes all launchers which have been developed and tested for launching ICBMs. If a launcher has been developed and tested for launching an ICBM, all launchers of that type shall be considered to have been developed and tested for launching ICBMs.

First Common Understanding. If a launcher contains or launches an ICBM, that launcher shall be considered to have been developed and tested for launching ICBMs.

Second Common Understanding. If a launcher has been developed and tested for launching an ICBM, all launchers of that type, except for ICBM test and training launchers, shall be included in the aggregate numbers of strategic offensive arms provided for in Article III of the Treaty, pursuant to the provisions of Article VI of the Treaty.

Third Common Understanding. The one hundred and seventy-seven former Atlas and Titan I ICBM launchers of the United States of America, which are no longer operational and are partially dismantled, shall not be considered as subject to the limitations provided for in the Treaty.

Second Agreed Statement. After the date on which the Protocol ceases to be in force, mobile ICBM launchers shall be subject to the relevant limitations provided for in the Treaty which are applicable to ICBM launchers, unless the Parties agree that mobile ICBM launchers shall not be deployed after that date.

---

2. Submarine-launched ballistic missile (SLBM) launchers are launchers of ballistic missiles installed on any nuclear-powered submarine or launchers of modern ballistic missiles installed on any submarine, regardless of its type.

Agreed Statement. Modern submarine-launched ballistic missiles are: for the United States of America, missiles installed in all nuclear-powered submarines; for the Union of Soviet Socialist Republics, missiles of the type installed in nuclear-powered submarines made operational since 1965; and for both Parties, submarine-launched ballistic missiles first flight-tested since 1965 and installed in any submarine, regardless of its type.

---

3. Heavy bombers are considered to be:

   (a) currently, for the United States of America, bombers of the B-52 and B-1 types, and for the Union of Soviet Socialist Republics, bombers of the Tupolev-95 and Myasishchev types;
   (b) in the future, types of bombers which can carry out the mission of a heavy bomber in a manner similar or superior to that of bombers listed in subparagraph (a) above;
   (c) types of bombers equipped for cruise missiles capable of a range in excess of 600 kilometers; and
   (d) types of bombers equipped for ASBMs.

First Agreed Statement. The term "bombers," as used in paragraph 3 of Article II and other provisions of the Treaty, means airplanes of types initially constructed to be equipped for bombs or missiles.

Second Agreed Statement. The Parties shall notify each other on a case-by-case basis in the Standing Consultative Commission of inclusion of types of bombers as heavy bombers pursuant to the provisions of paragraph 3 of Article II of the Treaty; in this connection the Parties shall hold consultations, as appropriate, consistent with the provisions of paragraph 2 of Article XVII of the Treaty.

Third Agreed Statement. The criteria the Parties shall use to make case-by-case determinations of which types of bombers in the future can carry out the mission of a heavy bomber in a manner similar or superior to that of current heavy bombers, as referred to in subparagraph 3(b) of Article II of the Treaty, shall be agreed upon in the Standing Consultative Commission.

Fourth Agreed Statement. Having agreed that every bomber of a type included in paragraph 3 of Article II of the Treaty is to be considered a heavy bomber, the Parties further agree that:

   (a) airplanes which otherwise would be bombers of a heavy bomber type shall not be considered to be bombers of a heavy bomber type if they have functionally related observable differences which indicate that they cannot perform the mission of a heavy bomber;
   (b) airplanes which otherwise would be bombers of a type equipped for cruise missiles capable of a range in excess of 600 kilometers shall not be considered to be bombers of a type equipped for cruise missiles capable of a range in excess of 600 kilometers if they have functionally related observable differences which indicate that they cannot perform the mission of a bomber equipped for cruise missiles capable of a range in excess of 600 kilometers, except that heavy bombers of current types, as designated in subparagraph 3(a) of Article II of the Treaty, which otherwise would be of a type equipped for cruise missiles capable of a range in excess of 600 kilometers shall not be considered to be heavy bombers of a type equipped for cruise missiles capable of a range in excess of 600 kilometers if they are distinguishable on the basis of externally observable differences from heavy bombers of a type equipped for cruise missiles capable of a range in excess of 600 kilometers; and
   (c) airplanes which otherwise would be bombers of a type equipped for ASBMs shall not be considered to be bombers of a type equipped for ASBMs if they have functionally related observable differences which indicate that they cannot perform the mission of a bomber equipped for ASBMs, except that heavy bombers of current types, as designated in subparagraph 3(a) of Article II of the Treaty, which otherwise would be of a type equipped for ASBMs shall not be considered to be heavy bombers of a type

equipped for ASBMs if they are distinguishable on the basis of externally observable differences from heavy bombers of a type equipped for ASBMs.

First Common Understanding. Functionally related observable differences are differences in the observable features of airplanes which indicate whether or not these airplanes can perform the mission of a heavy bomber, or whether or not they can perform the mission of a bomber equipped for cruise missiles capable of a range in excess of 600 kilometers or whether or not they can perform the mission of a bomber equipped for ASBMs. Functionally related observable differences shall be verifiable by national technical means. To this end, the Parties may take, as appropriate, cooperative measures contributing to the effectiveness of verification by national technical means.

Fifth Agreed Statement. Tupolev-142 airplanes in their current configuration, that is, in the configuration for anti-submarine warfare, are considered to be airplanes of a type different from types of heavy bombers referred to in subparagraph 3(a) of Article II of the Treaty and not subject to the Fourth Agreed Statement to paragraph 3 of Article II of the Treaty. This Agreed Statement does not preclude improvement of Tupolev-142 airplanes as an anti-submarine system, and does not prejudice or set a precedent for designation in the future of types of airplanes as heavy bombers pursuant to subparagraph 3(b) of Article II of the Treaty or for application of the Fourth Agreed Statement to paragraph 3 of Article II of the Treaty to such airplanes.

Second Common Understanding. Not later than six months after entry into force of the Treaty the Union of Soviet Socialist Republics will give its thirty-one Myasishchev airplanes used as tankers in existence as of the date of signature of the Treaty functionally related observable differences which indicate that they cannot perform the mission of a heavy bomber.

Third Common Understanding. The designations by the United States of America and by the Union of Soviet Socialist Republics for heavy bombers referred to in subparagraph 3(a) of Article II of the Treaty correspond in the following manner:

Heavy bombers of the types designated by the United States of America as the B-52 and the B-1 are known to the Union of Soviet Socialist Republics by the same designations;

Heavy bombers of the type designated by the Union of Socialist Republics as the Tupolev-95 are known to the United States of America as heavy bombers of the Bear type; and

Heavy bombers of the type designated by the Union of Soviet Socialist Republics as the Myasishchev are known to the United States of America as heavy bombers of the Bison type.

---

4. Air-to-surface ballistic missiles (ASBMs) are any such missiles capable of a range in excess of 600 kilometers and installed in an aircraft or on its external mountings.

5. Launchers of ICBMs and SLBMs equipped with multiple independently targetable reentry vehicles (MIRVs) are launchers of the types developed and tested for launching ICBMs or SLBMs equipped with MIRVs.

First Agreed Statement. If a launcher has been developed and tested for launching an ICBM or an SLBM equipped with MIRVs, all launchers of that type shall be considered to have been developed and tested for launching ICBMs or SLBMs equipped with MIRVs.

First Common Understanding. If a launcher contains or launches an ICBM or an SLBM equipped with MIRVs, that launcher shall be considered to have been developed and tested for launching ICBMs or SLBMs equipped with MIRVs.

Second Common Understanding. If a launcher has been developed and tested for launching an ICBM or an SLBM equipped with MIRVs, all launchers of that type, except for ICBM and SLBM test and training launchers, shall be included in the corresponding aggregate numbers provided for in Article V of the Treaty, pursuant to the provisions of Article VI of the Treaty.

Second Agreed Statement. ICBMs and SLBMs equipped with MIRVs are ICBMs and SLBMs of the types which have been flight-tested with two or more independently targetable reentry vehicles, regardless of whether or not they have also been flight-tested with a single reentry vehicle or with multiple reentry vehicles which are not independently targetable. As of the date of signature of the Treaty, such ICBMs and SLBMS are: for the United States of America, Minuteman III ICBMs, Poseidon C-3 SLBMs, and Trident C-4 SLBMs; and for the Union of Soviet Socialist Republics, RS-16, RS-18, RS-20 ICBMs and RSM-50 SLBMs.

Each Party will notify the other Party in the Standing Consultative Commission on a case-by-case basis of the designation of the one new type of light ICBM, if equipped with MIRVs, permitted pursuant to paragraph 9 of Article IV of the Treaty when first flight-tested; of designations of additional types of SLBMs equipped with MIRVs when first installed on a submarine; and of designations of types of ASBMs equipped with MIRVs when first flight-tested.

Third Common Understanding. The designations by the United States of America and by the Union of Soviet Socialist Republics for ICBMs and SLBMs equipped with MIRVs correspond in the following manner:

Missiles of the type designated by the United States of America as the Minuteman III and known to the Union of Soviet Socialist Republics by the same designation, a light ICBM that has been flight-tested with multiple independently targetable reentry vehicles;

Missiles of the type designated by the United States of America as the Poseiden C-3 and known to the Union of Soviet Socialist Republics by the same designation, an SLBM that was first flight-tested in 1968 and that has been flight-tested with multiple independently targetable reentry vehicles;

Missiles of the type designated by the United States of America as the Trident C-4 and known to the Union of Soviet Socialist Republics by the same designation, an SLBM that was first flight-tested in 1977 and that has been flight-tested with multiple independently targetable reentry vehicles;

Missiles of the type designated by the Union of Soviet Socialist Republics as the RS-16 and known to the United States of America as the SS-17, a light ICBM that has been flight-tested with a single reentry vehicle and with multiple independently targetable reentry vehicles;

Missiles of the type designated by the Union of Soviet Socialist Republics as the RS-18 and known to the United States of America as the SS-19, the heaviest in terms of launch-weight and throw-weight of light ICBMs, which has been flight-tested with a single reentry vehicle and with multiple independently targetable reentry vehicles;

Missiles of the type designated by the Union of Soviet Socialist Republics as the RS-20 and known to the United States of America as the SS-18, the heaviest in terms of launch-weight and throw-weight of heavy ICBMs, which has been flight-tested with a single reentry vehicle and with multiple independently targetable reentry vehicles;

Missiles of the type designated by the Union of Soviet Socialist Republics as the RSM-50 and known to the United States of America as the SS-N-18, an SLBM that has been flight-tested with a single reentry vehicle and with multiple independently targetable reentry vehicles.

Third Agreed Statement. Reentry vehicles are independently targetable:

(a) if, after separation from the booster, maneuvering and targeting of the reentry vehicles to separate aim points along trajectories which are unrelated to each other are accomplished by means of devices which are installed in a self-contained dispensing mechanism or on the reentry vehicles, and which are based on the use of electronic or other computers in combination with devices using jet engines, including rocket engines, or aerodynamic systems;

(b) if maneuvering and targeting of the reentry vehicles to separate aim points along trajectories which are unrelated to each other are accomplished by means of other devices which may be developed in the future.

Fourth Common Understanding. For the purposes of this Treaty, all ICBM launchers in the Derazhnya and Pervomaysk areas in the Union of Soviet Socialist Republics are included in the aggregate numbers provided for in Article V of the Treaty.

Fifth Common Understanding. If ICBM or SLBM launchers are converted, constructed or undergo significant changes to their principal observable structural design features after entry into force of the Treaty, any such launchers which are launchers of missiles equipped with MIRVs shall be distinguishable from launchers of missiles not equipped with MIRVs, and any such launchers which are launchers of missiles not equipped with MIRVs shall be distinguishable from launchers of missiles equipped with MIRVs, on the basis of externally observable design features of the launchers. Submarines with launchers of SLBMs equipped with MIRVs shall be distinguishable from submarines with launchers of SLBMs not equipped with MIRVs on the basis of externally observable design features of the submarines.

This Common Understanding does not require changes to launcher conversion or construction programs, or to programs including significant changes to the principal observable structural design features of launchers, underway as of the date of signature of the Treaty.

---

6. ASBMs equipped with MIRVs are ASBMs of the types which have been flight-tested with MIRVs.

First Agreed Statement. ASBMs of the types which have been flight-tested with MIRVs are all ASBMs of the types which have been flight-tested with two or more independently targetable reentry vehicles, regardless of whether or not they have also been flight-tested with a single reentry vehicle or with multiple reentry vehicles which are not independently targetable.

Second Agreed Statement. Reentry vehicles are independently targetable:

(a) if, after separation from the booster, maneuvering and targeting of the reentry vehicles to separate aim points along trajectories which are unrelated to each other are accomplished by means of devices which are installed in a self-contained dispensing mechanism or on the reentry vehicles, and which are based on the use of electronic or other computers in combination with devices using jet engines, including rocket engines, or aerodynamic systems;

(b) if maneuvering and targeting of the reentry vehicles to separate aim points along trajectories which are unrelated to each other are accomplished by means of other devices which may be developed in the future.

---

7. Heavy ICBMs are ICBMs which have a launch-weight greater or a throw-weight greater than that of the heaviest, in terms of either launch-weight or throw-weight, respectively, of the light ICBMs deployed by either Party as of the date of signature of this Treaty.

First Agreed Statement. The launch-weight of an ICBM is the weight of the fully loaded missile itself at the time of launch.

Second Agreed Statement. The throw-weight of an ICBM is the sum of the weight of:

(a) its reentry vehicle or reentry vehicles;
(b) any self-contained dispensing mechanisms or other appropriate devices for targeting one reentry vehicle, or for releasing or for dispensing and targeting two or more reentry vehicles; and
(c) its penetration aids, including devices for their release.

Common Understanding. The term "other appropriate devices," as used in the definition of the throw-weight of an ICBM in the Second Agreed Statement to paragraph 7 of Article II of the Treaty, means any devices for dispensing and targeting two or more reentry vehicles; and any devices for releasing two or more reentry vehicles or for targeting one reentry vehicle, which cannot provide their reentry vehicles or reentry vehicle with additional velocity of more than 1,000 meters per second.

---

8. Cruise missiles are unmanned, self-propelled, guided, weapon-delivery vehicles which sustain flight through the use of aerodynamic lift over most of their flight path and which are flight-tested from or deployed on aircraft, that is, air-launched cruise missiles, or such vehicles which are referred to as cruise missiles in subparagraph 1(b) of Article IX.

First Agreed Statement. If a cruise missile is capable of a range in excess of 600 kilometers, all cruise missiles of that type shall be considered to be cruise missiles capable of a range in excess of 600 kilometers.

First Common Understanding. If a cruise missile has been flight-tested to a range in excess of 600 kilometers, it shall be considered to be a cruise missile capable of a range in excess of 600 kilometers.

Second Common Understanding. Cruise missiles not capable of a range in excess of 600 kilometers shall not be considered to be of a type capable of a range in excess of 600 kilometers if they are distinguishable on the basis of externally observable design features from cruise missiles of types capable of a range in excess of 600 kilometers.

Second Agreed Statement. The range of which a cruise missile is capable is the maximum distance which can be covered by the missile in its standard design mode flying until fuel exhaustion, determined by projecting its flight path onto the Earth's sphere from the point of launch to the point of impact.

Third Agreed Statement. If an unmanned, self-propelled, guided vehicle which sustains flight through the use of aerodynamic lift over most of its flight path has been flight-tested or deployed for weapon delivery, all vehicles of that type shall be considered to be weapon-delivery vehicles.

Third Common Understanding. Unmanned, self-propelled, guided vehicles which sustain flight through the use of aerodynamic lift over most of their flight path and are not weapon-delivery vehicles, that is, unarmed, pilotless, guided vehicles, shall not be considered to be cruise missiles if such vehicles are distinguishable from cruise missiles on the basis of externally observable design features.

Fourth Common Understanding. Neither Party shall convert unarmed, pilotless, guided vehicles into cruise missiles capable of a range in excess of 600 kilometers, nor shall either Party convert cruise missiles capable of a range in excess of 600 kilometers into unarmed, pilotless, guided vehicles.

Fifth Common Understanding. Neither Party has plans during the term of the Treaty to flight-test from or deploy on aircraft unarmed, pilotless, guided vehicles which are capable of a range in excess of 600 kilometers. In the future, should a Party have such plans, that Party will provide notification thereof to the other Party well in advance of such flight-testing or deployment. This Common Understanding does not apply to target drones.

Article III

1. Upon entry into force of this Treaty, each Party undertakes to limit ICBM launchers, SLBM launchers, heavy bombers, and ASBMs to an aggregate number not to exceed 2,400.

2. Each Party undertakes to limit, from January 1, 1981, strategic offensive arms referred to in paragraph 1 of this Article to an aggregate number not to exceed 2,250, and to initiate reductions of those arms which as of that date would be in excess of this aggregate number.

3. Within the aggregate numbers provided for in paragraphs 1 and 2 of this Article and subject to the provisions of this Treaty, each Party has the right to determine the composition of these aggregates.

4. For each bomber of a type equipped for ASBMs, the aggregate numbers provided for in paragraphs 1 and 2 of this Article shall include the maximum number of such missiles for which a bomber of that type is equipped for one operational mission.

5. A heavy bomber equipped only for ASBMs shall not itself be included in the aggregate numbers provided for in paragraphs 1 and 2 of this Article.

6. Reductions of the numbers of strategic offensive arms required to comply with the provisions of paragraphs 1 and 2 of this Article shall be carried out as provided for in Article XI.

Article IV

1. Each Party undertakes not to start construction of additional fixed ICBM launchers.

2. Each Party undertakes not to relocate fixed ICBM launchers.

3. Each Party undertakes not to convert launchers of light ICBMs, or of ICBMs of older types deployed prior to 1964, into launchers of heavy ICBMs of types deployed after that time.

4. Each Party undertakes in the process of modernization and replacement of ICBM silo launchers not to increase the original internal volume of an ICBM silo launcher by

more than thirty-two percent. Within this limit each Party has the right to determine whether such an increase will be made through an increase in the original diameter or in the original depth of an ICBM silo launcher, or in both of these dimensions.

Agreed Statement. The word "original" in paragraph 4 of Article IV of the Treaty refers to the internal dimensions of an ICBM silo launcher, including its internal volume, as of May 26, 1972, or as of the date on which such launcher becomes operational, whichever is later.

Common Understanding. The obligations provided for in paragraph 4 of Article IV of the Treaty and in the Agreed Statement thereto mean that the original diameter or the original depth of an ICBM silo launcher may not be increased by an amount greater than that which would result in an increase in the original internal volume of the ICBM silo launcher by thirty-two percent solely through an increase in one of these dimensions.

---

5. Each Party undertakes:

   (a) not to supply ICBM launcher deployment areas with intercontinental ballistic missiles in excess of a number consistent with normal deployment, maintenance, training, and replacement requirements;
   (b) not to provide storage facilities for or to store ICBMs in excess of normal deployment requirements at launch sites of ICBM launchers;
   (c) not to develop, test, or deploy systems for rapid reload of ICBM launchers.

Agreed Statement. The term "normal deployment requirements," as used in paragraph 5 of Article IV of the Treaty, means the deployment of one missile at each ICBM launcher.

---

6. Subject to the provisions of this Treaty, each Party undertakes not to have under construction at any time strategic offensive arms referred to in paragraph 1 of Article III in excess of numbers consistent with a normal construction schedule.

Common Understanding. A normal construction schedule, in paragraph 6 of Article IV of the Treaty, is understood to be one consistent with the past or present construction practices of each Party.

---

7. Each Party undertakes not to develop, test, or deploy ICBMs which have a launch-weight greater or a throw-weight greater than that of the heaviest, in terms of either launch-weight or throw-weight, respectively, of the heavy ICBMs, deployed by either Party as of the date of signature of this Treaty.

First Agreed Statement. The launch-weight of an ICBM is the weight of the fully loaded missile itself at the time of launch.

Second Agreed Statement. The throw-weight of an ICBM is the sum of the weight of:

   (a) its reentry vehicle or reentry vehicles;

(b) any self-contained dispensing mechanisms or other appropriate devices for targeting one reentry vehicle, or for releasing or for dispensing and targeting two or more reentry vehicles; and

(c) its penetration aids, including devices for their release.

Common Understanding. The term "other appropriate devices," as used in the definition of the throw-weight of an ICBM in the Second Agreed Statement to paragraph 7 of Article IV of the Treaty, means any devices for dispensing and targeting two or more reentry vehicles; and any devices for releasing two or more reentry vehicles or for targeting one reentry vehicle, which cannot provide their reentry vehicles or reentry vehicle with additional velocity of more than 1,000 meters per second.

---

8. Each Party undertakes not to convert land-based launchers of ballistic missiles which are not ICBMs into launchers for launching ICBMs, and not to test them for this purpose.

Common Understanding. During the term of the Treaty, the Union of Soviet Socialist Republics will not produce, test, or deploy ICBMs of the type designated by the Union of Soviet Socialist Republics as the RS-14 and known to the United States of America as the SS-16, a light ICBM first flight-tested after 1970 and flight-tested only with a single reentry vehicle; this Common Understanding also means that the Union of Soviet Socialist Republics will not produce the third stage of that missile, the reentry vehicle of that missile, or the appropriate device for targeting the reentry vehicle of that missile.

---

9. Each Party undertakes not to flight-test or deploy new types of ICBMs, that is, types of ICBMs not flight-tested as of May 1, 1979, except that each Party may flight-test and deploy one new type of light ICBM.

First Agreed Statement. The term "new types of ICBMs," as used in paragraph 9 of Article IV of the Treaty, refers to any ICBM which is different from those ICBMs flight-tested as of May 1, 1979 in any one or more of the following respects:

(a) the number of stages, the length, the largest diameter, the launch-weight, or the throw-weight, of the missile;

(b) the type of propellant (that is, liquid or solid) of any of its stages.

First Common Understanding. As used in the First Agreed Statement to paragraph 9 of Article IV of the Treaty, the term "different," referring to the length, the diameter, the launch-weight, and the throw-weight, of the missile, means a difference in excess of five percent.

Second Agreed Statement. Every ICBM of the one new type of light ICBM permitted to each Party pursuant to paragraph 9 of Article IV of the Treaty shall have the same number of stages and the same type of propellant (that is, liquid or solid) of each stage as the first ICBM of the one new type of light ICBM launched by that Party. In addition, after the twenty-fifth launch of an ICBM of that type, or after the last launch before deployment begins of ICBMs of that type, whichever occurs earlier, ICBMs of the one new type of light ICBM permitted to that Party shall not be different in any one or more of the following respects: the length, the largest diameter, the launch-weight, or the throw-weight, of the missile.

A Party which launches ICBMs of the one new type of light ICBM permitted pursuant to paragraph 9 of Article IV of the Treaty shall promptly notify the other Party of the date of the first launch and of the date of either the twenty-fifth or the last launch before deployment begins of ICBMs of that type, whichever occurs earlier.

Second Common Understanding. As used in the Second Agreed Statement to paragraph 9 of Article IV of the Treaty, the term "different," referring to the length, the diameter, the launch-weight, and the throw-weight, of the missile, means a difference in excess of five percent from the value established for each of the above parameters as of the twenty-fifth launch or as of the last launch before deployment begins, whichever occurs earlier. The values demonstrated in each of the above parameters during the last twelve of the twenty-five launches or during the last twelve launches before deployment begins, whichever twelve launches occur earlier, shall not vary by more than ten percent from any other of the corresponding values demonstrated during those twelve launches.

Third Common Understanding. The limitations with respect to launch-weight and throw-weight, provided for in the First Agreed Statement and the First Common Understanding to paragraph 9 of Article IV of the Treaty, do not preclude the flight-testing or the deployment of ICBMs with fewer reentry vehicles, or fewer penetration aids, or both, than the maximum number of reentry vehicles and the maximum number of penetration aids with which ICBMs of that type have been flight-tested as of May 1, 1979, even if this results in a decrease in launch-weight or in throw-weight in excess of five percent.

In addition to the aforementioned cases, those limitations do not preclude a decrease in launch-weight or in throw-weight in excess of five percent, in the case of the flight-testing or the deployment of ICBMs with a lesser quantity of propellant, including the propellant of a self-contained dispensing mechanism or other appropriate device, than the maximum quantity of propellant, including the propellant of a self-contained dispensing mechanism or other appropriate device, with which ICBMs of that type have been flight-tested as of May 1, 1979, provided that such an ICBM is at the same time flight-tested or deployed with fewer reentry vehicles, or fewer penetration aids, or both, than the maximum number of reentry vehicles and the maximum number of penetration aids with which ICBMs of that type have been flight-tested as of May 1, 1979, and the decrease in launch-weight and throw-weight in such cases results only from the reduction in the number of reentry vehicles, or penetration aids, or both, and the reduction in the quantity of propellant.

Fourth Common Understanding. The limitations with respect to launch-weight and throw-weight, provided for in the Second Agreed Statement and the Second Common Understanding to paragraph 9 of Article IV of the Treaty, do not preclude the flight-testing or the deployment of ICBMs of the one new type of light ICBM permitted to each Party pursuant to paragraph 9 of Article IV of the Treaty with fewer reentry vehicles, or fewer penetration aids, or both, than the maximum number of reentry vehicles and the maximum number of penetration aids with which ICBMs of that type have been flight-tested, even if this results in a decrease in launch-weight or in throw-weight in excess of five percent.

In addition to the aforementioned cases, those limitations do not preclude a decrease in launch-weight or in throw-weight in excess of five percent, in the case of the flight-testing or the deployment of ICBMs of that type with a lesser quantity of propellant, including the propellant of a self-contained dispensing mechanism or other appropriate device, than the maximum quantity of propellant, including the propellant of a self-contained dispensing mechanism or other appropriate device, with which ICBMs of that type have been flight-tested, provided that such an ICBM is at the same time flight-tested or deployed with fewer reentry vehicles, or fewer penetration aids, or both, than the maximum number of reentry vehicles and the maximum number of penetration aids with which ICBMs of that type have been flight-tested, and the decrease in launch-weight and throw-weight in such cases results only from the reduction in the number of reentry vehicles, or penetration aids, or both, and the reduction in the quantity of propellant.

---

10. Each Party undertakes not to flight-test or deploy ICBMs of a type flight-tested as of May 1, 1979 with a number of reentry vehicles greater than the maximum number of reentry vehicles with which an ICBM of that type has been flight-tested as of that date.

First Agreed Statement. The following types of ICBMs and SLBMs equipped with MIRVs have been flight-tested with the maximum number of reentry vehicles set forth below:

For the United States of America

ICBMs of the Minuteman III type—seven reentry vehicles;
SLBMs of the Poseidon C-3 type—fourteen reentry vehicles;
SLBMs of the Trident C-4 type—seven reentry vehicles.

For the Union of Soviet Socialist Republics

ICBMs of the RS-16 type—four reentry vehicles;
ICBMs of the RS-18 type—six reentry vehicles;
ICBMs of the RS-20 type—ten reentry vehicles;
SLBMs of the RSM-50 type—seven reentry vehicles.

Common Understanding. Minuteman III ICBMs of the United States of America have been deployed with no more than three reentry vehicles. During the term of the Treaty, the United States of America has no plans to and will not flight-test or deploy missiles of this type with more than three reentry vehicles.

Second Agreed Statement. During the flight-testing of any ICBM, SLBM, or ASBM after May 1, 1979, the number of procedures for releasing or for dispensing may not exceed the maximum number of reentry vehicles established for missiles of corresponding types as provided for in paragraphs 10, 11, 12, and 13 of Article IV of the Treaty. In this Agreed Statement "procedures for releasing or for dispensing" are understood to mean maneuvers of a missile associated with targeting and releasing or dispensing its reentry vehicles to aim points, whether or not a reentry vehicle is actually released or dispensed. Procedures for releasing anti-missile defense penetration aids will not be considered to be procedures for releasing or for dispensing a reentry vehicle so long as the procedures for releasing anti-missile defense penetration aids differ from those for releasing or for dispensing reentry vehicles.

Third Agreed Statement. Each Party undertakes:

(a) not to flight-test or deploy ICBMs equipped with multiple reentry vehicles, of a type flight-tested as of May 1, 1979, with reentry vehicles the weight of any of which is less than the weight of the lightest of those reentry vehicles with which an ICBM of that type has been flight-tested as of that date;

(b) not to flight-test or deploy ICBMs equipped with a single reentry vehicle and without an appropriate device for targeting a reentry vehicle, of a type flight-tested as of May 1, 1979, with a reentry vehicle the weight of which is less than the weight of the lightest reentry vehicle on an ICBM of a type equipped with MIRVs and flight-tested by that Party as of May 1, 1979; and

(c) not to flight-test or deploy ICBMs equipped with a single reentry vehicle and with an appropriate device for targeting a reentry vehicle, of a type flight-tested as of May 1, 1979, with a reentry vehicle the weight of which is less than fifty percent of the throw-weight of that ICBM.

---

11. Each Party undertakes not to flight-test or deploy ICBMs of the one new type permitted pursuant to paragraph 9 of this Article with a number of reentry vehicles greater than the maximum number of reentry vehicles with which an ICBM of either Party has been flight-tested as of May 1, 1979, that is, ten.

First Agreed Statement. Each Party undertakes not to flight-test or deploy the one new type of light ICBM permitted to each Party pursuant to paragraph 9 of Article IV of the Treaty with a number of reentry vehicles greater than the maximum number of reentry vehicles with which an ICBM of that type has been flight-tested as of the twenty-fifth launch or the last launch before deployment begins of ICBMs of that type, whichever occurs earlier.

Second Agreed Statement. During the flight-testing of any ICBM, SLBM, or ASBM after May 1, 1979 the number of procedures for releasing or for dispensing may not exceed the maximum number of reentry vehicles established for missiles of corresponding types as provided for in paragraphs 10, 11, 12, and 13 of Article IV of the Treaty. In this Agreed Statement "procedures for releasing or for dispensing" are understood to mean maneuvers of a missile associated with targeting and releasing or dispensing its reentry vehicles to aim points, whether or not a reentry vehicle is actually released or dispensed. Procedures for releasing anti-missile defense penetration aids will not be considered to be procedures for releasing or for dispensing a reentry vehicle so long as the procedures for releasing anti-missile defense penetration aids differ from those for releasing or for dispensing reentry vehicles.

---

12. Each Party undertakes not to flight-test or deploy SLBMs with a number of reentry vehicles greater than the maximum number of reentry vehicles with which an SLBM of either Party has been flight-tested as of May 1, 1979, that is, fourteen.

First Agreed Statement. The following types of ICBMs and SLBMs equipped with MIRVs have been flight-tested with the maximum number of reentry vehicles set forth below:

For the United States of America

ICBMs of the Minuteman III type—seven reentry vehicles;
SLBMs of the Poseidon C-3 type—fourteen reentry vehicles;
SLBMs of the Trident C-4 type—seven reentry vehicles.

For the Union of Soviet Socialist Republics

ICBMs of the RS-16 type—four reentry vehicles;
ICBMs of the RS-18 type—six reentry vehicles;
ICBMs of the RS-20 type—ten reentry vehicles;
SLBMs of the RSM-50 type—seven reentry vehicles.

Second Agreed Statement. During the flight-testing of any ICBM, SLBM, or ASBM after May 1, 1979 the number of procedures for releasing or for dispensing may not exceed the maximum number of reentry vehicles established for missiles of corresponding types as provided for in paragraphs 10, 11, 12, and 13 of Article IV of the Treaty. In this Agreed Statement "procedures for releasing or dispensing" are understood to mean maneuvers of a missile associated with targeting and releasing or dispensing its reentry vehicles to aim points, whether or not a reentry vehicle is actually released or dispensed. Procedures for releasing anti-missile defense penetration aids will not be considered to be procedures for releasing or for dispensing a reentry vehicle so long as the procedures for releasing anti-missile defense penetration aids differ from those for releasing or for dispensing reentry vehicles.

---

13. Each Party undertakes not to flight-test or deploy ASBMs with a number of reentry vehicles greater than the maximum number of reentry vehicles with which an ICBM of either Party has been flight-tested as of May 1, 1979, that is, ten.

Agreed Statement. During the flight-testing of any ICBM, SLBM, or ASBM after May 1, 1979 the number of procedures for releasing or for dispensing may not exceed the maximum number of reentry vehicles established for missiles of corresponding types as provided for in paragraphs 10, 11, 12, and 13 of Article IV of the Treaty. In this Agreed Statement "procedures for releasing or for dispensing" are understood to mean maneuvers of a missile associated with targeting and releasing or dispensing its reentry vehicles to aim points, whether or not a reentry vehicle is actually released or dispensed. Procedures for releasing anti-missile defense penetration aids will not be considered to be procedures for releasing or for dispensing a reentry vehicle so long as the procedures for releasing anti-missile defense penetration aids differ from those for releasing or for dispensing reentry vehicles.

---

14. Each Party undertakes not to deploy at any one time on heavy bombers equipped for cruise missiles capable of a range in excess of 600 kilometers a number of such cruise missiles which exceeds the product of 28 and the number of such heavy bombers.

First Agreed Statement. For the purposes of the limitation provided for in paragraph 14 of Article IV of the Treaty, there shall be considered to be deployed on each heavy bomber of a type equipped for cruise missiles capable of a range in excess of 600 kilometers the maximum number of such missiles for which any bomber of that type is equipped for one operational mission.

Second Agreed Statement. During the term of the Treaty no bomber of the B-52 or B-1 types of the United States of America and no bomber of the Tupolev-95 or Myasishchev types of the Union of Soviet Socialist Republics will be equipped for more than twenty cruise missiles capable of a range in excess of 600 kilometers.

Article V

1. Within the aggregate numbers provided for in paragraphs 1 and 2 of Article III, each Party undertakes to limit launchers of ICBMs and SLBMs equipped with MIRVs, ASBMs equipped with MIRVs, and heavy bombers equipped for cruise missiles

capable of a range in excess of 600 kilometers to an aggregate number not to exceed 1,320.

2. Within the aggregate number provided for in paragraph 1 of this Article, each Party undertakes to limit launchers of ICBMs and SLBMs equipped with MIRVs, and ASBMs equipped with MIRVs to an aggregate number not to exceed 1,200.

3. Within the aggregate number provided for in paragraph 2 of this Article, each Party undertakes to limit launchers of ICBMs equipped with MIRVs to an aggregate number not to exceed 820.

4. For each bomber of a type equipped for ASBMs equipped with MIRVs, the aggregate numbers provided for in paragraphs 1 and 2 of this Article shall include the maximum number of ASBMs for which a bomber of that type is equipped for one operational mission.

Agreed Statement. If a bomber is equipped for ASBMs equipped with MIRVs, all bombers of that type shall be considered to be equipped for ASBMs equipped with MIRVs.

---

5. Within the aggregate numbers provided for in paragraphs 1, 2, and 3 of this Article and subject to the provisions of this Treaty, each Party has the right to determine the composition of these aggregates.

Article VI

1. The limitations provided for in this Treaty shall apply to those arms which are:

(a) operational;
(b) in the final stage of construction;
(c) in reserve, in storage, or mothballed;
(d) undergoing overhaul, repair, modernization, or conversion.

2. Those arms in the final stage of construction are:

(a) SLBM launchers on submarines which have begun sea trials;
(b) ASBMs after a bomber of a type equipped for such missiles has been brought out of the shop, plant, or other facility where its final assembly or conversion for the purpose of equipping it for such missiles has been performed;
(c) other strategic offensive arms which are finally assembled in a shop, plant, or other facility after they have been brought out of the shop, plant, or other facility where their final assembly has been performed.

3. ICBM and SLBM launchers of a type not subject to the limitation provided for in Article V, which undergo conversion into launchers of a type subject to that limitation, shall become subject to that limitation as follows:

(a) fixed ICBM launchers when work on their conversion reaches the stage which first definitely indicates that they are being so converted;
(b) SLBM launchers on a submarine when that submarine first goes to sea after their conversion has been performed.

Agreed Statement. The procedures referred to in paragraph 7 of Article VI of the Treaty shall include procedures determining the manner in which mobile ICBM launchers of a type not subject to the limitation provided for in Article V of the Treaty, which undergo conversion into launchers of a type subject to that limitation, shall become subject to that limitation, unless the Parties agree that mobile ICBM launchers shall not be deployed after the date on which the Protocol ceases to be in force.

---

4. ASBMs on a bomber which undergoes conversion from a bomber of a type equipped for ASBMs which are not subject to the limitation provided for in Article V into a bomber of a type equipped for ASBMs which are subject to that limitation shall become subject to that limitation when the bomber is brought out of the shop, plant, or other facility where such conversion has been performed.

5. A heavy bomber of a type not subject to the limitation provided for in paragraph 1 of Article V shall become subject to that limitation when it is brought out of the shop, plant, or other facility where it has been converted into a heavy bomber of a type equipped for cruise missiles capable of a range in excess of 600 kilometers. A bomber of a type not subject to the limitation provided for in paragraph 1 or 2 of Article III shall become subject to that limitation and to the limitation provided for in paragraph 1 of Article V when it is brought out of the shop, plant, or other facility where it has been converted into a bomber of a type equipped for cruise missiles capable of a range in excess of 600 kilometers.

6. The arms subject to the limitations provided for in this Treaty shall continue to be subject to these limitations until they are dismantled, are destroyed, or otherwise cease to be subject to these limitations under procedures to be agreed upon.

Agreed Statement. The procedures for removal of strategic offensive arms from the aggregate numbers provided for in the Treaty, which are referred to in paragraph 6 of Article VI of the Treaty, and which are to be agreed upon in the Standing Consultative Commission, shall include:

(a) procedures for removal from the aggregate numbers, provided for in Article V of the Treaty, of ICBM and SLBM launchers which are being converted from launchers of a type subject to the limitation provided for in Article V of the Treaty, into launchers of a type not subject to that limitation;

(b) procedures for removal from the aggregate numbers, provided for in Articles III and V of the Treaty, of bombers which are being converted from bombers of a type subject to the limitations provided for in Article III of the Treaty or in Articles III and V of the Treaty into airplanes or bombers of a type not so subject.

Common Understanding. The procedures referred to in subparagraph (b) of the Agreed Statement to paragraph 6 of Article VI of the Treaty for removal of bombers from the aggregate numbers provided for in Articles III and V of the Treaty shall be based upon the existence of functionally related observable differences which indicate whether or not they can perform the mission of a heavy bomber, or whether or not they can perform the mission of a bomber equipped for cruise missiles capable of a range in excess of 600 kilometers.

---

7. In accordance with the provisions of Article XVII, the Parties will agree in the Standing Consultative Commission upon procedures to implement the provisions of this Article.

Article VII

1. The limitations provided for in Article III shall not apply to ICBM and SLBM test and training launchers or to space vehicle launchers for exploration and use of outer

space. ICBM and SLBM test and training launchers are ICBM and SLBM launchers used only for testing or training.

Common Understanding. The term "testing," as used in Article VII of the Treaty, includes research and development.

---

2. The Parties agree that:

(a) there shall be no significant increase in the number of ICBM or SLBM test and training launchers or in the number of such launchers of heavy ICBMs; .

(b) construction or conversion of ICBM launchers at test ranges shall be undertaken only for purposes of testing and training;

(c) there shall be no conversion of ICBM test and training launchers or of space vehicle launchers into ICBM launchers subject to the limitations provided for in Article III.

First Agreed Statement. The term "significant increase," as used in subparagraph 2(a) of Article VII of the Treaty, means an increase of fifteen percent or more. Any new ICBM test and training launchers which replace ICBM test and training launchers at test ranges will be located only at test ranges.

Second Agreed Statement. Current test ranges where ICBMs are tested are located: for the United States of America, near Santa Maria, California, and at Cape Canaveral, Florida; and for the Union of Soviet Socialist Republics, in the areas of Tyura-Tam and Plesetskaya. In the future, each Party shall provide notification in the Standing Consultative Commission of the location of any other test range used by that Party to test ICBMs.

First Common Understanding. At test ranges where ICBMs are tested, other arms, including those not limited by the Treaty, may also be tested.

Second Common Understanding. Of the eighteen launchers of fractional orbital missiles at the test range where ICBMs are tested in the area of Tyura-Tam, twelve launchers shall be dismantled or destroyed and six launchers may be converted to launchers for testing missiles undergoing modernization.

Dismantling or destruction of the twelve launchers shall begin upon entry into force of the Treaty and shall be completed within eight months, under procedures for dismantling or destruction of these launchers to be agreed upon in the Standing Consultative Commission. These twelve launchers shall not be replaced.

Conversion of the six launchers may be carried out after entry into force of the Treaty. After entry into force of the Treaty, fractional orbital missiles shall be removed and shall be destroyed pursuant to the provisions of subparagraph 1(c) of Article IX and of Article XI of the Treaty and shall not be replaced by other missiles, except in the case of conversion of these six launchers for testing missiles undergoing modernization. After removal of the fractional orbital missiles, and prior to such conversion, any activities associated with these launchers shall be limited to normal maintenance requirements for launchers in which missiles are not deployed. These six launchers shall be subject to the provisions of Article VII of the Treaty and, if converted, to the provisions of the Fifth Common Understanding to paragraph 5 of Article II of the Treaty.

Article VIII

1. Each Party undertakes not to flight-test cruise missiles capable of a range in excess of 600 kilometers or ASBMs from aircraft other than bombers or to convert such aircraft into aircraft equipped for such missiles.

Agreed Statement. For purposes of testing only, each Party has the right, through initial construction or, as an exception to the provisions of paragraph 1 of Article VIII of the Treaty, by conversion, to equip for cruise missiles capable of a range in excess of 600 kilometers or for ASBMs no more than sixteen airplanes, including airplanes which are prototypes of bombers equipped for such missiles. Each Party also has the right, as an exception to the provisions of paragraph 1 of Article VIII of the Treaty, to flight-test from such airplanes cruise missiles capable of a range in excess of 600 kilometers and, after the date on which the Protocol ceases to be in force, to flight-test ASBMs from such airplanes as well, unless the Parties agree that they will not flight-test ASBMs after that date. The limitations provided for in Article III of the Treaty shall not apply to such airplanes.

The aforementioned airplanes may include only:

(a) airplanes other than bombers which, as an exception to the provisions of paragraph 1 of Article VIII of the Treaty, have been converted into airplanes equipped for cruise missiles capable of a range in excess of 600 kilometers or for ASBMs;

(b) airplanes considered to be heavy bombers pursuant to subparagraph 3(c) or 3(d) of Article II of the Treaty; and

(c) airplanes other than heavy bombers which, prior to March 7, 1979, were used for testing cruise missiles capable of a range in excess of 600 kilometers.

The airplanes referred to in subparagraphs (a) and (b) of this Agreed Statement shall be distinguishable on the basis of functionally related observable differences from airplanes which otherwise would be of the same type but cannot perform the mission of a bomber equipped for cruise missiles capable of a range in excess of 600 kilometers or for ASBMs.

The airplanes referred to in subparagraph (c) of this Agreed Statement shall not be used for testing cruise missiles capable of a range in excess of 600 kilometers after the expiration of a six-month period from the date of entry into force of the Treaty, unless by the expiration of that period they are distinguishable on the basis of functionally related observable differences from airplanes which otherwise would be of the same type but cannot perform the mission of a bomber equipped for cruise missiles capable of a range in excess of 600 kilometers.

First Common Understanding. The term "testing," as used in the Agreed Statement to paragraph 1 of Article VIII of the Treaty, includes research and development.

Second Common Understanding. The Parties shall notify each other in the Standing Consultative Commission of the number of airplanes, according to type, used for testing pursuant to the Agreed Statement to paragraph 1 of Article VIII of the Treaty. Such notification shall be provided at the first regular session of the Standing Consultative Commission held after an airplane has been used for such testing.

Third Common Understanding. None of the sixteen airplanes referred to in the Agreed Statement to paragraph 1 of Article VIII of the Treaty may be replaced, except in the event of the involuntary destruction of any such airplane or in the case of the dismantling or destruction of any such airplane. The procedures for such replacement and for removal of any such airplane from that number, in case of its conversion, shall be agreed upon in the Standing Consultative Commission.

---

2. Each Party undertakes not to convert aircraft other than bombers into aircraft which can carry out the mission of a heavy bomber as referred to in subparagraph 3(b) of Article II.

Article IX

1. Each Party undertakes not to develop, test, or deploy:

(a) ballistic missiles capable of a range in excess of 600 kilometers for installation on waterborne vehicles other than submarines, or launchers of such missiles;

Common Understanding to subparagraph (a). The obligations provided for in subparagraph 1(a) of Article IX of the Treaty do not affect current practices for transporting ballistic missiles.

(b) fixed ballistic or cruise missile launchers for emplacement on the ocean floor, on the seabed, or on the beds of internal waters and inland waters, or in the subsoil thereof, or mobile launchers of such missiles, which move only in contact with the ocean floor, the seabed, or the beds of internal waters and inland waters, or missiles for such launchers;

Agreed Statement to subparagraph (b). The obligations provided for in subparagraph 1(b) of Article IX of the Treaty shall apply to all areas of the ocean floor and the seabed, including the seabed zone referred to in Articles I and II of the 1971 Treaty on the Prohibition of the Emplacement of Nuclear Weapons and Other Weapons of Mass Destruction on the Seabed and the Ocean Floor and in the Subsoil Thereof.

(c) systems for placing into Earth orbit nuclear weapons or any other kind of weapons of mass destruction, including fractional orbital missiles;

Common Understanding to subparagraph (c). The provisions of subparagraph 1(c) of Article IX of the Treaty do not require the dismantling or destruction of any existing launchers of either Party.

(d) mobile launchers of heavy ICBMs;
(e) SLBMs which have a launch-weight greater or a throw-weight greater than that of the heaviest, in terms of either launch-weight or throw-weight, respectively, of the light ICBMs deployed by either Party as of the date of signature of this Treaty, or launchers of such SLBMs; or
(f) ASBMs which have a launch-weight greater or a throw-weight greater than that of the heaviest, in terms of either launch-weight or throw-weight, respectively, of the light ICBMs deployed by either Party as of the date of signature of this Treaty.

First Agreed Statement to subparagraphs (e) and (f). The launch-weight of an SLBM or of an ASBM is the weight of the fully loaded missile itself at the time of launch.

Second Agreed Statement to subparagraphs (e) and (f). The throw-weight of an SLBM or of an ASBM is the sum of the weight of:

(a) its reentry vehicle or reentry vehicles;
(b) any self-contained dispensing mechanisms or other appropriate devices for targeting one reentry vehicle, or for releasing or for dispensing and targeting two or more reentry vehicles; and
(c) its penetration aids, including devices for their release.

Common Understanding to subparagraphs (e) and (f). The term "other appropriate devices," as used in the definition of the throw-weight of an SLBM or of an ASBM in the Second Agreed Statement to subparagraphs 1(e) and 1(f) of Article IX of the Treaty, means any devices for dispensing and targeting two or more reentry vehicles; and any devices for releasing two or more reentry vehicles or for targeting one reentry vehicle, which cannot provide their reentry vehicles or reentry vehicle with additional velocity of more than 1,000 meters per second.

---

2. Each Party undertakes not to flight-test from aircraft cruise missiles capable of a range in excess of 600 kilometers which are equipped with multiple independently targetable warheads and not to deploy such cruise missiles on aircraft.

Agreed Statement. Warheads of a cruise missile are independently targetable if maneuvering or targeting of the warheads to separate aim points along ballistic trajectories or any other flight paths, which are unrelated to each other, is accomplished during a flight of a cruise missile.

Article X

Subject to the provisions of this Treaty, modernization and replacement of strategic offensive arms may be carried out.

Article XI

1. Strategic offensive arms which would be in excess of the aggregate numbers provided for in this Treaty as well as strategic offensive arms prohibited by this Treaty shall be dismantled or destroyed under procedures to be agreed upon in the Standing Consultative Commission.

2. Dismantling or destruction of strategic offensive arms which would be in excess of the aggregate number provided for in paragraph 1 of Article III shall begin on the date of the entry into force of this Treaty and shall be completed within the following periods from that date: four months for ICBM launchers; six months for SLBM launchers; and three months for heavy bombers.

3. Dismantling or destruction of strategic offensive arms which would be in excess of the aggregate number provided for in paragraph 2 of Article III shall be initiated no later than January 1, 1981, shall be carried out throughout the ensuing twelve-month period, and shall be completed no later than December 31, 1981.

4. Dismantling or destruction of strategic offensive arms prohibited by this Treaty shall be completed within the shortest possible agreed period of time, but not later than six months after the entry into force of this Treaty.

Article XII

1. In order to ensure the viability and effectiveness of this Treaty, each Party undertakes not to circumvent the provisions of this Treaty, through any other state or states, or in any other manner.

Article XIII

1. Each Party undertakes not to assume any international obligations which would conflict with this Treaty.

Article XIV

The Parties undertake to begin, promptly after the entry into force of this Treaty, active negotiations with the objective of achieving, as soon as possible, agreement on further measures for the limitation and reduction of strategic arms. It is also the objective of the Parties to conclude well in advance of 1985 an agreement limiting strategic offensive arms to replace this Treaty upon its expiration.

Article XV

1. For the purpose of providing assurance of compliance with the provisions of this Treaty, each Party shall use national technical means of verification at its disposal in a manner consistent with generally recognized principles of international law.

2. Each party undertakes not to interfere with the national technical means of verification of the other Party operating in accordance with paragraph 1 of this Article.

3. Each Party undertakes not to use deliberate concealment measures which impede verification by national technical means of compliance with the provisions of this Treaty. This obligation shall not require changes in current construction, assembly, conversion, or overhaul practices.

First Agreed Statement. Deliberate concealment measures, as referred to in paragraph 3 of Article XV of the Treaty, are measures carried out deliberately to hinder or deliberately to impede verification by national technical means of compliance with the provisions of the Treaty.

Second Agreed Statement. The obligation not to use deliberate concealment measures, provided for in paragraph 3 of Article XV of the Treaty, does not preclude the testing of anti-missile defense penetration aids.

First Common Understanding. The provisions of paragraph 3 of Article XV of the Treaty and the First Agreed Statement thereto apply to all provisions of the Treaty, including provisions associated with testing. In this connection, the obligation not to use deliberate concealment measures includes the obligation not to use deliberate concealment measures associated with testing, including those measures aimed at concealing the association between ICBMs and launchers during testing.

Second Common Understanding. Each Party is free to use various methods of transmitting telemetric information during testing, including its encryption, except that, in accordance with the provisions of paragraph 3 of Article XV of the Treaty, neither Party shall engage in deliberate denial of telemetric information, such as through the use of telemetry encryption, whenever such denial impedes verfication of compliance with the provisions of the Treaty.

Third Common Understanding. In addition to the obligations provided for in paragraph 3 of Article XV of the Treaty, no shelters which impede verification by national technical means of compliance with the provisions of the Treaty shall be used over ICBM silo launchers.

Article XVI

1. Each Party undertakes, before conducting each planned ICBM launch, to notify the other Party well in advance on a case-by-case basis that such a launch will occur, except for single ICBM launches from test ranges or from ICBM launcher deployment areas, which are not planned to extend beyond its national territory.

First Common Understanding. ICBM launches to which the obligations provided for in Article XVI of the Treaty apply, include, among others, those ICBM launches for which advance notification is required pursuant to the provisions of the Agreement on Measures to Reduce the Risk of Outbreak of Nuclear War Between the United States of America and the Union of Soviet Socialist Republics, signed September 30, 1971, and the Agreement Between the Government of the United States of America and the Government of the Union of Soviet Socialist Republics on the Prevention of Incidents On and Over the High Seas, signed May 25, 1972. Nothing in Article XVI of the Treaty is intended to inhibit advance notification, on a voluntary basis, of any ICBM launches not subject to its provisions, the advance notification of which would enhance confidence between the Parties.

Second Common Understanding. A multiple ICBM launch conducted by a Party, as distinct from single ICBM launches referred to in Article XVI of the Treaty, is a launch which would result in two or more of its ICBMs being in flight at the same time.

Third Common Understanding. The test ranges referred to in Article XVI of the Treaty are those covered by the Second Agreed Statement to paragraph 2 of Article VII of the Treaty.

---

2. The Parties shall agree in the Standing Consultative Commission upon procedures to implement the provisions of this Article.

Article XVII

1. To promote the objectives and implementation of the provisions of this Treaty, the Parties shall use the Standing Consultative Commission established by the Memorandum of Understanding Between the Government of the United States of America and the Government of the Union of Soviet Socialist Republics Regarding the Establishment of a Standing Consultative Commission of December 21, 1972.

2. Within the framework of the Standing Consultative Commission, with respect to this Treaty, the Parties will:

    (a) consider questions concerning compliance with the obligations assumed and related situations which may be considered ambiguous;

    (b) provide on a voluntary basis such information as either Party considers necessary to assure confidence in compliance with the obligations assumed;

(c) consider questions involving unintended interference with national technical means of verification, and questions involving unintended impeding of verification by national technical means of compliance with the provisions of this Treaty;

(d) consider possible changes in the strategic situation which have a bearing on the provisions of this Treaty;

(e) agree upon procedures for replacement, conversion, and dismantling or destruction, of strategic offensive arms in cases provided for in the provisions of this Treaty and upon procedures for removal of such arms from the aggregate numbers when they otherwise cease to be subject to the limitations provided for in this Treaty, and at regular sessions of the Standing Consultative Commission, notify each other in accordance with the aforementioned procedures, at least twice annually, of actions completed and those in process;

(f) consider, as appropriate, possible proposals for further increasing the viability of this Treaty, including proposals for amendments in accordance with the provisions of this Treaty;

(g) consider, as appropriate, proposals for further measures limiting strategic offensive arms.

3. In the Standing Consultative Commission the Parties shall maintain by category the agreed data base on the numbers of strategic offensive arms established by the Memorandum of Understanding Between the United States of America and the Union of Soviet Socialist Republics Regarding the Establishment of a Data Base on the Numbers of Strategic Offensive Arms of June 18, 1979.

Agreed Statement. In order to maintain the agreed data base on the numbers of strategic offensive arms subject to the limitations provided for in the Treaty in accordance with paragraph 3 of Article XVII of the Treaty, at each regular session of the Standing Consultative Commission the Parties will notify each other of and consider changes in those numbers in the following categories: launchers of ICBMs; fixed launchers of ICBMs; launchers of ICBMs equipped with MIRVs; launchers of SLBMs; launchers of SLBMs equipped with MIRVs; heavy bombers; heavy bombers equipped for cruise missiles capable of a range in excess of 600 kilometers; heavy bombers equipped only for ASBMs; ASBMs; and ASBMs equipped with MIRVs.

Article XVIII

Each Party may propose amendments to this Treaty. Agreed amendments shall enter into force in accordance with the procedures governing the entry into force of this Treaty.

Article XIX

1. This Treaty shall be subject to ratification in accordance with the constitutional procedures of each Party. This Treaty shall enter into force on the day of the exchange of instruments of ratification and shall remain in force through December 31, 1985, unless replaced earlier by an agreement further limiting strategic offensive arms.

2. This Treaty shall be registered pursuant to Article 102 of the Charter of the United Nations.

3. Each Party shall, in exercising its national sovereignty, have the right to withdraw from this Treaty if it decides that extraordinary events related to the subject matter of this Treaty have jeopardized its supreme interests. It shall give notice of its decision to the other Party six months prior to withdrawal from the Treaty. Such notice shall

include a statement of the extraordinary events the notifying Party regards as having jeopardized its supreme interests.

DONE at Vienna on June 18, 1979, in two copies, each in the English and Russian languages, both texts being equally authentic.

For the United States of America:

JIMMY CARTER

President of the United States of America

For the Union of Soviet Socialist Republics:

L. BREZHNEV,

General Secretary of the CPSU, Chairman of the Presidium of the Supreme Soviet of the U.S.S.R.

Protocol to the Treaty Between the United States of America and the Union of Soviet Socialist Republics on the Limitation of Strategic Offensive Arms

The United States of America and the Union of Soviet Socialist Republics, hereinafter referred to as the Parties,

Having agreed on limitations on strategic offensive arms in the Treaty,

Have agreed on additional limitations for the period during which this Protocol remains in force, as follows:

Article I

Each Party undertakes not to deploy mobile ICBM launchers or to flight-test ICBMs from such launchers.

Article II

1. Each Party undertakes not to deploy cruise missiles capable of a range in excess of 600 kilometers on sea-based launchers or on land-based launchers.

2. Each Party undertakes not to flight-test cruise missiles capable of a range in excess of 600 kilometers which are equipped with multiple independently targetable warheads from sea-based launchers or from land-based launchers.

Agreed Statement. Warheads of a cruise missile are independently targetable if maneuvering or targeting of the warheads to separate aim points along ballistic trajectories or any other flight paths, which are unrelated to each other, is accomplished during a flight of a cruise missile.

---

3. For the purposes of this Protocol, cruise missiles are unmanned, self-propelled, guided, weapon-delivery vehicles which sustain flight through the use of aerodynamic lift over most of their flight path and which are flight-tested from or deployed on sea-based or land-based launchers, that is, sea-launched cruise missiles and ground-launched cruise missiles, respectively.

First Agreed Statement. If a cruise missile is capable of a range in excess of 600 kilometers, all cruise missiles of that type shall be considered to be cruise missiles capable of a range in excess of 600 kilometers.

First Common Understanding. If a cruise missile has been flight-tested to a range in excess of 600 kilometers, it shall be considered to be a cruise missile capable of a range in excess of 600 kilometers.

Second Common Understanding. Cruise missiles not capable of a range in excess of 600 kilometers shall not be considered to be of a type capable of a range in excess of 600 kilometers if they are distinguishable on the basis of externally observable design features from cruise missiles of types capable of a range in excess of 600 kilometers.

314

Second Agreed Statement. The range of which a cruise missile is capable is the maximum distance which can be covered by the missile in its standard design mode flying until fuel exhaustion, determined by projecting its flight path onto the Earth's sphere from the point of launch to the point of impact.

Third Agreed Statement. If an unmanned, self-propelled, guided vehicle which sustains flight through the use of aerodynamic lift over most of its flight path has been flight-tested or deployed for weapon delivery, all vehicles of that type shall be considered to be weapon-delivery vehicles.

Third Common Understanding. Unmanned, self-propelled, guided vehicles which sustain flight through the use of aerodynamic lift over most of their flight path and are not weapon-delivery vehicles, that is, unarmed, pilotless, guided vehicles, shall not be considered to be cruise missiles if such vehicles are distinguishable from cruise missiles on the basis of externally observable design features.

Fourth Common Understanding. Neither Party shall convert unarmed, pilotless, guided vehicles into cruise missiles capable of a range in excess of 600 kilometers, nor shall either Party convert cruise missiles capable of a range in excess of 600 kilometers into unarmed, pilotless, guided vehicles.

Fifth Common Understanding. Neither Party has plans during the term of the Protocol to flight-test from or deploy on sea-based or land-based launchers unarmed, pilotless, guided vehicles which are capable of a range in excess of 600 kilometers. In the future, should a Party have such plans, that Party will provide notification thereof to the other Party well in advance of such flight-testing or deployment. This Common Understanding does not apply to target drones.

Article III

Each Party undertakes not to flight-test or deploy ASBMs.

Article IV

This Protocol shall be considered an integral part of the Treaty. It shall enter into force on the day of the entry into force of the Treaty and shall remain in force through December 31, 1981, unless replaced earlier by an agreement on further measures limiting strategic offensive arms.

DONE at Vienna on June 18, 1979, in two copies, each in the English and Russian languages, both texts being equally authentic.

For the United States of America:

JIMMY CARTER

President of the United States of America

For the Union of Soviet Socialist Republics:

L. BREZHNEV

General Secretary of the CPSU, Chairman of the Presidium of the Supreme Soviet of the U.S.S.R.

Memorandum of Understanding Between the United States of America and the Union of Soviet Socialist Republics Regarding the Establishment of a Data Base on the Numbers of Strategic Offensive Arms

For the purposes of the Treaty Between the United States of America and the Union of Soviet Socialist Republics on the Limitation of Strategic Offensive Arms, the Parties have considered data on numbers of strategic offensive arms and agree that as of November 1, 1978 there existed the following numbers of strategic offensive arms subject to the limitations provided for in the Treaty which is being signed today.

	U.S.A.	U.S.S.R.
Launchers of ICBMs	1,054	1,398
Fixed launchers of ICBMs	1,054	1,398
Launchers of ICBMs equipped with MIRVs	550	576
Launchers of SLBMs	656	950
Launchers of SLBMs equipped with MIRVs	496	128
Heavy bombers	574	156
Heavy bombers equipped for cruise missiles capable of a range in excess of 600 kilometers	0	0
Heavy bombers equipped only for ASBMs	0	0
ASBMs	0	0
ASBMs equipped with MIRVs	0	0

At the time of entry into force of the Treaty the Parties will update the above agreed data in the categories listed in this Memorandum.

DONE at Vienna on June 18, 1979, in two copies, each in the English and Russian languages, both texts being equally authentic.

For the United States of America

RALPH EARLE II

Chief of the United States Delegation to the Strategic Arms Limitation Talks

For the Union of Soviet Socialist Republics

V. KARPOV

Chief of the U.S.S.R. Delegation to the Strategic Arms Limitation Talks

Statement of Data on the Numbers of Strategic Offensive Arms as of the Date of Signature of the Treaty

The United States of America declares that as of June 18, 1979 it possesses the following numbers of strategic offensive arms subject to the limitations provided for in the Treaty which is being signed today:

Launchers of ICBMs	1,054
Fixed launchers of ICBMs	1,054
Launchers of ICBMs equipped with MIRVs	550
Launchers of SLBMs	656
Launchers of SLBMs equipped with MIRVs	496
Heavy bombers	573
Heavy bombers equipped for cruise missiles capable of a range in excess of 600 kilometers	3
Heavy bombers equipped only for ASBMs	0
ASBMs	0
ASBMs equipped with MIRVs	0

June 18, 1979

RALPH EARLE II

Chief of the United States Delegation to the Strategic Arms Limitation Talks

I certify that this is a true copy of the document signed by Ambassador Ralph Earle II entiled "Statement of Data on the Numbers of Strategic Offensive Arms as of the Date of Signature of the Treaty" and given to Ambassador V. Karpov on June 18, 1979 in Vienna, Austria.

THOMAS GRAHAM, JR.

General Counsel
United States Arms Control and Disarmament Agency

Statement of Data on the Numbers of Strategic Offensive Arms as of the Date of Signature of the Treaty

The Union of Soviet Socialist Republics declares that as of June 18, 1979, it possesses the following numbers of strategic offensive arms subject to the limitations provided for in the Treaty which is being signed today:

Launchers of ICBMs	1,398
Fixed launchers of ICBMs	1,398

Launchers of ICBMs equipped with MIRVs	608
Launchers of SLBMs	950
Launchers of SLBMs equipped with MIRVs	144
Heavy bombers	156
Heavy bombers equipped for cruise missiles capable of a range in excess of 600 kilometers	0
Heavy bombers equipped only for ASBMs	0
ASBMs	0
ASBMs equipped with MIRVs	0

June 18, 1979

V. KARPOV

Chief of the U.S.S.R. Delegation to the Strategic Arms Limitation Talks

Translation certified by:
   W.D. Krimer,
      Senior Language Officer,
      Division of Language Services, U.S. Department of State

WILLIAM D. KRIMER

Joint Statement of Principles and Basic Guidelines for Subsequent Negotiations on the Limitation of Strategic Arms

The United States of America and the Union of Soviet Socialist Republics, hereinafter referred to as the Parties,

Having concluded the Treaty on the Limitation of Strategic Offensive Arms,

Reaffirming that the strengthening of strategic stability meets the interests of the Parties and the interests of international security,

Convinced that early agreement on the further limitation and further reduction of strategic arms would serve to strengthen international peace and security and to reduce the risk of outbreak of nuclear war,

Have agreed as follows:

First, The Parties will continue to pursue negotiations, in accordance with the principle of equality and equal security, on measures for the further limitation and reduction in the numbers of strategic arms, as well as for their further qualitative limitation.

In furtherance of existing agreements between the Parties on the limitation and reduction of strategic arms, the Parties will continue, for the purposes of reducing and averting the risk of outbreak of nuclear war, to seek measures to strengthen strategic stability by, among other things, limitations on strategic offensive arms most destabilizing to the strategic balance and by measures to reduce and to avert the risk of surprise attack.

Second. Further limitations and reductions of strategic arms must be subject to adequate verification by national technical means, using additionally, as appropriate, cooperative measures contributing to the effectiveness of verification by national technical means. The Parties will seek to strengthen verification and to perfect the operation of the Standing Consultative Commission in order to promote assurance of compliance with the obligations assumed by the Parties.

Third. The Parties shall pursue in the course of these negotiations, taking into consideration factors that determine the strategic situation, the following objectives:

1) significant and substantial reductions in the numbers of strategic offensive arms;

2) qualitative limitations on strategic offensive arms, including restrictions on the development, testing, and deployment of new types of strategic offensive arms and on the modernization of existing strategic offensive arms;

3) resolution of the issues included in the Protocol to the Treaty Between the United States of America and the Union of Soviet Socialist Republics on the Limitation of Strategic Offensive Arms in the context of the negotiations relating to the implementation of the principles and objectives set out herein.

Fourth. The Parties will consider other steps to ensure and enhance strategic stability, to ensure the equality and equal security of the Parties, and to implement the above principles and objectives. Each Party will be free to raise any issue relative to the further limitation of strategic arms. The Parties will also consider further joint measures, as appropriate, to strengthen international peace and security and to reduce the risk of outbreak of nuclear war.

Vienna, June 18, 1979

For the United States of America

JIMMY CARTER

President of the United States of America

For the Union of Soviet Socialist Republics

L. BREZHNEV

General Secretary of the CPSU, Chairman of the Presidium of the Supreme Soviet of the U.S.S.R.

Soviet Backfire Statement

On June 16, 1979, President Brezhnev handed President Carter the following written statement [original Russian text was attached]:

"The Soviet side informs the US side that the Soviet 'Tu-22M' airplane, called 'Backfire' in the USA, is a medium-range bomber, and that it does not intend to give this airplane the capability of operating at intercontinental distances. In this connection, the Soviet side states that it will not increase the radius of action of this airplane in such a way as to enable it to strike targets on the territory of the USA. Nor does it intend to give it such a capability in any other manner, including by in-flight refueling. At the same time, the Soviet side states that it will not increase the production rate of this airplane as compared to the present rate."

President Brezhnev confirmed that the Soviet Backfire production rate would not exceed 30 per year.

President Carter stated that the United States enters into the SALT II Agreement on the basis of the commitments contained in the Soviet statement and that it considers the carrying out of these commitments to be essential to the obligations assumed under the Treaty.

CYRUS VANCE

Appendix C

Treaty Between the United States of America and the Union of Soviet Socialist Republics on the Limitation of Anti-Ballistic Missile Systems

Treaty Between the United States of America and the Union of Soviet Socialist Republics on the Limitation of Anti-Ballistic Missile Systems

Signed at Moscow May 26, 1972
Ratification advised by U.S. Senate August 3, 1972
Ratified by U.S. President September 30, 1972
Proclaimed by U.S. President October 3, 1972
Instruments of ratification exchanged October 3, 1972
Entered into force October 3, 1972

The United States of America and the Union of Soviet Socialist Republics, hereinafter referred to as the Parties,

Proceeding from the premise that nuclear war would have devastating consequences for all mankind,

Considering that effective measures to limit anti-ballistic missile systems would be a substantial factor in curbing the race in strategic offensive arms and would lead to a decrease in the risk of outbreak of war involving nuclear weapons,

Proceeding from the premise that the limitation of anti-ballistic missile systems, as well as certain agreed measures with respect to the limitation of strategic offensive arms, would contribute to the creation of more favorable conditions for further negotiations on limiting strategic arms,

Mindful of their obligations under Article VI of the Treaty on the Non-Proliferation of Nuclear Weapons,

Declaring their intention to achieve at the earliest possible date the cessation of the nuclear arms race and to take effective measures toward reductions in strategic arms, nuclear disarmament, and general and complete disarmament,

Desiring to contribute to the relaxation of international tension and the strengthening of trust between States,

Have agreed as follows:

Article I

1. Each party undertakes to limit anti-ballistic missile (ABM) systems and to adopt other measures in accordance with the provisions of this Treaty.

2. Each Party undertakes not to deploy ABM systems for a defense of the territory of its country and not to provide a base for such a defense, and not to deploy ABM systems for defense of an individual region except as provided for in Article III of this Treaty.

Article II

1. For the purpose of this Treaty an ABM system is a system to counter strategic ballistic missiles or their elements in flight trajectory, currently consisting of:

(a) ABM interceptor missiles, which are interceptor missiles constructed and deployed for an ABM role, or of a type tested in an ABM mode;

(b) ABM launchers, which are launchers constructed and deployed for launching ABM interceptor missiles; and

(c) ABM radars, which are radars constructed and deployed for an ABM role, or of a type tested in an ABM mode.

2. The ABM system components listed in paragraph 1 of this Article include those which are:

(a) operational;
(b) under construction;
(c) undergoing testing;
(d) undergoing overhaul, repair or conversion; or
(e) mothballed.

Article III

Each Party undertakes not to deploy ABM systems or their components except that:

(a) within one ABM system deployment area having a radius of one hundred and fifty kilometers and centered on the Party's national capital, a Party may deploy: (1) no more than one hundred ABM launchers and no more than one hundred ABM interceptor missiles at launch sites, and (2) ABM radars within no more than six ABM radar complexes, the area of each complex being circular and having a diameter of no more than three kilometers; and

(b) within one ABM system deployment area having a radius of one hundred and fifty kilometers and containing ICBM silo launchers, a Party may deploy: (1) no more than one hundred ABM launchers and no more than one hundred ABM interceptor missiles at launch sites, (2) two large phased-array ABM radars comparable in potential to corresponding ABM radars operational or under construction on the date of signature of the Treaty in an ABM system deployment area containing ICBM silo launchers, and (3) no more than eighteen ABM radars each having a potential less than the potential of the smaller of the above-mentioned two large phased-array ABM radars.

Article IV

The limitations provided for in Article III shall not apply to ABM systems or their components used for development or testing, and located within current or additionally agreed test ranges. Each Party may have no more than a total of fifteen ABM launchers at test ranges.

Article V

1. Each Party undertakes not to develop, test, or deploy ABM systems or components which are sea-based, air-based, space-based, or mobile land-based.

2. Each Party undertakes not to develop, test, or deploy ABM launchers for launching more than one ABM interceptor missile at a time from each launcher, not to modify deployed launchers to provide them with such a capability, not to develop, test, or deploy automatic or semi-automatic or other similar systems for rapid reload of ABM launchers.

Article VI

To enhance assurance of the effectiveness of the limitations on ABM systems and their components provided by the Treaty, each Party undertakes:

(a) not to give missiles, launchers, or radars, other than ABM interceptor missiles, ABM launchers, or ABM radars, capabilities to counter strategic ballistic missiles or their elements in flight trajectory, and not to test them in an ABM mode; and

(b) not to deploy in the future radars for early warning of strategic ballistic missile attack except at locations along the periphery of its national territory and oriented outward.

Article VII

Subject to the provisions of this Treaty, modernization and replacement of ABM systems or their components may be carried out.

Article VIII

ABM systems or their components in excess of the numbers or outside the areas specified in this Treaty, as well as ABM systems or their components prohibited by this Treaty, shall be destroyed or dismantled under agreed procedures within the shortest possible agreed period of time.

Article IX

To assure the viability and effectiveness of this Treaty, each Party undertakes not to transfer to other States, and not to deploy outside its national territory, ABM systems or their components limited by this Treaty.

Article X

Each Party undertakes not to assume any international obligations which would conflict with this Treaty.

Article XI

The Parties undertake to continue active negotiations for limitations on strategic offensive arms.

Article XII

1. For the purpose of providing assurance of compliance with the provisions of this Treaty, each Party shall use national technical means of verification at its disposal in a manner consistent with generally recognized principles of international law.

2. Each Party undertakes not to interfere with the national technical means of verification of the other Party operating in accordance with paragraph 1 of this Article.

3. Each Party undertakes not to use deliberate concealment measures which impede verification by national technical means of compliance with the provisions of this Treaty. This obligation shall not require changes in current construction, assembly, conversion, or overhaul practices.

Article XIII

1. To promote the objectives and implementation of the provisions of this Treaty, the Parties shall establish promptly a Standing Consultative Commission, within the framework of which they will:

(a) consider questions concerning compliance with the obligations assumed and related situations which may be considered ambiguous;

(b) provide on a voluntary basis such information as either Party considers necessary to assure confidence in compliance with the obligations assumed;

(c) consider questions involving unintended interference with national technical means of verification;

(d) consider possible changes in the strategic situation which have a bearing on the provisions of this Treaty;

(e) agree upon procedures and dates for destruction or dismantling of ABM systems or their components in cases provided for by the provisions of this Treaty;

(f) consider, as appropriate, possible proposals for further increasing the viability of this Treaty; including proposals for amendments in accordance with the provisions of this Treaty;

(g) consider, as appropriate, proposals for further measures aimed at limiting strategic arms.

2. The Parties through consultation shall establish, and may amend as appropriate, Regulations for the Standing Consultative Commission governing procedures, composition and other relevant matters.

Article XIV

1. Each Party may propose amendments to this Treaty. Agreed amendments shall enter into force in accordance with the procedures governing the entry into force of this Treaty.

2. Five years after entry into force of this Treaty, and at five-year intervals thereafter, the Parties shall together conduct a review of this Treaty.

Article XV

1. This Treaty shall be of unlimited duration.

2. Each Party shall, in exercising its national sovereignty, have the right to withdraw from this Treaty if it decides that extraordinary events related to the subject matter of this Treaty have jeopardized its supreme interests. It shall give notice of its decision to the other Party six months prior to withdrawal from the Treaty. Such notice shall include a statement of the extraordinary events the notifying Party regards as having jeopardized its supreme interests.

Article XVI

1. This Treaty shall be subject to ratification in accordance with the constitutional procedures of each Party. The Treaty shall enter into force on the day of the exchange of instruments of ratification.

2. This Treaty shall be registered pursuant to Article 102 of the Charter of the United Nations.

DONE at Moscow on May 26, 1972, in two copies, each in the English and Russian languages, both texts being equally authentic.

FOR THE UNITED STATES OF AMERICA

FOR THE UNION OF SOVIET SOCIALIST REPUBLICS

RICHARD NIXON

L. I. BREZHNEV

President of the United States of America

General Secretary of the Central Committee of the CPSU

Agreed Statements, Common Understandings, and Unilateral Statements Regarding the Treaty Between the United States of America and the Union of Soviet Socialist Republics on the Limitation of Anti-Ballistic Missiles

1. Agreed Statements

The document set forth below was agreed upon and initialed by the Heads of the Delegations on May 26, 1972 (letter designations added);

AGREED STATEMENTS REGARDING THE TREATY BETWEEN THE UNITED STATES OF AMERICA AND THE UNION OF SOVIET SOCIALIST REPUBLICS ON THE LIMITATION OF ANTI-BALLISTIC MISSILE SYTEMS

[A]

The Parties understand that, in addition to the ABM radars which may be deployed in accordance with subparagraph (a) of Article III of the Treaty, those non-phased- array ABM radars operational on the date of signature of the Treaty within the ABM system deployment area for defense of the national capital may be retained.

[B]

The Parties understand that the potential (the product of mean emitted power in watts and antenna area in square meters) of the smaller of the two large phased-array ABM radars referred to in subparagraph (b) of Article III of the Treaty is considered for purposes of the Treaty to be three million.

[C]

The Parties understand that the center of the ABM system deployment area centered on the national capital and the center of the ABM system deployment area containing ICBM silo launchers for each Party shall be separated by no less than thirteen hundred kilometers.

[D]

In order to insure fulfillment of the obligation not to deploy ABM systems and their components except as provided in Article III of the Treaty, the Parties agree that in the event ABM systems based on other physical principles and including components capable of substituting for ABM interceptor missiles, ABM launchers, or ABM radars are created in the future, specific limitations on such systems and their components would be subject to discussion in accordance with Article XIII and agreement in accordance with Article XIV of the Treaty.

[E]

The Parties understand that Article V of the Treaty includes obligations not to develop, test or deploy ABM interceptor missiles for the delivery by each ABM interceptor missile of more than one independently guided warhead.

[F]

The Parties agree not to deploy phased-array radars having a potential (the product of mean emitted power in watts and antenna area in square meters) exceeding three million, except as provided for in Articles III, IV and VI of the Treaty, or except for the purposes of tracking objects in outer space or for use as national technical means of verification.

[G]

The Parties understand that Article IX of the Treaty includes the obligation of the US and the USSR not to provide to other States technical descriptions or blue prints specially worked out for the construction of ABM systems and their components limited by the Treaty.

2. Common Understandings

Common understanding of the Parties on the following matters was reached during the negotiations:

A. Location of ICBM Defenses

The U.S. Delegation made the following statement on May 26, 1972:

Article III of the ABM Treaty provides for each side one ABM system deployment area centered on its national capital and one ABM system deployment area containing ICBM silo launchers. The two sides have registered agreement on the following statement: "The Parties understand that the center of the ABM system deployment area centered on the national capital and the center of the ABM system deployment area containing ICBM silo launchers for each Party shall be separated by no less than thirteen hundred kilometers." In this connection, the U.S. side notes that its ABM system deployment area for defense of ICBM silo launchers, located west of the Mississippi River, will be centered in the Grand Forks ICBM silo launcher deployment area. (See Agreed Statement [C].)

B. ABM Test Ranges

The U.S. Delegation made the following statement on April 26, 1972:

Article IV of the ABM Treaty provides that "the limitations provided for in Article III shall not apply to ABM systems or their components used for development or testing, and located within current or additionally agreed test ranges." We believe it would be useful to assure that there is no misunderstanding as to current ABM test ranges. It is our understanding that ABM test ranges encompass the area within which ABM components are located for test purposes. The current U.S. ABM test ranges are at White Sands, New Mexico, and at Kwajalein Atoll, and the current Soviet ABM test range is near Sary Shagan in Kazakhstan. We consider that non-phased array radars of types used for range safety or instrumentation purposes may be located outside of ABM test ranges. We interpret the reference in Article IV to "additionally agreed test ranges" to mean that ABM components will not be located at any other test ranges without prior agreement between our Governments that there will be such additional ABM test ranges.

On May 5, 1972, the Soviet Delegation stated that there was a common understanding on what ABM test ranges were, that the use of the types of non-ABM radars for range safety or instrumentation was not limited under the Treaty, that the reference in Article IV to "additionally agreed" test ranges was sufficiently clear, and that national means permitted identifying current test ranges.

C. Mobile ABM Systems

On January 29, 1972, the U.S. Delegation made the following statement:

Article V(1) of the Joint Draft Text of the ABM Treaty includes an undertaking not to develop, test, or deploy mobile land-based ABM systems and their components. On May 5, 1971, the U.S. side indicated that, in its view, a prohibition on deployment of mobile ABM systems and components would rule out the deployment of ABM launchers and radars which were not permanent fixed types. At that time, we asked for the Soviet view of this interpretation. Does the Soviet side agree with the U.S. side's interpretation put forward on May 5, 1971?

On April 13, 1972, the Soviet Delegation said there is a general common understanding on this matter.

D. Standing Consultative Commission

Ambassador Smith made the following statement on May 22, 1972:

The United States proposes that the sides agree that, with regard to initial implementation of the ABM Treaty's Article XIII on the Standing Consultative Commission (SCC) and of the consultation Articles to the Interim Agreement on offensive arms and the Accidents Agreement,[1] agreement establishing the SCC will be worked out early in the follow-on SALT negotiations; until that is completed, the following arrangements will prevail: when SALT is in session, any consultation desired by either side under these Articles can be carried out by the two SALT Delegations; when SALT is not in session, *ad hoc* arrangements for any desired consultations under these Articles may be made through diplomatic channels.

Minister Semenov replied that, on an *ad referendum* basis, he could agree that the U.S. statement corresponded to the Soviet understanding.

E. Standstill

On May 6, 1972, Minister Semenov made the following statement:

In an effort to accommodate the wishes of the U.S. side, the Soviet Delegation is prepared to proceed on the basis that the two sides will in fact observe the obligations of both the Interim Agreement and the ABM Treaty beginning from the date of signature of these two documents.

In reply, the U.S. Delegation made the following statement on May 20, 1972:

---

[1]See Article 7 of Agreement to Reduce the Risk of Outbreak of Nuclear War Between the United States of America and the Union of Soviet Socialist Republics, signed Sept. 30, 1971.

The U.S. agrees in principle with the Soviet statement made on May 6 concerning observance of obligations beginning from date of signature but we would like to make clear our understanding that this means that, pending ratification and acceptance, neither side would take any action prohibited by the agreements after they had entered into force. This understanding would continue to apply in the absence of notification by either signatory of its intention not to proceed with ratification or approval.

The Soviet Delegation indicated agreement with the U.S. statement.

3. Unilateral Statements

The following noteworthy unilateral statements were made during the negotiations by the United States Delegation:

A. Withdrawal from the ABM Treaty

On May 9, 1972, Ambassador Smith made the following statement:

The U.S. Delegation has stressed the importance the U.S. Government attaches to achieving agreement on more complete limitations on strategic offensive arms, following agreement on an ABM Treaty and on an Interim Agreement on certain measures with respect to the limitation of strategic offensive arms. The U.S. Delegation believes that an objective of the follow-on negotiations should be to constrain and reduce on a long-term basis threats to the survivability of our respective strategic retaliatory forces. The USSR Delegation has also indicated that the objectives of SALT would remain unfulfilled without the achievement of an agreement providing for more complete limitations on strategic offensive arms. Both sides recognize that the initial agreements would be steps toward the achievement of more complete limitations on strategic arms. If an agreement providing for more complete strategic offensive arms limitations were not achieved within five years, U.S. supreme interests could be jeopardized. Should that occur, it would constitute a basis for withdrawal from the ABM Treaty. The U.S. does not wish to see such a situation occur, nor do we believe that the USSR does. It is because we wish to prevent such a situation that we emphasize the importance the U.S. Government attaches to achievement of more complete limitations on strategic offensive arms. The U.S. Executive will inform the Congress, in connection with Congressional consideration of the ABM Treaty and the Interim Agreement, of this statement of the U.S. position.

B. Tested in ABM Mode

On April 7, 1972, the U.S. Delegation made the following statement:

Article II of the Joint Text Draft uses the term "tested in an ABM mode," in defining ABM components, and Article VI includes certain obligations concerning such testing. We believe that the sides should have a common understanding of this phrase. First, we would note that the testing provisions of the ABM Treaty are intended to apply to testing which occurs after the date of signature of the Treaty, and not to any testing which may have occurred in the past. Next, we would amplify the remarks we have made on this subject during the previous Helsinki phase by setting forth the objectives which govern the U.S. view on the subject, namely, while prohibiting testing of non-ABM components for ABM purposes: not to prevent testing of ABM components, and not to prevent testing of non-ABM components for non-ABM purposes. To clarify our interpretation of "tested in an ABM mode," we note that we would consider a launcher, missile or radar to be "tested in an ABM mode" if, for example, any of the following events occur: (1) a launcher is used to launch an ABM interceptor missile, (2) an interceptor missile is flight tested against a target vehicle which has a flight trajectory with characteristics of a strategic ballistic missile flight trajectory, or is flight tested in conjunction with the test of an ABM interceptor missile or an ABM radar at the same test range, or is flight tested to an altitude inconsistent with interception of targets against which air defenses are deployed, (3) a radar makes measurements on a cooperative target vehicle of the kind referred to in item (2) above during the reentry portion of its trajectory or makes measurements in conjunction with the test of an ABM interceptor missile or an ABM radar at the same test range. Radars used for purposes such as range safety or instrumentation would be exempt from application of these criteria.

C.   No-Transfer Article of ABM Treaty

On April 18, 1972, the U.S. Delegation made the following statement:

In regard to this Article [IX], I have a brief and I believe self-explanatory statement to make. The U.S. side wishes to make clear that the provisions of this Article do not set a precedent for whatever provision may be considered for a Treaty on Limiting Strategic Offensive Arms. The question of transfer of strategic offensive arms is a far more complex issue, which may require a different solution.

D.   No Increase in Defense of Early Warning Radars

On July 28, 1970, the U.S. Delegation made the following statement:

Since Hen House radars [Soviet ballistic missile early warning radars] can detect and track ballistic missile warheads at great distances, they have a significant ABM potential. Accordingly, the U.S. would regard any increase in the defenses of such radars by surface-to-air missiles as inconsistent with an agreement.

Protocol to the Treaty Between the United States of America and the Union of Soviet Socialist Republics on the Limitation of Anti-Ballistic Missile Systems

Signed at Moscow July 3, 1974
Ratification advised by U.S. Senate November 10, 1975
Ratified by U.S. President March 19, 1976
Instruments of ratification exchanged May 24, 1976
Proclaimed by U.S. President July 6, 1976
Entered into force May 24, 1976

The United States of America and the Union of Soviet Socialist Republics, hereinafter referred to as the Parties,

Proceeding from the Basic Principles of Relations between the United States of America and the Union of Soviet Socialist Republics signed on May 29, 1972,

Desiring to further the objectives of the Treaty between the United States of America and the Union of Soviet Socialist Republics on the Limitation of Anti-Ballistic Missile Systems signed on May 26, 1972, hereinafter referred to as the Treaty,

Reaffirming their conviction that the adoption of further measures for the limitation of strategic arms would contribute to strengthening international peace and security,

Proceeding from the premise that further limitation of anti-ballistic missile systems will create more favorable conditions for the completion of work on a permanent agreement on more complete measures for the limitation of strategic offensive arms,

Have agreed as follows:

Article I

1. Each Party shall be limited at any one time to a single area out of the two provided in Article III of the Treaty for deployment of anti-ballistic missile (ABM) systems or their components and accordingly shall not exercise its right to deploy an ABM system or its components in the second of the two ABM system deployment areas permitted by Article III of the Treaty, except as an exchange of one permitted area for the other in accordance with Article II of this Protocol.

2. Accordingly, except as permitted by Article II of this Protocol: the United States of America shall not deploy an ABM system or its components in the area centered on its capital, as permitted by Article III(a) of the Treaty, and the Soviet Union shall not deploy an ABM system or its components in the deployment area of intercontinental ballistic missile (ICBM) silo launchers as permitted by Article III(b) of the Treaty.

Article II

1. Each Party shall have the right to dismantle or destroy its ABM system and the components thereof in the area where they are presently deployed and to deploy an ABM system or its components in the alternative area permitted by Article III of the Treaty, provided that prior to initiation of construction, notification is given in accord

with the procedure agreed to in the Standing Consultative Commission, during the year beginning October 3, 1977 and ending October 2, 1978, or during any year which commences at five year intervals thereafter, those being the years for periodic review of the Treaty, as provided in Article XIV of the Treaty. This right may be exercised only once.

2. Accordingly, in the event of such notice, the United States would have the right to dismantle or destroy the ABM system and its components in the deployment area of ICBM silo launchers and to deploy an ABM system or its components in an area centered on its capital, as permitted by Article III(a) of the Treaty, and the Soviet Union would have the right to dismantle or destroy the ABM system and its components in the area centered on its capital and to deploy an ABM system or its components in an area containing ICBM silo launchers, as permitted by Article III(b) of the Treaty.

3. Dismantling or destruction and deployment of ABM systems or their components and the notification thereof shall be carried out in accordance with Article VIII of the ABM Treaty and procedures agreed to in the Standing Consultative Commission.

Article III

The rights and obligations established by the Treaty remain in force and shall be complied with by the Parties except to the extent modified by this Protocol. In particular, the deployment of an ABM system or its components within the area selected shall remain limited by the levels and other requirements established by the Treaty.

Article IV

This Protocol shall be subject to ratification in accordance with the constitutional procedures of each Party. It shall enter into force on the day of the exchange of instruments of ratification and shall thereafter be considered an integral part of the Treaty.

DONE at Moscow on July 3, 1974, in duplicate, in the English and Russian languages, both texts being equally authentic.

For the United States of America:

RICHARD NIXON

President of the United States of America

For the Union of Soviet Socialist Republics:

L. I. BREZHNEV

General Secretary of the Central Committee of the CPSU

Appendix D

Treaty Banning Nuclear Weapon Tests in the Atmosphere, in Outer Space and Under Water

Treaty Banning Nuclear Weapon Tests in the Atmosphere, in Outer Space and Under Water

Signed at Moscow August 5, 1963
Ratification advised by U.S. Senate September 24, 1963
Ratified by U.S. President October 7, 1963
U.S. ratification deposited at Washington, London, and Moscow October 10, 1963
Proclaimed by U.S. President October 10, 1963
Entered into force October 10, 1963

The Governments of the United States of America, the United Kingdom of Great Britain and Northern Ireland, and the Union of Soviet Socialist Republics, hereinafter referred to as the "Original Parties,"

Proclaiming as their principal aim the speediest possible achievement of an agreement on general and complete disarmament under strict international control in accordance with the objectives of the United Nations which would put an end to the armaments race and eliminate the incentive to the production and testing of all kinds of weapons, including nuclear weapons.

Seeking to achieve the discontinuance of all test explosions of nuclear weapons for all time, determined to continue negotiations to this end, and desiring to put an end to the contamination of man's environment by radioactive substances,

Have agreed as follows:

Article I

1. Each of the Parties to this Treaty undertakes to prohibit, to prevent, and not to carry out any nuclear weapon test explosion, or any other nuclear explosion, at any place under its jurisdiction or control:

(a) in the atmosphere; beyond its limits, including outer space; or under water, including territorial waters or high seas; or

(b) in any other environment if such explosion causes radioactive debris to be present outside the territorial limits of the State under whose jurisdiction or control such explosion is conducted. It is understood in this connection that the provisions of this subparagraph are without prejudice to the conclusion of a treaty resulting in the permanent banning of all nuclear test explosions, including all such explosions underground, the conclusion of which, as the Parties have stated in the Preamble to this Treaty, they seek to achieve.

2. Each of the Parties to this Treaty undertakes furthermore to refrain from causing, encouraging, or in any way participating in, the carrying out of any nuclear weapon test explosion, or any other nuclear explosion, anywhere which would take place in any of the environments described, or have the effect referred to, in paragraph 1 of this Article.

Article II

1. Any Party may propose amendments to this Treaty. The text of any proposed amendment shall be submitted to the Depositary Governments which shall circulate it to all Parties to this Treaty. Thereafter, if requested to do so by one-third or more of the

Parties, the Depositary Governments shall convene a conference, to which they shall invite all the Parties, to consider such amendment.

2. Any amendment to this Treaty must be approved by a majority of the votes of all the Parties to this Treaty, including the votes of all of the Original Parties. The amendment shall enter into force for all Parties upon the deposit of instruments of ratification by a majority of all the Parties, including the instruments of ratification of all of the Original Parties.

Article III

1. This Treaty shall be open to all States for signature. Any State which does not sign this Treaty before its entry into force in accordance with paragraph 3 of this Article may accede to it at any time.

2. This Treaty shall be subject to ratification by signatory States. Instruments of ratification and instruments of accession shall be deposited with the Governments of the Original Parties—the United States of America, the United Kingdom of Great Britain and Northern Ireland, and the Union of Soviet Socialist Republics—which are hereby designated the Depositary Governments.

3. This Treaty shall enter into force after its ratification by all the Original Parties and the deposit of their instruments of ratification.

4. For States whose instruments of ratification or accession are deposited subsequent to the entry into force of this Treaty, it shall enter into force on the date of the deposit of their instruments of ratification or accession.

5. The Depositary Governments shall promptly inform all signatory and acceding States of the date of each signature, the date of deposit of each instrument of ratification of and accession to this Treaty, the date of its entry into force, and the date of receipt of any requests for conferences or other notices.

6. This Treaty shall be registered by the Depositary Governments pursuant to Article 102 of the Charter of the United Nations.

Article IV

This Treaty shall be of unlimited duration.

Each Party shall in exercising its national sovereignty have the right to withdraw from the Treaty if it decides that extraordinary events, related to the subject matter of this Treaty, have jeopardized the supreme interests of its country. It shall give notice of such withdrawal to all other Parties to the Treaty three months in advance.

Article V

This Treaty, of which the English and Russian texts are equally authentic, shall be deposited in the archives of the Depositary Governments. Duly certified copies of this Treaty shall be transmitted by the Depositary Governments to the Governments of the signatory and acceding States.

IN WITNESS WHEREOF the undersigned, duly authorized, have signed this Treaty.

DONE in triplicate at the city of Moscow the fifth day of August, one thousand nine hundred and sixty-three.

For the Government of the United States of America	For the Government of the United Kingdom of Great Britain and Northern Ireland	For the Government of the Union of Soviet Socialist Republics
DEAN RUSK	**HOME**	**A. GROMYKO**

Appendix *E*

Treaty Between the United States of America and the Union of Soviet Socialist Republics on the Limitation of Underground Nuclear Weapon Tests

Treaty Between the United States of America and the Union of Soviet Socialist Republics on the Limitation of Underground Nuclear Weapon Tests

Signed at Moscow July 3, 1974

The United States of America and the Union of Soviet Socialist Republics, hereinafter referred to as the Parties,

Declaring their intention to achieve at the earliest possible date the cessation of the nuclear arms race and to take effective measures toward reductions in strategic arms, nuclear disarmament, and general and complete disarmament under strict and effective international control,

Recalling the determination expressed by the Parties to the 1963 Treaty Banning Nuclear Weapon Tests in the Atmosphere, in Outer Space and Under Water in its Preamble to seek to achieve the discontinuance of all test explosions of nuclear weapons for all time, and to continue negotiations to this end,

Noting that the adoption of measures for the further limitation of underground nuclear weapon tests would contribute to the achievement of these objectives and would meet the interests of strengthening peace and the further relaxation of international tension,

Reaffirming their adherence to the objectives and principles of the Treaty Banning Nuclear Weapon Tests in the Atmosphere, in Outer Space and Under Water and of the Treaty on the Non-Proliferation of Nuclear Weapons,

Have agreed as follows:

Article I

1. Each Party undertakes to prohibit, to prevent, and not to carry out any underground nuclear weapon test having a yield exceeding 150 kilotons at any place under its jurisdiction or control, beginning March 31, 1976.

2. Each Party shall limit the number of its underground nuclear weapon tests to a minimum.

3. The Parties shall continue their negotiations with a view toward achieving a solution to the problem of the cessation of all underground nuclear weapon tests.

Article II

1. For the purpose of providing assurance of compliance with the provisions of this Treaty, each Party shall use national technical means of verification at its disposal in a manner consistent with the generally recognized principles of international law.

2. Each Party undertakes not to interfere with the national technical means of verification of the other Party operating in accordance with paragraph 1 of this Article.

3. To promote the objectives and implementation of the provisions of this Treaty the Parties shall, as necessary, consult with each other, make inquiries and furnish information in response to such inquiries.

Article III

The provisions of this Treaty do not extend to underground nuclear explosions carried out by the Parties for peaceful purposes. Underground nuclear explosions for peaceful purposes shall be governed by an agreement which is to be negotiated and concluded by the Parties at the earliest possible time.

Article IV

This Treaty shall be subject to ratification in accordance with the constitutional procedures of each Party. This Treaty shall enter into force on the day of the exchange of instruments of ratification.

Article V

1. This Treaty shall remain in force for a period of five years. Unless replaced earlier by an agreement in implementation of the objectives specified in paragraph 3 of Article I of this Treaty, it shall be extended for successive five-year periods unless either Party notifies the other of its termination no later than six months prior to the expiration of the Treaty. Before the expiration of this period the Parties may, as necessary, hold consultations to consider the situation relevant to the substance of this Treaty and to introduce possible amendments to the text of the Treaty.

2. Each Party shall, in exercising its national sovereignty, have the right to withdraw from this Treaty if it decides that extraordinary events related to the subject matter of this Treaty have jeopardized its supreme interests. It shall give notice of its decision to the other Party six months prior to withdrawal from this Treaty. Such notice shall include a statement of the extraordinary events the notifying Party regards as having jeopardized its supreme interests.

3. This Treaty shall be registered pursuant to Article 102 of the Charter of the United Nations.

DONE at Moscow on July 3, 1974, in duplicate, in the English and Russian languages, both texts being equally authentic.

For the United States of America:

RICHARD NIXON,

The President of the United States of America

For the Union of Soviet Socialist Republics:

L. BREZHNEV,

General Secretary of the Central Committee of the CPSU.

Protocol to the Treaty Between the United States of America and the Union of Soviet Socialist Republics on the Limitation of Underground Nuclear Weapon Tests

The United States of America and the Union of Soviet Socialist Republics, hereinafter referred to as the Parties,
Having agreed to limit underground nuclear weapon tests,

Have agreed as follows:

1. For the Purpose of ensuring verification of compliance with the obligations of the Parties under the Treaty by national technical means, the Parties shall, on the basis of reciprocity, exchange the following data:

　a. The geographic coordinates of the boundaries of each test site and of the boundaries of the geophysically distinct testing areas therein.

　b. Information on the geology of the testing areas of the sites (the rock characteristics of geological formations and the basic physical properties of the rock, i.e., density, seismic velocity, water saturation, porosity and the depth of water table).

　c. The geographic coordinates of underground nuclear weapon tests, after they have been conducted.

　d. Yield, date, time, depth and coordinates for two nuclear weapon tests for calibration purposes from each geophysically distinct testing area where underground nuclear weapon tests have been and are to be conducted. In this connection the yield of such explosions for calibration purposes should be as near as possible to the limit defined in Article I of the Treaty and not less than one-tenth of that limit. In the case of testing areas where data are not available on two tests for calibration purposes, the data pertaining to one such test shall be exchanged, if available, and the data pertaining to the second test shall be exchanged as soon as possible after the second test having a yield in the above-mentioned range. The provisions of this Protocol shall not require the Parties to conduct tests solely for calibration purposes.

2. The Parties agree that the exchange of data pursuant to subparagraphs a, b, and d of paragraph 1 shall be carried out simultaneously with the exchange of instruments of ratification of the Treaty, as provided in Article IV of the Treaty, having in mind that the Parties shall, on the basis of reciprocity, afford each other the opportunity to familiarize themselves with these data before the exchange of instruments of ratification.

3. Should a Party specify a new test site or testing area after the entry into force of the Treaty, the data called for by subparagraphs a and b of paragraph 1 shall be transmitted to the other Party in advance of use of that site or area. The data called for by subparagraph d of paragraph 1 shall also be transmitted in advance of use of that site or area if they are available; if they are not available, they shall be transmitted as soon as possible after they have been obtained by the transmitting Party.

4. The Parties agree that the test sites of each Party shall be located at places under its jurisdiction or control and that all nuclear weapon tests shall be conducted solely within the testing areas specified in accordance with paragraph 1.

5. For the purposes of the Treaty, all underground nuclear explosions at the specified test sites shall be considered nuclear weapon tests and shall be subject to all the provisions of the Treaty relating to nuclear weapon tests. The provisions of Article III of the Treaty apply to all underground nuclear explosions conducted outside of the specified test sites, and only to such explosions.

This Protocol shall be considered an integral part of the Treaty.

DONE at Moscow on July 3, 1974.

For the United States of America:

RICHARD M. NIXON,

The President of the United States of America

For the Union of Soviet Socialist Republics:

L. BREZHNEV,

General Secretary of the Central Committee of the CPSU.

Appendix *F*

Treaty Between the United States of America and the Union of Soviet Socialist Republics on Underground Nuclear Explosions for Peaceful Purposes

Treaty Between the United States of America and the Union of Soviet Socialist Republics on Underground Nuclear Explosions for Peaceful Purposes

Signed at Washington and Moscow May 28, 1976

The United States of America and the Union of Soviet Socialist Republics, hereinafter referred to as the Parties,

Proceeding from a desire to implement Article III of the Treaty between the United States of America and the Union of Soviet Socialist Republics on the Limitation of Underground Nuclear Weapon Tests, which calls for the earliest possible conclusion of an agreement on underground nuclear explosions for peaceful purposes,

Reaffirming their adherence to the objectives and principles of the Treaty Banning Nuclear Weapon Tests in the Atmosphere, in Outer Space and Under Water, the Treaty on Non-Proliferation of Nuclear Weapons, and the Treaty on the Limitation of Underground Nuclear Weapon Tests, and their determination to observe strictly the provisions of these international agreements,

Desiring to assure that underground nuclear explosions for peaceful purposes shall not be used for purposes related to nuclear weapons,

Desiring that utilization of nuclear energy be directed only toward peaceful purposes,

Desiring to develop appropriately cooperation in the field of underground nuclear explosions for peaceful purposes,

Have agreed as follows:

Article I

1. The Parties enter into this Treaty to satisfy the obligations in Article III of the Treaty on the Limitation of Underground Nuclear Weapon Tests, and assume additional obligations in accordance with the provisions of this Treaty.

2. This Treaty shall govern all underground nuclear explosions for peaceful purposes conducted by the Parties after March 31, 1976.

Article II

For the purposes of this Treaty:

(a) "explosion" means any individual or group underground nuclear explosion for peaceful purposes;

(b) "explosive" means any device, mechanism or system for producing an individual explosion;

(c) "group explosion" means two or more individual explosions for which the time interval between successive individual explosions does not exceed five seconds and for which the emplacement points of all explosives can be interconnected by straight line segments, each of which joins two emplacement points and each of which does not exceed 40 kilometers.

346

Article III

1. Each Party, subject to the obligations assumed under this Treaty and other international agreements, reserves the right to:

(a) carry out explosions at any place under its jurisdiction or control outside the geographical boundaries of test sites specified under the provisions of the Treaty on the Limitation of Underground Nuclear Weapon Tests; and

(b) carry out, participate or assist in carrying out explosions in the territory of another State at the request of such other State.

2. Each Party undertakes to prohibit, to prevent and not to carry out at any place under its jurisdiction or control, and further undertakes not to carry out, participate or assist in carrying out anywhere:

(a) any individual explosion having a yield exceeding 150 kilotons;

(b) any group explosion:

(1) having an aggregate yield exceeding 150 kilotons except in ways that will permit identification of each individual explosion and determination of the yield of each individual explosion in the group in accordance with the provisions of Article IV of and the Protocol to this Treaty;

(2) having an aggregate yield exceeding one and one-half megatons;

(c) any explosion which does not carry out a peaceful application;

(d) any explosion except in compliance with the provisions of the Treaty Banning Nuclear Weapon Tests in the Atmosphere, in Outer Space and Under Water, the Treaty on the Non-Proliferation of Nuclear Weapons, and other international agreements entered into by that Party.

3. The question of carrying out any individual explosion having a yield exceeding the yield specified in paragraph 2(a) of this article will be considered by the Parties at an appropriate time to be agreed.

Article IV

1. For the purpose of providing assurance of compliance with the provisions of this Treaty, each Party shall:

(a) use national technical means of verification at its disposal in a manner consistent with generally recognized principles of international law; and

(b) provide to the other Party information and access to sites of explosions and furnish assistance in accordance with the provisions set forth in the Protocol to this Treaty.

2. Each Party undertakes not to interfere with the national technical means of verification of the other Party operating in accordance with paragraph 1(a) of this article, or with the implementation of the provisions of paragraph 1(b) of this article.

Article V

1. To promote the objectives and implementation of the provisions of this Treaty, the Parties shall establish promptly a Joint Consultative Commission within the framework of which they will:

(a) consult with each other, make inquiries and furnish information in response to such inquiries, to assure confidence in compliance with the obligations assumed;

(b) consider questions concerning compliance with the obligations assumed and related situations which may be considered ambiguous;

(c) consider questions involving unintended interference with the means for assuring compliance with the provisions of this Treaty;

(d) consider changes in technology or other new circumstances which have a bearing on the provisions of this Treaty; and

(e) consider possible amendments to provisions governing underground nuclear explosions for peaceful purposes.

2. The Parties through consultation shall establish, and may amend as appropriate, Regulations for the Joint Consultative Commission governing procedures, composition and other relevant matters.

Article VI

1. The Parties will develop cooperation on the basis of mutual benefit, equality, and reciprocity in various areas related to carrying out underground nuclear explosions for peaceful purposes.

2. The Joint Consultative Commission will facilitate this cooperation by considering specific areas and forms of cooperation which shall be determined by agreement between the Parties in accordance with their constitutional procedures.

3. The Parties will appropriately inform the International Atomic Energy Agency of results of their cooperation in the field of underground nuclear explosions for peaceful purposes.

Article VII

1. Each Party shall continue to promote the development of the international agreement or agreements and procedures provided for in Article V of the Treaty on the Non-Proliferation of Nuclear Weapons, and shall provide appropriate assistance to the International Atomic Energy Agency in this regard.

2. Each Party undertakes not to carry out, participate or assist in the carrying out of any explosion in the territory of another State unless that State agrees to the implementation in its territory of the international observation and procedures contemplated by Article V of the Treaty on the Non-Proliferation of Nuclear Weapons and the provisions of Article IV of and the Protocol to this Treaty, including the provision by that State of the assistance necessary for such implementation and of the privileges and immunities specified in the Protocol.

Article VIII

1. This Treaty shall remain in force for a period of five years, and it shall be extended for successive five-year periods unless either Party notifies the other of its termination no later than six months prior to its expiration. Before the expiration of this period the Parties may, as necessary, hold consultations to consider the situation relevant to the substance of this Treaty. However, under no circumstances shall either Party be entitled to terminate this Treaty while the Treaty on the Limitation of Underground Nuclear Weapon Tests remains in force.

2. Termination of the Treaty on the Limitation of Underground Nuclear Weapon Tests shall entitle either Party to withdraw from this Treaty at any time.

3. Each Party may propose amendments to this Treaty. Amendments shall enter into force on the day of the exchange of instruments of ratification of such amendments.

Article IX

1. This Treaty including the Protocol which forms an integral part hereof, shall be subject to ratification in accordance with the constitutional procedures of each Party. This Treaty shall enter into force on the day of the exchange of instruments of ratification which exchange shall take place simultaneously with the exchange of instruments of ratification of the Treaty on the Limitation of Underground Nuclear Weapon Tests.

2. This Treaty shall be registered pursuant to Article 102 of the Charter of the United Nations.

DONE at Washington and Moscow, on May 28, 1976, in duplicate, in the English and Russian languages, both texts being equally authentic.

For the United States of America:

GERALD R. FORD,

The President of the United States of America.

For the Union of Soviet Socialist Republics:

L. BREZHNEV,

General Secretary of the Central Committee of the CPSU.

Protocol to the Treaty Between the United States of America and the Union of Soviet Socialist Republics on Underground Nuclear Explosions for Peaceful Purposes

The United States of America and the Union of Soviet Socialist Republics, hereinafter referred to as the Parties,

Having agreed to the provisions in the Treaty on Underground Nuclear Explosions for Peaceful Purposes, hereinafter referred to as the Treaty,

Have agreed as follows:

Article I

1. No individual explosion shall take place at a distance, in meters, from the ground surface which is less than 30 times the 3.4 root of its planned yield in kilotons.

2. Any group explosion with a planned aggregate yield exceeding 500 kilotons shall not include more than five individual explosions, each of which has a planned yield not exceeding 50 kilotons.

Article II

1. For each explosion, the Party carrying out the explosion shall provide the other Party:

(a) not later than 90 days before the beginning of emplacement of the explosives when the planned aggregate yield of the explosion does not exceed 100 kilotons, or not later than 180 days before the beginning of emplacement of the explosives when the planned aggregate yield of the explosion exceeds 100 kilotons, with the following information to the extent and degree of precision available when it is conveyed:

(1) the purpose of the planned explosion;

(2) the location of the explosion expressed in geographical coordinates with a precision of four or less kilometers, planned date and aggregate yield of the explosion;

(3) the type or types of rock in which the explosion will be carried out, including the degree of liquid saturation of the rock at the point of emplacement of each explosive; and

(4) a description of specific technological features of the project, of which the explosion is a part, that could influence the determination of its yield and confirmation of purpose; and

(b) not later than 60 days before the beginning of emplacement of the explosives the information specified in subparagraph 1(a) of this article to the full extent and with the precision indicated in that subparagraph.

2. For each explosion with a planned aggregate yield exceeding 50 kilotons, the Party carrying out the explosion shall provide the other Party, not later than 60 days before the beginning of emplacement of the explosives, with the following information:

(a) the number of explosives, the planned yield of each explosive, the location of each explosive to be used in a group explosion relative to all other explosives in the group with a precision of 100 or less meters, the depth of emplacement of each explosive with a precision of one meter and the time intervals between individual explosions in any group explosion with a precision of one-tenth second; and

(b) a description of specific features of geological structure or other local conditions that could influence the determination of the yield.

3. For each explosion with a planned aggregate yield exceeding 75 kilotons, the Party carrying out the explosion shall provide the other Party, not later than 60 days before the beginning of emplacement of the explosives, with a description of the geological and geophysical characteristics of the site of each explosion which could influence determination of the yield, which shall include: the depth of the water table; a stratigraphic column above each emplacement point; the position of each emplacement point relative to nearby geological and other features which influenced the design of the project of which the explosion is a part; and the physical parameters of the rock, including density, seismic velocity, porosity, degree of liquid saturation, and rock strength, within the sphere centered on each emplacement point and having a radius, in meters, equal to 30 times the cube root of the planned yield in kilotons of the explosive emplaced at that point.

4. For each explosion with a planned aggregate yield exceeding 100 kilotons, the Party carrying out the explosion shall provide the other Party, not later than 60 days before the beginning of emplacement of the explosives, with:

(a) information on locations and purposes of facilities and installations which are associated with the conduct of the explosion;

(b) information regarding the planned date of the beginning of emplacement of each explosive; and

(c) a topographic plan in local coordinates of the areas specified in paragraph 7 of Article IV, at a scale of 1 : 24,000 or 1 : 25,000 with a contour interval of 10 meters or less.

5. For application of an explosion to alleviate the consequences of an emergency situation involving an unforeseen combination of circumstances which calls for immediate action for which it would not be practicable to observe the timing requirements of paragraphs 1, 2 and 3 of this article, the following conditions shall be met:

(a) the Party deciding to carry out an explosion for such purposes shall inform the other Party of that decision immediately after it has been made and describe such circumstances;

(b) the planned aggregate yield of an explosion for such purpose shall not exceed 100 kilotons; and

(c) the Party carrying out an explosion for such purpose shall provide to the other Party the information specified in paragraph 1 of this article, and the information specified in paragraphs 2 and 3 of this article if applicable, after the decision to conduct the explosion is taken, but not later than 30 days before the beginning of emplacement of the explosives.

6. For each explosion, the Party carrying out the explosion shall inform the other Party, not later than two days before the explosion, of the planned time of detonation of each explosive with a precision of one second.

7. Prior to the explosion, the Party carrying out the explosion shall provide the other Party with timely notification of changes in the information provided in accordance with this article.

8. The explosion shall not be carried out earlier than 90 days after notification of any change in the information provided in accordance with this article which requires more extensive verification procedures than those required on the basis of the original information, unless an earlier time for carrying out the explosion is agreed between the Parties.

9. Not later than 90 days after each explosion the Party carrying out the explosion shall provide the other Party with the following information:

(a) the actual time of the explosion with a precision of one-tenth second and its aggregate yield;

(b) when the planned aggregate yield of a group explosion exceeds 50 kilotons, the actual time of the first individual explosion with a precision of one-tenth second, the time interval between individual explosions with a precision of one milli-second and the yield of each individual explosion; and

(c) confirmation of other information provided in accordance with paragraphs 1, 2, 3 and 4 of this article and explanation of any changes or corrections based on the results of the explosion.

10. At any time, but not later than one year after the explosion, the other Party may request the Party carrying out the explosion to clarify any item of the information provided in accordance with this article. Such clarification shall be provided as soon as practicable, but not later than 30 days after the request is made.

Article III

1. For the purposes of this Protocol:

(a) "designated personnel" means those nationals of the other Party identified to the Party carrying out an explosion as the persons who will exercise the rights and functions provided for in the Treaty and this Protocol; and

(b) "emplacement hole" means the entire interior of any drill-hole, shaft, adit or tunnel in which an explosive and associated cables and other equipment are to be installed.

2. For any explosion with a planned aggregate yield exceeding 100 kilotons but not exceeding 150 kilotons if the Parties, in consultation based on information provided in accordance with Article II and other information that may be introduced by either Party, deem it appropriate for the confirmation of the yield of the explosion, and for any explosion with a planned aggregate yield exceeding 150 kilotons, the Party carrying out the explosion shall allow designated personnel within the areas and at the locations described in Article V to exercise the following rights and functions:

(a) confirmation that the local circumstances, including facilities and installations associated with the project, are consistent with the stated peaceful purposes;

(b) confirmation of the validity of the geological and geophysical information provided in accordance with Article II through the following procedures:

(1) examination by designated personnel of research and measurement data of the Party carrying out the explosion and of rock core or rock fragments removed from each emplacement hole, and of any logs and drill core from existing exploratory holes which shall be provided to designated personnel upon their arrival at the site of the explosion;

(2) examination by designated personnel of rock core or rock fragments as they become available in accordance with the procedures specified in subparagraph 2(b)(3) of this article; and

(3) observation by designated personnel of implementation by the Party carrying out the explosion of one of the following four procedures, unless this right is waived by the other Party:

(i) construction of that portion of each emplacement hole starting from a point nearest the entrance of the emplacement hole which is at a distance, in meters, from the nearest emplacement point equal to 30 times the cube root of the planned yield in kilotons of the explosive to be emplaced at that point and continuing to the completion of the emplacement hole; or

(ii) construction of that portion of each emplacement hole starting from a point nearest the entrance of the emplacement hole which is at a distance, in meters, from the nearest emplacement point equal to six times the cube root of the planned yield in kilotons of the explosive to be emplaced at that point and continuing to the completion of the emplacement hole as well as the removal of rock core or rock fragments from the wall of an existing exploratory hole, which is substantially parallel with and at no point more than 100 meters from the emplacement hole, at locations specified by designated personnel which lie within a distance, in meters, from the same horizon as each emplacement point of 30 times the cube root of the planned yield in kilotons of the explosive to be emplaced at that point; or

(iii) removal of rock core or rock fragments from the wall of each emplacement hole at locations specified by designated personnel which lie within a distance, in meters, from each emplacement point of 30 times the cube root of the planned yield in kilotons of the explosive to be emplaced at each such point; or

(iv) construction of one or more new exploratory holes so that for each emplacement hole there will be a new exploratory hole to the same depth as that of the emplacement of the explosive, substantially parallel with and at no point more than 100 meters from each emplacement hole, from which rock cores would be removed at locations specified by designated personnel which lie within a distance, in meters, from the same horizon as each emplacement point of 30 times the cube root of the planned yield in kilotons of the explosive to be emplaced at each such point:

(c) observation of the emplacement of each explosive, confirmation of the depth of its emplacement and observation of the stemming of each emplacement hole;

(d) unobstructed visual observation of the area of the entrance to each emplacement hole at any time from the time of emplacement of each explosive until all personnel have been withdrawn from the site for the detonation of the explosion; and

(e) observation of each explosion.

3. Designated personnel, using equipment provided in accordance with paragraph 1 of Article IV, shall have the right, for any explosion with a planned aggregate yield exceeding 150 kilotons, to determine the yield of each individual explosion in a group explosion in accordance with the provisions of Article VI.

4. Designated personnel, when using their equipment in accordance with paragraph 1 of Article IV, shall have the right, for any explosion with planned aggregate yield exceeding 500 kilotons, to emplace, install and operate under the observation and with the assistance of personnel of the Party carrying out the explosion, if such assistance is requested by designated personnel, a local seismic network in accordance with the provisions of paragraph 7 of Article IV. Radio links may be used for the transmission of data and control signals between the seismic stations and the control center. Frequencies, maximum power output of radio transmitters, directivity of antennas and times of operation of the local seismic network radio transmitters before the explosion shall be agreed between the Parties in accordance with Article X and time of operation after the explosion shall conform to the time specified in paragraph 7 of Article IV.

5. Designated personnel shall have the right to:

(a) acquire photographs under the following conditions:

(1) the Party carrying out the explosion shall identify to the other Party those personnel of the Party carrying out the explosion who shall take photographs as requested by designated personnel;

(2) photographs shall be taken by personnel of the Party carrying out the explosion in the presence of designated personnel and at the time requested by designated personnel for taking such photographs. Designated personnel shall determine whether these photographs are in conformity with their requests and, if not, additional photographs shall be taken immediately;

(3) photographs shall be taken with cameras provided by the other Party having built-in, rapid developing capability and a copy of each photograph shall be provided at the completion of the development process to both Parties;

(4) cameras provided by designated personnel shall be kept in agreed secure storage when not in use; and

(5) the request for photographs can be made, at any time, of the following:

(i) exterior views of facilities and installations associated with the conduct of the explosion as described in subparagraph 4(a) of Article II;

(ii) geological samples used for confirmation of geological and geophysical information, as provided for in subparagraph 2(b) of this article and the equipment utilized in the acquisition of such samples;

(iii) emplacement and installation of equipment and associated cables used by designated personnel for yield determination;

(iv) emplacement and installation of the local seismic network used by designated personnel;

(v) emplacement of the explosives and the stemming of the emplacement hole; and

(vi) containers, facilities and installations for storage and operation of equipment used by designated personnel;

(b) photographs of visual displays and records produced by the equipment used by designated personnel and photographs within the control centers taken by cameras which are component parts of such equipment; and

(c) receive at the request of designated personnel and with the agreement of the Party carrying out the explosion supplementary photographs taken by the Party carrying out the explosion.

Article IV

1. Designated personnel in exercising their rights and functions may choose to use the following equipment of either Party, of which choice the Party carrying out the explosion shall be informed not later than 150 days before the beginning of emplacement of the explosives:

(a) electrical equipment for yield determination and equipment for a local seismic network as described in paragraphs 3, 4 and 7 of this article; and

(b) geologist's field tools and kits and equipment for recording of field notes.

2. Designated personnel shall have the right in exercising their rights and functions to utilize the following additional equipment which shall be provided by the Party carrying out the explosion, under procedures to be established in accordance with Article X to ensure that the equipment meets the specifications of the other Party: portable short-range communication equipment, field glasses, optical equipment for

surveying and other items which may be specified by the other Party. A description of such equipment and operating instructions shall be provided to the other Party not later than 90 days before the beginning of emplacement of the explosives in connection with which such equipment is to be used.

3. A complete set of electrical equipment for yield determination shall consist of:

(a) sensing elements and associated cables for transmission of electrical power, control signals and data;

(b) equipment of the control center, electrical power supplies and cables for transmission of electrical power, control signals and data; and

(c) measuring and calibration instruments, maintenance equipment and spare parts necessary for ensuring the functioning of sensing elements, cables and equipment of the control center.

4. A complete set of equipment for the local seismic network shall consist of:

(a) seismic stations each of which contains a seismic instrument, electrical power supply and associated cables and radio equipment for receiving and transmission of control signals and data or equipment for recording control signals and data;

(b) equipment of the control center and electrical power supplies; and

(c) measuring and calibration instruments, maintenance equipment and spare parts necessary for ensuring the functioning of the complete network.

5. In case designated personnel, in accordance with paragraph 1 of this article, choose to use equipment of the Party carrying out the explosion for yield determination or for a local seismic network, a description of such equipment and installation and operating instructions shall be provided to the other Party not later than 90 days before the beginning of emplacement of the explosives in connection with which such equipment is to be used. Personnel of the Party carrying out the explosion shall emplace, install and operate the equipment in the presence of designated personnel. After the explosion, designated personnel shall receive duplicate copies of the recorded data. Equipment for yield determination shall be emplaced in accordance with Article VI. Equipment for a local seismic network shall be emplaced in accordance with paragraph 7 of this article.

6. In case designated personnel, in accordance with paragraph 1 of this article, choose to use their own equipment for yield determination and their own equipment for a local seismic network, the following procedures shall apply:

(a) the Party carrying out the explosion shall be provided by the other Party with the equipment and information specified in subparagraphs (a)(1) and (a)(2) of this paragraph not later than 150 days prior to the beginning of emplacement of the explosives in connection with which such equipment is to be used in order to permit the Party carrying out the explosion to familiarize itself with such equipment, if such equipment and information has not been previously provided, which equipment shall be returned to the other Party not later than 90 days before the beginning of emplacement of the explosives. The equipment and information to be provided are:

(1) one complete set of electrical equipment for yield determination as described in paragraph 3 of this aritcle, electrical and mechanical design information, specifications and installation and operating instructions concerning this equipment; and

(2) one complete set of equipment for the local seismic network described in paragraph 4 of this article, including one seismic station, electrical and mechanical design information, specifications and installation and operating instructions concerning this equipment;

(b) not later than 35 days prior to the beginning of emplacement of the explosives in connection with which the following equipment is to be used, two complete sets of electrical equipment for yield determination as described in paragraph 3 of this article and specific installation instructions for the emplacement of the sensing elements based on information provided in accordance with subparagraph 2(a) of Article VI and two complete sets of equipment for the local seismic network as described in paragraph 4 of this article, which sets of equipment shall have the same components and technical characteristics as the corresponding equipment specified in subparagraph 6(a) of this article, shall be delivered in sealed containers to the port of entry;

(c) The Party carrying out the explosion shall choose one of each of the two sets of equipment described above which shall be used by designated personnel in connection with the explosions;

(d) the set or sets of equipment not chosen for use in connection with the explosion shall be at the disposal of the Party carrying out the explosion for a period that may be as long as 30 days after the explosion at which time such equipment shall be returned to the other Party;

(e) the set or sets of equipment chosen for use shall be transported by the Party carrying out the explosion in the sealed containers in which this equipment arrived, after seals of the Party carrying out the explosion have been affixed to them, to the site of the explosion, so that this equipment is delivered to designated personnel for emplacement, installation and operation not later than 20 days before the beginning of emplacement of the explosives. This equipment shall remain in the custody of designated personnel in accordance with paragraph 7 of Article V or in agreed secure storage. Personnel of the Party carrying out the explosion shall have the right to observe the use of this equipment by designated personnel during the time the equipment is at the site of the explosion. Before the beginning of emplacement of the explosives, designated personnel shall demonstrate to personnel of the Party carrying out the explosion that this equipment is in working order;

(f) each set of equipment shall include two sets of components for recording data and associated calibration equipment. Both of these sets of components in the equipment chosen for use shall simultaneously record data. After the explosion, and after duplicate copies of all data have been obtained by designated personnel and the Party carrying out the explosion, one of each of the two sets of components for recording data and associated calibration equipment shall be selected, by an agreed process of chance, to be retained by designated personnel. Designated personnel shall pack and seal such components for recording data and associated calibration equipment which shall accompany them from the site of the explosion to the port of exit; and

(g) all remaining equipment may be retained by the Party carrying out the explosion for a period that may be as long as 30 days, after which time this equipment shall be returned to the other Party.

7. For any explosion with a planned aggregate yield exceeding 500 kilotons, a local seismic network, the number of stations of which shall be determined by designated personnel but shall not exceed the number of explosives in the group plus five, shall be emplaced, installed and operated at agreed sites of emplacement within an area circumscribed by circles of 15 kilometers in radius centered on points on the surface of the earth above the points of emplacement of the explosives during a period beginning not later than 20 days before the beginning of emplacement of the explosives and continuing after the explosion not later than three days unless otherwise agreed between the Parties.

8. The Party carrying out the explosion shall have the right to examine in the presence of designated personnel all equipment, instruments and tools of designated personnel specified in subparagraph 1(b) of this article.

9. The Joint Consultative Commission will consider proposals that either Party may put forward for the joint development of standardized equipment for verification purposes.

Article V

1. Except as limited by the provisions of paragraph 5 of this article, designated personnel in the exercise of their rights and functions shall have access along agreed routes:

(a) for an explosion with a planned aggregate yield exceeding 100 kilotons in accordance with paragraph 2 of Article III:

(1) to the locations of facilities and installations associated with the conduct of the explosion provided in accordance with subparagraph 4(a) of Article II; and

(2) to the locations of activities described in paragraph 2 of Article III; and

(b) for any explosion with a planned aggregate yield exceeding 150 kilotons, in addition to the access described in subparagraph 1(a) of this article:

(1) to other locations within the area circumscribed by circles of 10 kilometers in radius centered on points on the surface of the earth above the points of emplacement of the explosives in order to confirm that the local circumstances are consistent with the stated peaceful purposes;

(2) to the locations of the components of the electrical equipment for yield determination to be used for recording data when, by agreement between the Parties, such equipment is located outside the area described in subparagraph 1(b)(1) of this article; and

(3) to the sites of emplacement of the equipment of the local seismic network provided for in paragraph 7 of Article IV.

2. The Party carrying out the explosion shall notify the other Party of the procedure it has chosen from among those specified in subparagraph 2(b)(3) of Article III not later than 30 days before beginning the implementation of such procedure. Designated personnel shall have the right to be present at the site of the explosion to exercise their rights and functions in the areas and at the locations described in paragraph 1 of this article for a period of time beginning two days before the beginning of the implementation of the procedure and continuing for a period of three days after the completion of this procedure.

3. Except as specified in paragraph 5 of this article, designated personnel shall have the right to be present in the areas and at the locations described in paragraph 1 of this article:

(a) for an explosion with a planned aggregate yield exceeding 100 kilotons but not exceeding 150 kilotons, in accordance with paragraph 2 of Article III, at any time beginning five days before the beginning of emplacement of the explosives and continuing after the explosion and after safe access to evacuated areas has been established according to standards determined by the Party carrying out the explosion for a period of two days; and

(b) for any explosion with a planned aggregate yield exceeding 150 kilotons, at any time beginning 20 days before the beginning of emplacement of the explosives and continuing after the explosion and after safe access to evacuated areas has been established according to standards determined by the Party carrying out the explosion for a period of:

(1) five days in the case of an explosion with a planned aggregate yield exceeding 150 kilotons but not exceeding 500 kilotons; or

(2) eight days in the case of an explosion with a planned aggregate yield exceeding 500 kilotons.

4. Designated personnel shall not have the right to be present in those areas from which all personnel have been evacuated in connection with carrying out an explosion, but shall have the right to re-enter those areas at the same time as personnel of the Party carrying out the explosion.

5. Designated personnel shall not have or seek access by physical, visual or technical means to the interior of the canister containing an explosive, to documentary or other information descriptive of the design of an explosive nor to equipment for control and firing of explosives. The Party carrying out the explosion shall not locate documentary or other information descriptive of the design of an explosive in such ways as to impede the designated personnel in the exercise of their rights and functions.

6. The number of designated personnel present at the site of an explosion shall not exceed:

(a) for the exercise of their rights and functions in connection with the confirmation of the geological and geophysical information in accordance with the provisions of subparagraph 2(b) and applicable provisions of paragraph 5 of Article III—the number of emplacement holes plus three;

(b) for the exercise of their rights and functions in connection with confirming that the local circumstances are consistent with the information provided and with the stated peaceful purposes in accordance with the provisions in subparagraphs 2(a), 2(c), 2(d) and 2(e) and applicable provisions of paragraph 5 of Article III—the number of explosives plus two;

(c) for the exercise of their rights and functions in connection with confirming that the local circumstances are consistent with the information provided and with the stated peaceful purposes in accordance with the provisions in subparagraphs 2(a), 2(c), 2(d) and 2(e) and applicable provisions of paragraph 5 of Article III and in connection with the use of electrical equipment for determination of the yield in accordance with paragraph 3 of Article III—the number of explosives plus seven; and

(d) for the exercise of their rights and functions in connection with confirming that the local circumstances are consistent with the information provided and with the stated peaceful purposes in accordance with the provisions in subparagraph 2(a), 2(c), 2(d) and 2(e) and applicable provisions of paragraph 5 of Article III and in connection with the use of electrical equipment for determination of the yield in accordance with paragraph 3 of Article III and with the use of the local seismic network in accordance with paragraph 4 of Article III—the number of explosives plus 10.

7. The Party carrying out the explosion shall have the right to assign its personnel to accompany designated personnel while the latter exercise their rights and functions.

8. The Party carrying out an explosion shall assure for designated personnel telecommunications with their authorities, transportation and other services appropriate to their presence and to the exercise of their rights and functions at the site of the explosion.

9. The expenses incurred for the transportation of designated personnel and their equipment to and from the site of the explosion, telecommunications provided for in paragraph 8 of this article, their living and working quarters, subsistence and all other personal expenses shall be the responsibility of the Party other than the Party carrying out the explosion.

10. Designated personnel shall consult with the Party carrying out the explosion in order to coordinate the planned program and schedule of activities of designated personnel with the program of the Party carrying out the explosion for the conduct of the project so as to ensure that designated personnel are able to conduct their activities in an orderly and timely way that is compatible with the implementation of the project. Procedures for such consultations shall be established in accordance with Article X.

Article VI

For any explosion with a planned aggregate yield exceeding 150 kilotons, determination of the yield of each explosive used shall be carried out in accordance with the following provisions:

1. Determination of the yield of each individual explosion in the group shall be based on measurements of the velocity of propagation, as a function of time, of the hydrodynamic shock wave generated by the explosion, taken by means of electrical equipment described in paragraph 3 of Article IV.

2. The Party carrying out the explosion shall provide the other Party with the following information:

(a) not later than 60 days before the beginning of emplacement of the explosives, the length of each canister in which the explosive will be contained in the corresponding emplacement hole, the dimensions of the tube or other device used to emplace the canister and the cross-sectional dimensions of the emplacement hole to a distance, in meters, from the emplacement point of 10 times the cube root of its yield in kilotons;

(b) not later than 60 days before the beginning of emplacement of the explosives, a description of materials, including their densities, to be used to stem each emplacement hole; and

(c) not later than 30 days before the beginning of emplacement of the explosives, for each emplacement hole of a group explosion, the local coordinates of the point of emplacement of the explosive, the entrance of the emplacement hole, the point of the emplacement hole most distant from the entrance, the location of the emplacement hole at each 200 meters distance from the entrance and the configuration of any known voids larger than one cubic meter located within the distance, in meters, of 10 times the cube root of the planned yield in kilotons measured from the bottom of the canister containing the explosive. The error in these coordinates shall not exceed one percent of the distance between the emplacement hole and the nearest other emplacement hole or one percent of the distance between the point of measurement and the entrance of the emplacement hole, whichever is smaller, but in no case shall the error be required to be less than one meter.

3. The Party carrying out the explosion shall emplace for each explosive that portion of the electrical equipment for yield determination described in subparagraph 3(a) of Article IV, supplied in accordance with paragraph 1 of Article IV, in the same emplacement hole as the explosive in accordance with the installation instructions supplied under the provisions of paragraph 5 or 6 of Article IV. Such emplacement shall be carried out under the observation of designated personnel. Other equipment specified in subparagraph 3(b) of Article IV shall be emplaced and installed:

(a) by designated personnel under the observation and with the assistance of personnel of the Party carrying out the explosion, if such assistance is requested by designated personnel; or

(b) in accordance with paragraph 5 of Article IV.

4. That portion of the electrical equipment for yield determination described in subparagraph 3(a) of Article IV that is to be emplaced in each emplacement hole shall be located so that the end of the electrical equipment which is farthest from the entrance to the emplacement hole is at a distance, in meters, from the bottom of the canister containing the explosive equal to 3.5 times the cube root of the planned yield in kilotons of the explosive when the planned yield is less than 20 kilotons and three times the cube root of the planned yield in kilotons of the explosive when the planned yield is 20 kilotons or more. Canisters longer than 10 meters containing the explosive shall only be utilized if there is prior agreement between the Parties establishing provisions for their use. The Party carrying out the explosion shall provide the other Party with data on the distribution of density inside any other canister in the emplacement hole with a transverse cross-sectional area exceeding 10 square centimeters located within a distance, in meters, of 10 times the cube root of the planned yield in kilotons of the explosion from the bottom of the canister containing the explosive. The Party carrying out the explosion shall provide the other Party with access to confirm such data on density distribution within any such canister.

5. The Party carrying out an explosion shall fill each emplacement hole, including all pipes and tubes contained therein which have at any transverse section an aggregate cross-sectional area exceeding 10 square centimeters in the region containing the electrical equipment for yield determination and to a distance, in meters, of six times the cube root of the planned yield in kilotons of the explosive from the explosive emplacement point, with material having a density not less than seven-tenths of the average density of the surrounding rock, and from that point to a distance of not less than 60 meters from the explosive emplacement point with material having a density greater than one gram per cubic centimeter.

6. Designated personnel shall have the right to:

(a) confirm information provided in accordance with subparagraph 2(a) of this article;

(b) confirm information provided in accordance with subparagraph 2(b) of this article and be provided, upon request, with a sample of each batch of stemming material as that material is put into the emplacement hole; and

(c) confirm the information provided in accordance with subparagraph 2(c) of this article by having access to the data acquired and by observing, upon their request, the making of measurements.

7. For those explosives which are emplaced in separate holes, the emplacement shall be such that the distance D, in meters, between any explosive and any portion of the electrical equipment for determination of the yield of any other explosive in the group shall be not less than 10 times the cube root of the planned yield in kilotons of the larger explosive of such a pair of explosives. Individual explosions shall be separated by time intervals, in milliseconds, not greater than one-sixth the amount by which the distance D, in meters, exceeds 10 times the cube root of the planned yield in kilotons of the larger explosive of such a pair of explosives.

8. For those explosives in a group which are emplaced in a common emplacement hole, the distance, in meters, between each explosive and any other explosive in that emplacement hole shall be not less than 10 times the cube root of the planned yield in kilotons of the larger explosive of such a pair of explosives, and the explosives shall be detonated in sequential order, beginning with the explosive farthest from the entrance to the emplacement hole, with the individual detonations separated by time intervals, in milliseconds, of not less than one times the cube root of the planned yield in kilotons of the largest explosive in this emplacement hole.

Article VII

1. Designated personnel with their personal baggage and their equipment as provided in Article IV shall be permitted to enter the territory of the Party carrying out the explosion at an entry port to be agreed upon by the Parties, to remain in the territory of the Party carrying out the explosion for the purpose of fulfilling their rights and functions provided for in the Treaty and this Protocol, and to depart from an exit port to be agreed upon by the Parties.

2. At all times while designated personnel are in the territory of the Party carrying out the explosion, their persons, property, personal baggage, archives and documents as well as their temporary official and living quarters shall be accorded the same privileges and immunities as provided in Articles 22, 23, 24, 29, 30, 31, 34 and 36 of the Vienna Convention on Diplomatic Relations of 1961 to the persons, property, personal baggage, archives and documents of diplomatic agents as well as to the premises of diplomatic missions and private residences of diplomatic agents.

3. Without prejudice to their privileges and immunities it shall be the duty of designated personnel to respect the laws and regulations of the State in whose territory the explosion is to be carried out insofar as they do not impede in any way whatsoever the proper exercising of their rights and functions provided for by the Treaty and this Protocol.

Article VIII

The Party carrying out an explosion shall have sole and exclusive control over and full responsibility for the conduct of the explosion.

Article IX

1. Nothing in the Treaty and this Protocol shall affect proprietary rights in information made available under the Treaty and this Protocol and in information which may be disclosed in preparation for and carrying out of explosions; however, claims to such proprietary rights shall not impede implementation of the provisions of the Treaty and this Protocol.

2. Public release of the information provided in accordance with Article II or publication of material using such information, as well as public release of the results of observation and measurements obtained by designated personnel, may take place only by agreement with the Party carrying out an explosion; however, the other Party shall have the right to issue statements after the explosion that do not divulge information in which the Party carrying out the explosion has rights which are referred to in paragraph 1 of this article.

Article X

The Joint Consultative Commission shall establish procedures through which the Parties will, as appropriate, consult with each other for the purpose of ensuring efficient implementation of this Protocol.

DONE at Washington and Moscow, on May 28, 1976.

For the United States of America:

GERALD R. FORD,

The President of the United States of America.

For the Union of Soviet Socialist Republics:

L. BREZHNEV,

General Secretary of the Central Committee of the CPSU.

Agreed Statement

May 13, 1976

The Parties to the Treaty Between the United States of America and the Union of Soviet Socialist Republics on Underground Nuclear Explosions for Peaceful Purposes, hereinafter referred to as the Treaty, agree that under subparagraph 2(c) of Article III of the Treaty:

(a) Development testing of nuclear explosives does not constitute a "peaceful application" and any such development tests shall be carried out only within the boundaries of nuclear weapon test sites specified in accordance with the Treaty between the United States of America and the Union of Soviet Socialist Republics on the Limitation of Underground Nuclear Weapon Tests;

(b) Associating test facilities, instrumentation or procedures related only to testing of nuclear weapons or their effects with any explosion carried out in accordance with the Treaty does not constitute a "peaceful application."

Appendix G

Treaty on the Non-Profileration of Nuclear Weapons

Treaty on the Non-Proliferation of Nuclear Weapons

Signed at Washington, London, and Moscow July 1, 1968
Ratification advised by U.S. Senate March 13, 1969
Ratified by U.S. President November 24, 1969
U.S. ratification deposited at Washington, London, and Moscow March 5, 1970
Proclaimed by U.S. President March 5, 1970
Entered into force March 5, 1970

The States concluding this Treaty, hereinafter referred to as the "Parties to the Treaty",

Considering the devastation that would be visited upon all mankind by a nuclear war and the consequent need to make every effort to avert the danger of such a war and to take measures to safeguard the security of peoples,

Believing that the proliferation of nuclear weapons would seriously enhance the danger of nuclear war,

In conformity with resolutions of the United Nations General Assembly calling for the conclusion of an agreement on the prevention of wider dissemination of nuclear weapons,

Undertaking to cooperate in facilitating the application of International Atomic Energy Agency safeguards on peaceful nuclear activities,

Expressing their support for research, development and other efforts to further the application, within the framework of the International Atomic Energy Agency safeguards system, of the principle of safeguarding effectively the flow of source and special fissionable materials by use of instruments and other techniques at certain strategic points,

Affirming the principle that the benefits of peaceful applications of nuclear technology, including any technological by-products which may be derived by nuclear-weapon States from the development of nuclear explosive devices, should be available for peaceful purposes to all Parties of the Treaty, whether nuclear-weapon or non-nuclear weapon States,

Convinced that, in furtherance of this principle, all Parties to the Treaty are entitled to participate in the fullest possible exchange of scientific information for, and to contribute alone or in cooperation with other States to, the further development of the applications of atomic energy for peaceful purposes,

Declaring their intention to achieve at the earliest possible date the cessation of the nuclear arms race and to undertake effective measures in the direction of nuclear disarmament,

Urging the cooperation of all States in the attainment of this objective,

Recalling the determination expressed by the Parties to the 1963 Treaty banning nuclear weapon tests in the atmosphere, in outer space and under water in its Preamble to seek to achieve the discontinuance of all test explosions of nuclear weapons for all time and to continue negotiations to this end,

Desiring to further the easing of international tension and the strengthening of trust between States in order to facilitate the cessation of the manufacture of nuclear weapons, the liquidation of all their existing stockpiles, and the elimination from national arsenals of nuclear weapons and the means of their delivery pursuant to a treaty on general and complete disarmament under strict and effective international control,

Recalling that, in accordance with the Charter of the United Nations, States must refrain in their international relations from the threat or use of force against the territorial integrity or political independence of any State, or in any other manner inconsistent with the Purposes of the United Nations, and that the establishment and maintenance of international peace and security are to be promoted with the least diversion for armaments of the world's human and economic resources,

Have agreed as follows:

Article I

Each nuclear-weapon State Party to the Treaty undertakes not to transfer to any recipient whatsoever nuclear weapons or other nuclear explosive devices or control over such weapons or explosive devices directly, or indirectly; and not in any way to assist, encourage, or induce any non-nuclear-weapon State to manufacture or otherwise acquire nuclear weapons or other nuclear explosive devices, or control over such weapons or explosive devices.

Article II

Each non-nuclear-weapon State Party to the Treaty undertakes not to receive the transfer from any transferor whatsoever of nuclear weapons or other nuclear explosive devices or of control over such weapons or explosive devices directly, or indirectly; not to manufacture or otherwise acquire nuclear weapons or other nuclear explosive devices; and not to seek or receive any assistance in the manufacture of nuclear weapons or other nuclear explosive devices.

Article III

1. Each non-nuclear-weapon State Party to the Treaty undertakes to accept safeguards, as set forth in an agreement to be negotiated and concluded with the International Atomic Energy Agency in accordance with the Statute of the International Atomic Energy Agency and the Agency's safeguards system, for the exclusive purpose of verification of the fulfillment of its obligations assumed under this Treaty with a view to preventing diversion of nuclear energy from peaceful uses to nuclear weapons or other nuclear explosive devices. Procedures for the safeguards required by this article shall be followed with respect to source or special fissionable material whether it is being produced, processed or used in any principal nuclear facility or is outside any such facility. The safeguards required by this article shall be applied to all source or special fissionable material in all peaceful nuclear activities within the territory of such State, under its jurisdiction, or carried out under its control anywhere.

2. Each State Party to the Treaty undertakes not to provide: (a) source or special fissionable material, or (b) equipment or material especially designed or prepared for the processing, use or production of special fissionable material, to any non-nuclear-weapon State for peaceful purposes, unless the source or special fissionable material shall be subject to the safeguards required by this article.

3. The safeguards required by this article shall be implemented in a manner designed to comply with article IV of this Treaty, and to avoid hampering the economic or technological development of the Parties or international cooperation in the field of peaceful nuclear activities, including the international exchange of nuclear material and equipment for the processing, use or production of nuclear material for peaceful purposes in accordance with the provisions of this article and the principle of safeguarding set forth in the Preamble of the Treaty.

4. Non-nuclear-weapon States Party to the Treaty shall conclude agreements with the International Atomic Energy Agency to meet the requirements of this article either

individually or together with other States in accordance with the Statute of the International Atomic Energy Agency. Negotiation of such agreements shall commence within 180 days from the original entry into force of this Treaty. For States depositing their instruments of ratification or accession after the 180-day period, negotiation of such agreements shall commence not later than the date of such deposit. Such agreements shall enter into force not later than eighteen months after the date of initiation of negotiations.

Article IV

1. Nothing in this Treaty shall be interpreted as affecting the inalienable right of all the Parties to the Treaty to develop research, production and use of nuclear energy for peaceful purposes without discrimination and in conformity with articles I and II of this Treaty.

2. All the Parties to the Treaty undertake to facilitate, and have the right to participate in, the fullest possible exchange of equipment, materials and scientific and technological information for the peaceful uses of nuclear energy. Parties to the Treaty in a position to do so shall also cooperate in contributing alone or together with other States or international organizations to the further development of the applications of nuclear energy for peaceful purposes, especially in the territories of non-nuclear-weapon States Party to the Treaty, with due consideration for the needs of the developing areas of the world.

Article V

Each party to the Treaty undertakes to take appropriate measures to ensure that, in accordance with this Treaty, under appropriate international observation and through appropriate international procedures, potential benefits from any peaceful applications of nuclear explosions will be made available to non-nuclear-weapon States Party to the Treaty on a nondiscriminatory basis and that the charge to such Parties for the explosive devices used will be as low as possible and exclude any charge for research and development. Non-nuclear-weapon States Party to the Treaty shall be able to obtain such benefits, pursuant to a special international agreement or agreements, through an appropriate international body with adequate representation of non-nuclear-weapon States. Negotiations on this subject shall commence as soon as possible after the Treaty enters into force. Non-nuclear-weapon States Party to the Treaty so desiring may also obtain such benefits pursuant to bilateral agreements.

Article VI

Each of the Parties to the Treaty undertakes to pursue negotiations in good faith on effective measures relating to cessation of the nuclear arms race at an early date and to nuclear disarmament, and on a treaty on general and complete disarmament under strict and effective international control.

Article VII

Nothing in this Treaty affects the right of any group of States to conclude regional treaties in order to assure the total absence of nuclear weapons in their respective territories.

Article VIII

1. Any Party to the Treaty may propose amendments to this Treaty. The text of any proposed amendment shall be submitted to the Depositary Governments which shall

circulate it to all Parties to the Treaty. Thereupon, if requested to do so by one-third or more of the Parties to the Treaty, the Depositary Governments shall convene a conference, to which they shall invite all the Parties to the Treaty, to consider such an amendment.

2. Any amendment to this Treaty must be approved by a majority of the votes of all the Parties to the Treaty, including the votes of all nuclear-weapon States Party to the Treaty and all other Parties which, on the date the amendment is circulated, are members of the Board of Governors of the International Atomic Energy Agency. The amendment shall enter into force for each Party that deposits its instrument of ratification of the amendment upon the deposit of such instruments of ratification by a majority of all the Parties, including the instruments of ratification of all nuclear-weapon States Party to the Treaty and all other Parties which, on the date the amendment is circulated, are members of the Board of Governors of the International Atomic Energy Agency. Thereafter, it shall enter into force for any other Party upon the deposit of its instrument of ratification of the amendment.

3. Five years after the entry into force of this Treaty, a conference of Parties to the Treaty shall be held in Geneva, Switzerland, in order to review the operation of this Treaty with a view to assuring that the purposes of the Preamble and the provisions of the Treaty are being realized. At intervals of five years thereafter, a majority of the Parties to the Treaty may obtain, by submitting a proposal to this effect to the Depositary Governments, the convening of further conferences with the same objective of reviewing the operation of the Treaty.

Article IX

1. This Treaty shall be open to all States for signature. Any State which does not sign the Treaty before its entry into force in accordance with paragraph 3 of this article may accede to it at any time.

2. This Treaty shall be subject to ratification by signatory States. Instruments of ratification and instruments of accession shall be deposited with the Governments of the United States of America, the United Kingdom of Great Britain and Northern Ireland and the Union of Soviet Socialist Republics, which are hereby designated the Depositary Governments.

3. This Treaty shall enter into force after its ratification by the States, the Governments of which are designated Depositaries of the Treaty, and forty other States signatory to this Treaty and the deposit of their instruments of ratification. For the purposes of this Treaty, a nuclear-weapon State is one which has manufactured and exploded a nuclear weapon or other nuclear explosive device prior to January 1, 1967.

4. For States whose instruments of ratification or accession are deposited subsequent to the entry into force of this Treaty, it shall enter into force on the date of the deposit of their instruments of ratification or accession.

5. The Depositary Governments shall promptly inform all signatory and acceding States of the date of each signature, the date of deposit of each instrument of ratification or of accession, the date of the entry into force of this Treaty, and the date of receipt of any requests for convening a conference or other notices.

6. This Treaty shall be registered by the Depositary Governments pursuant to article 102 of the Charter of the United Nations.

Article X

1. Each Party shall in exercising its national sovereignty have the right to withdraw from the Treaty if it decides that extraordinary events, related to the subject matter of this Treaty, have jeopardized the supreme interests of its country. It shall give notice of

such withdrawal to all other Parties to the Treaty and to the United Nations Security Council three months in advance. Such notice shall include a statement of the extraordinary events it regards as having jeopardized its supreme interests.

2. Twenty-five years after the entry into force of the Treaty, a conference shall be convened to decide whether the Treaty shall continue in force indefinitely, or shall be extended for an additional fixed period or periods. This decision shall be taken by a majority of the Parties to the Treaty.

Article XI

This Treaty, the English, Russian, French, Spanish and Chinese texts of which are equally authentic, shall be deposited in the archives of the Depositary Governments. Duly certified copies of this Treaty shall be transmitted by the Depositary Governments to the Governments of the signatory and acceding States.

Index

369